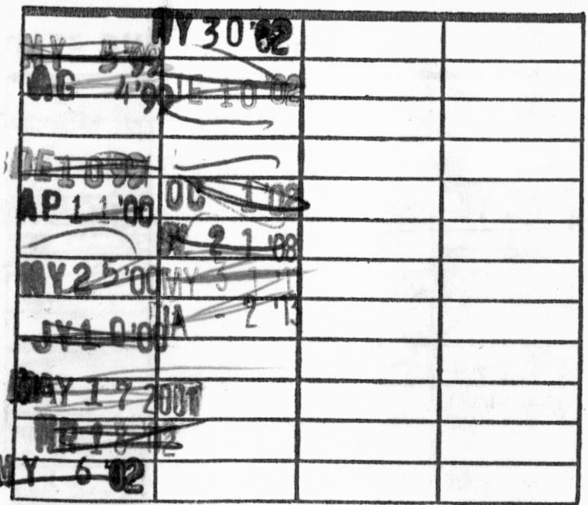

Science and creationism

QH
371
S397
1984

RIVERSIDE CITY COLLEGE
LIBRARY
Riverside, California

OC '85

DEMCO

Science and Creationism

SOCRATES:	Did you say that you believe in the separation of Church and State?
[W. J.] BRYAN:	I did. It is a fundamental principle.
SOCRATES:	Is the right of the majority to rule a fundamental principle?
BRYAN:	It is.
SOCRATES:	Is freedom of thought a fundamental principle, Mr. Jefferson?
[T.] JEFFERSON:	It is.
SOCRATES:	Well, how would you gentlemen compose your fundamental principles if a majority exercising its fundamental right to rule, ordained that only Buddhism should be taught in public schools?
BRYAN:	I'd move to a Christian country.
JEFFERSON:	I'd exercise the sacred right to revolution. What would you do, Socrates?
SOCRATES:	I'd re-examine my fundamental principles.

Walter Lippmann, *Four Dialogues*

SCIENCE AND CREATIONISM

EDITED BY

Ashley Montagu

OXFORD UNIVERSITY PRESS
Oxford New York Toronto Melbourne
1984

Riverside Community College
Library
4800 Magnolia Avenue
Riverside, CA 92506

Oxford University Press
Oxford London Glasgow
New York Toronto Melbourne Auckland
Delhi Bombay Calcutta Madras Karachi
Kuala Lumpur Singapore Hong Kong Tokyo
Nairobi Dar es Salaam Cape Town

and associate companies in
Beirut Berlin Ibadan Mexico City Nicosia

Copyright © 1984 by Ashley Montagu

First published by Oxford University Press, New York, 1984

First issued as an Oxford University Press paperback, 1984

Library of Congress Cataloging in Publication Data

Main entry under title:

Science and creationism.

1. Evolution—Addresses, essays, lectures.
2. Evolution and christianity—Addresses, essays, lectures. 3. Bible and evolution—Addresses, essays, lectures. I. Montagu, Ashley, 1905– .
QH371.S397 1984 575 82–14173
ISBN 0–19–503252–7
ISBN 0–19–503253–5 (pbk.)

Printing: 9 8 7 6 5 4 3 2

Printed in the United States of America

Contents

ROGER LEWIN

A TALE WITH MANY CONNECTIONS

When the "Balanced Treatment for Creation Science and Evolution Science Act" successfully passed through the Arkansas legislature and onto Governor Frank White's desk in March 1981, the one common reaction was pure astonishment. The bill's promoters were astonished that it had slid through the legislative process with such expedition. And many Arkansans were astonished that such a bill was passed at all. But passed it was, and readers of one Little Rock newspaper voted the creationism controversy as the number one news story of the year.

For a measure with such far-reaching implications for the nature and content of public school education, the Balanced Treatment bill had a legislative history that can only be described as bizarre. James Holsted, a senator for North Little Rock, introduced the bill on 24 February in the closing days of the session, and within less than a month the governor's signature had already dried on what was to become known as Act 590. Holsted not only had not penned a single word of the bill, but he also had no idea who had. The bill had slipped through the senate with no committee hearings. Consideration of the measure in the house was barely more probing, with all of 15 minutes being devoted to testimony on the bill in the education committee. White signed it with much ostentatious flourish,

From *Science,* Vol. 215 (29 January 1982), pp. 484–88. Copyright © 1982 American Association for the Advancement of Science. Reprinted by permission of Roger Lewin and the AAAS.

without first reading it and against the advice of a legislative assistant who had.

Once the initial shock had lifted, the obvious question was, how did such an extraordinary thing happen? One explanation is that the measure was so laden with such virtues as fair play, expansion of academic freedom, and embodiment of Christian ideals that the legislature had displayed unusual perspicacity in passing it at top speed. Another is that Act 590 is the cherished prize of a "conspiracy" between fundamentalists and conservative organizations operating as a closely coordinated network throughout the state.

Neither account is correct, although, if anything, the truth lies closer to the second. The story of Act 590 is long and meandering, but its denouement does involve the effective commitment to the measure by certain key organizations, specifically the Moral Majority.

The tale begins in May 1979 with the publication by the Institute for Creation Research of a model "Resolution for Balanced Presentation of Evolution and Scientific Creationism." The resolution, which had been prepared by Wendell Bird, who has recently become general counsel for the institute, covered four pages of ICR's *Impact* series and went to subscribers throughout the country. "Please note that this is a suggested *resolution,* to be adopted by boards of education, not *legislation* proposed for enactment as law," stated *Impact*'s editor.

Through a curious set of circumstances the Bird resolution was eventually to become both an adopted school board resolution and an enacted piece of legislation (albeit modified from the original) within three months of each other in the state of Arkansas. And, as the unfolding story of Act 590 reveals, controversy over the first was a trigger in promoting the second.

From its inception the Institute for Creation Research has favored action at the local school board level rather than aiming for state legislation. Other organizations, by contrast, see state legislation as the primary route by which to get creationism taught in schools. The most effective group treading this path is the Citizens for Fairness in Education, founded and run almost singlehandedly by Paul Ellwanger in Anderson, South Carolina.

Ellwanger says he is associated with no religious, political, or creationist organization. "I am beholden to no one," he asserts. "I can do my own thing and I don't have to answer to anyone." He has in fact been highly effective, and has poured tremendous energy and his own funds into promoting a model creationist bill, which has gone to sympathizers drawn from the ICR mailing list in all 50 states. The bills that became law in Arkansas and subsequently in Louisiana were the work of Ellwanger's hand.

When Ellwanger drafted the first version of the model bill in 1979 he drew heavily on the resolution published by the ICR. The draft bill's sections on "Clarifications" and "Findings of fact" were almost word for word taken from Bird's resolution. While the bulk of the model bill derived from Bird's draftsmanship, Ellwanger says that, in preparing a constitutionally strong version to be distributed around the country, he consulted with many people in many states, specifically attorneys and legislators.

One copy of this first version landed on the desk of the Reverend W. A. Blount, pastor of the Sylvan Hills Community Church in Little Rock. Blount's church makes small monthly contributions to the Institute for Creation Research, and so he regularly receives creationist literature. Blount was not particularly surprised to see the draft bill when it arrived toward the end of 1979, although he had no idea where it had come from. He read it, and thought it interesting; but he put it aside, where it was to remain for 18 months.

Meanwhile, Ellwanger was improving his product, and he came up with a second version in 1980. The wording was tighter and less vulnerable to interpretation as religious apologetics. The phrase "scientific creationism" was replaced by "creation science," for instance. In the first 2 years of promoting legislative efforts throughout the nation. Ellwanger had the gratification of watching his bill being introduced into more than 20 legislatures. All, however, were either blocked in committee or expired as legislative sessions came to an end with no action taken on them. And then the Arkansas phenomenon happened.

Larry Fisher, a mathematics teacher at a school in North Little Rock, was a catalyst in the developments in Arkansas. He has been on the Institute for Creation Research's mailing list ever since the early 1970s and has built up a comprehensive library of creationist material. Although a somewhat diffident man, Fisher nurtured a passion for getting what he viewed as balanced treatment for ideas on creation into the schools. "Over the years I had read the creation arguments and had become convinced of their validity," he says. "I came to believe that it was in the interests of good education that students should be presented with both models, the scientific evidence only."

During 1980, junior high schools in the area where Fisher teaches were adopting new science books for the next five years. "For the previous five years the schools had been using the Intermediate Science Curriculum Study, a kind of student-oriented discovery approach," explains Fisher. "Teachers and supervisors weren't very happy with this technique, and it was decided to return to the more traditional teacher-oriented material." Fisher was looking through the biology text that was to be adopted for 1981 through 1986 when he came across a section on evolution. "I knew from my reading of creation science literature that there were a lot of problems with the idea of evolution and so I decided I would try to get something done about it."

Fisher went through his file on ICR literature and pulled out the *Impact* issue of May 1979. He extracted the resolution on Balanced Presentation, appended an explanatory letter, and on 12 December 1980 sent a copy to Tom Hardin, superintendent of the Pulaski County Special School District.

"School districts across the country are beginning to pass similar resolutions," Fisher explained. "Our district would be promoting good public relations by adopting this resolution since surveys across the country indicate that about 80 percent of the patrons support it. By adopting this resolution, I feel our district would be providing a leadership role by promoting academic integrity and responsibility on this issue." Fisher sent copies of the resolution to all the school board members.

The school board had recently been the target of vigorous lobbying by Fundamentalist Christians, the Moral Majority, and a group known as Family Life America under God (FLAG), with issues such as sex education and adoption of certain "liberal" textbooks coming under fire. One local educator describes the board as being "very much under the influence of a southwest Little Rock fundamentalist church." It was against this background that Fisher's proposal was to be heard.

Fisher claims he did not do any active lobbying for the resolution, but he acknowledges that his spell of teaching with the board's chairman, Bob Moore, might have smoothed the passage of his proposal. When the January board meeting came around, Fisher said that there was good scientific evidence for creationism and surely it was fair to give students all the evidence relating to origins rather than censoring some of it because some scientists do not care for it. The board was quickly persuaded of the apparent good sense of the resolution and directed that a committee be established to come up with a curriculum on scientific creationism.

Fisher was delighted, although mildly surprised. Local residents wrote to Little Rock newspapers declaring that evolution was at last going to be kicked out of the public schools. And some science teachers commented that, if Fisher were really interested in education rather than politics, he would have taken the resolution to the county's science supervisor rather than straight to the school board.

A committee was set up, with the county's curriculum director as its chairperson, and was due to meet on 6 January. Fisher took a selection of creationist books and other material to this meeting, and these were distributed among committee members. Most of the committee's members, made up of country educators, teachers, two school board members, and a parent, were totally unfamiliar with scientific creationism as described in the resolution. Fisher took the lead in discussions at the first meeting, and he immediately detected opposition to the proposal. "Even some people who had not read any of the material were predisposed against it," he recalls.

The second meeting was an emotional affair. "I could find no science at all in the material that I had been given," says Bill Wood, a Little Rock science teacher. "The things I read were full of religious references." Wood's views were echoed by most people at the meeting, except Fisher. Nevertheless, Fisher insisted that the committee push ahead with developing a curriculum.

"We were surprised," says Wood, "because at the first meeting we said that if we could find no science in the books we would want to drop it. Fisher seemed happy with that suggestion, but at the second meeting he was adamant that we should continue. Something must have happened to him between those two meetings."

Wood had been the most vocal participant at the committee's second meeting, and therefore he was chosen to present its conclusion to the school board on 10 March. "The hall was packed," remembers Wood, "and we realized that a lot of people had come along to support Fisher. When he came in, he shook hands with 15 or 20 people. It was quite a reception for him."

Wood explained to the board what the committee had done and what conclusions it had arrived at. "The committee did not support implementing instruction in creationism in the district's classrooms," notes the official record of the 10 March board meeting, "nor would the committee endorse the materials submitted by Mr. Fisher." That, one might have anticipated, would be the end of it. But, no. "After discussion," the official record continued, "Gene Jones, director of secondary education, assured the board that the committee would continue to work on a sample curriculum which would offer several alternative theories to evolution."

Behind those bland words had been an uproar. "I was told that we had not been asked our opinion," says Wood, "but that we had been instructed to produce a curriculum and that's what we should have done." Members of the public booed and jeered when Wood tried to explain why the committee had recommended that the proposal be dropped. Cheers and bursts of applause encouraged Fisher when he rose to defend the resolution. Public opposition to the committee's findings had apparently been thoroughly well organized.

"We were bitter and angry," says Wood. "Larry admitted to me after the meeting that the books we had reviewed couldn't be used in public schools because of their religious content. He wouldn't say that in the meeting." The upshot was that a two-person working party was set up to draft a curriculum. One member was Marianne Wilson, science coordinator for the district. The other was Fisher.

Wilson describes the efforts she and Fisher went to in trying to draw up a teaching unit that did not make use of literature from creationist organizations. "I met with teachers and professors at the University of Arkansas, Little Rock (UALR), and asked advice," she recalls. "Some people tried to help us, although others all but asked us to leave." One person who did offer support was Ed Gran, a UALR physics professor who later was to be influential in establishing creationist organizations in the state.

Richard Bliss, curriculum director for the Institute for Creation

Research, visited Little Rock in April, and Wilson met with him to discuss potential teaching material. "What he had was trash," says Wilson. "It was just full of religious references, and the science was awful."

Eventually Wilson and Fisher did put together a curriculum that has noncreationist material as references. However, very little of the material comes from conventional scientific sources, and one article referred to is in *Reader's Digest.* "We had to produce something," says Wilson, "but it really isn't in a teachable form."

By now, however, this little farce at the school board level had been overtaken by weightier events at the state level, and the curriculum was temporarily shelved.

When news of Pulaski County school board's adoption of the creationist resolution hit the local papers in January the Reverend Blount was jerked into action. He searched through his papers and found the draft creationist bill that had arrived some 18 months earlier. "I believe that this is an idea whose time has come," he mused to himself.

"For more than 20 years I have been . . . trying to stop the teaching of evolution in public schools of Arkansas as a scientific fact," Blount stated when being deposed by ACLU lawyers before the recent trial of Act 590. Blount and a number of other ministers held seminars on scientific creationism (as it was termed at the time), and they put copies of *Twilight of Evolution* and *The Genesis Flood,* two creationist texts, into the libraries of every junior and senior high school in Pulaski County, with the approval of the school authorities. "We also donated a set of these books to every science and biology teacher in these schools."

Blount's long commitment to creationism and to promotion of its acceptance has been a quiet effort, and no one recalls any undue pressure on teachers and schools. Fisher's successful initiative was therefore a catalyst for further and different action.

Blount is president of the Greater Little Rock Evangelical Fellowship, which he describes as a loose alliance of ministers and others who believe in a literal interpretation of the Bible. The group meets regularly, and it happened that a meeting was imminent when reports of the Pulaski County resolution appeared in the newspapers. Blount suggested that the time had come for an initiative at the state level. The Arkansas legislature meets for just two months every two years, and by yet another of the many coincidences in this saga, it just happened to be in session during January and February of 1981.

Blount and his associates realized that they would have to act quickly if they were not to miss the current legislative session, and a two-man committee was set up to see what could be achieved. The committee was composed of the Reverend Curtis Thomas, of the Sovereign Baptist Church, and the Reverend W. A. Young of Bethel Chapel, vice president and secretary-treasurer respectively of the Evangelical Fellowship.

Thomas was the prime actor of the two, and the first thing he did was to contact Paul Ellwanger. "The material that Brother Blount had was a couple of years old and I didn't know . . . whether it had been changed," Thomas said during his deposition. "I didn't want to take to anybody a bill that had been declared unconstitutional." Ellwanger sent him the new version of the draft bill.

As he was an innocent in matters of legislation, Thomas consulted his friend Bill Simmons, an Associated Press reporter who works at the state capitol. "Simmons told me that the bill didn't have a chance, especially coming so late in the session," remembers Thomas. Nevertheless, Simmons gave Thomas the names of some legislators who might be willing to sponsor the bill. One of them was James Holsted. "I knew someone who was a business associate of Holsted," says Thomas, "so I contacted him and explained what I was trying to do." This was Carl Hunt, a businessman who knew his way around the legislative process and, more significantly as things turned out, a close friend of Frank White, the governor.

Hunt and Holsted met with White before the bill was introduced into the legislature, to "encourage him to sign such a bill," says Hunt. White, who describes himself as a "born-again" Christian, owed political debts to the Moral Majority for their efforts in helping him get elected, and he saw his endorsement of the bill as a way of paying some of these. He told Hunt and Holsted that he would sign the bill if it came to his desk.

Meanwhile, Thomas had met with Fisher to talk about the experience with the school board and to discuss prospects for state legislation. "He said we were wasting our time," recalls Thomas.

"He didn't encourage me." Although Fisher was pessimistic at that point, he gained strength for his own efforts, and it was this encounter that encouraged him to push the school board resolution in spite of the curriculum committee's negative reaction.

Holsted said he would sponsor the bill because it reflected his deep religious convictions, and he introduced it into the senate on 24 February. The bill was read a first and second time and then referred to the judiciary committee, on which Holsted serves. The committee's chairman, Max Howell, is the senior legislator in the senate and happens to be Holsted's neighbor in the chamber (such things are important in political spheres of influence). Like White, Howell is a born-again Christian, and he was happy to encourage the bill's progress from the committee with a "Do pass" recommendation, which happened on 3 March.

By now both the Moral Majority and FLAG were beginning to organize their forces in support of the bill, but it was not until the measure reached the house that their lobbying efforts became clearly overt. Members of both groups were in the house on the day that the bill reached the education committee, 13 March, and also when it received its third and final reading on the floor, 17 March. "There were six or seven of us on that last day," says the Reverend Roy McLaughlin, leader of Arkansas' Moral Majority. "We each had a roster of representatives so we could call them off the floor to ask them to vote for a suspension of the rules that would allow the bill to be read. It was tremendously effective."

Fisher testified in favor of the bill in the education committee hearings, as did Holsted. Two people spoke in opposition. The bill went out of committee with a "Do pass" recommendation on a voice vote.

Representatives had faced a barrage of telephone calls, particularly those who serve on the education committee. "I must have had 60 to 70 calls," says Representative Bill Sherman. "In the end I stopped returning the calls. The messages simply said, 'vote for bill 482.' " According to Representative Mike Wilson, "the calls were clearly orchestrated." It was, he says, a classic example of the activity of a single issue group.

At least part of that orchestration was directed by Thomas, who wrote to members of the Evangelical Fellowship and to other sympathizers and handed notices about the impending bill to his congregation. "We think the governor will sign the bill should it reach his desk," he noted in one letter.

The effort worked, with votes of 20 to 2 in the senate and 69 to 18 in the house. "When you get a mass of phone calls in favor of a bill and none against, and when it appears to be in support of motherhood, apple pie, and the American way of life, it is hard to vote against it," says Wilson. He did, but most of his colleagues took the easier route of "voting for God," as many of them put it.

With time for reflection since those frantic last few days of the session, a large number of legislators have indicated that they might have acted in haste. Ben Allen, president pro tem of the senate, has publicly stated that it was a mistake. "It looks fair and right on the surface," he told this reporter, "but when you probe into it, it begins to look wrong." For his public recantation Allen earned himself an ominous rebuke from Mary Ann Miller, a leading figure in FLAG: "He has marked himself for obscurity," she said in her deposition to the ACLU lawyers.

As soon as the bill became law, Ed Gran and Roy McLaughlin set up a group that came to be called the Arkansas Citizens for Balanced Education in Origins. The group was to promote teaching of creation science, and to this end it invited Bliss, from the Institute for Creation Research, to give a seminar on 22 April.

By this time it was clear that the American Civil Liberties Union was going to challenge the law in the courts and therefore the citizens group budded a second organization: the Creation Science Legal Defense Fund. The fund was to raise money and organize support for the defense of the statute, specifically by engaging the services of creationist lawyers Wendell Bird and John Whitehead.

Not only did the ACLU challenge promote the proliferation of creationist groups, but it also caused the membership to shake down in an intriguing way. "The ACLU was looking for a religious connection in this bill, so I dropped out of the citizens group," says McLaughlin. "I insisted on this," says Gran. "The citizens group is principally interested in science, while the defense fund has more religious connections."

Bird was particularly influential in shaping the response to the ACLU challenge in Arkansas, not only in helping establish the legal defense fund but in ensuring that Blount's Evangelical Fellowship became invisible. "He talked to us about our position as a group of ministers," said Blount in his deposition. "We agreed at that time to withdraw from any public action because it was not our purpose to inject religion into this. . . . We did not want to prejudice the case."

In the event the legal defense fund was thwarted in its attempts to take part in the creation trial in December. The attorney general, Steve Clark, said that he did not need help and Judge William Overton would not allow any outside intervention in any case. Toward the end of the second week of the trial, when prospects looked bleak for the state, the defense fund issued a blistering attack on the attorney general for his alleged poor handling of the case. But the organization had lost interest in Arkansas and had already focused on Louisiana.

Just before the Arkansas trial began, Bird and Whitehead, in conjunction with the Louisiana attorney general, filed suit in Baton Rouge, asking for declaratory judgment on the constitutionality of the state's Balanced Treatment law (the law is very similar to Arkansas' Act 590). Bird will be on leave from the Institute for Creation Research while he fights the case and will be supported by the legal defense fund, of which there will be a local chapter in Louisiana. Meanwhile, the original Creation Science Legal Defense Fund takes on the look of a national organization, with such notable figures as Duane Gish, Henry Morris, and Tim La Haye serving on the board. The organization clearly anticipates more battles elsewhere.

The last twist in this saga is Whitehead's contacts with the national Moral Majority. Jerry Falwell, the movement's national leader, asked Whitehead for help in establishing a group that will be called the Moral Majority Legal Defense Organization. Whitehead was asked to head the organization, but he declined because he did not wish to be seen to be too closely associated with the Moral Majority. The organization's function will be to counter what are perceived as assaults on the Christian viewpoint by the ACLU.

Although the passage of Arkansas' Act 590 may not have been a true conspiracy—"It all just came together," insists Thomas—there are an awful lot of interesting connections.

ASHLEY MONTAGU

INTRODUCTION

> . . . But man, proud man!
> Drest in a little brief authority,
> Most ignorant of what he's most assur'd,
> His glassy essence, like an angry ape,
> Plays such fantastic tricks before high heaven,
> As make the angels weep.
> Shakespeare, *Measure for Measure,* II, 2.

> The god who is reputed to have created fleas to keep dogs from moping over their situation must also have created fundamentalists to keep rationalists from getting flabby. Let us be duly thankful for our blessings.
> Garrett Hardin

There has arisen, in recent years, in the United States, a group of people who call themselves "Scientific Creationists." They have an Institute in San Diego, California, which serves them as headquarters, from which they issue a variety of publications, and send out lecturers to debate scientists who are naïve enough to believe that they can briefly state the case for evolution and convincingly dispose of the criticisms raised by these self-styled "Scientific Creationists." Several such evolutionists, including the writer, have lived to regret their folly. The conditions of such a debate essentially present the scientist with a "no-win" situation. It is quite impossible within the half-hour or forty-five minutes available to respond adequately to the criticisms of "Scientific Creationists" relating to the gigantic complex jigsaw puzzle that is evolution. It has taken generations of devoted workers to put the pieces of that puzzle together, involving such fields as astronomy, biology, chemistry, biochemistry, physics, paleontology, geology, genetics, anthropology, geochemistry, geophysics, dating methods, and many others, to help in the solution of that puzzle. There remains, and will always remain, a great deal of tidying-up to do. That tidying-up consists of filling in gaps, not so much for which there are no pieces, but in finding out the ways in which, in the state of nature, the pieces were actually put together. Scientists have gone a long way toward unraveling that mystery, but the work to be done before anything resembling a complete answer is found—if, indeed, that will ever become possible—is prodigious. That we shall ever achieve a complete answer is doubtful. It is too much to expect. What we do have is incontrovertible proof of the fact of evolution, namely, that genetic changes have come about in populations which have resulted in the great variety of plant and animal forms on this Earth, or put more simply, the transformation of the form and mode of existence of organisms in such a way that the descendants differ from their predecessors.

It is because it is not possible to respond adequately to the claims and to the criticisms made by "Scientific Creationists," of evolutionary facts and theories within the brief compass of a verbal debate, that it has been thought desirable to make those responses and clarifications in a more adequate and permanent form. The responses and clarifications that constitute the present volume may not quiet the "Scientific Creationists," but the hope is that they will be of help to all who seek a clarification of the issues involved.

A brief account of the rise of so-called "Scientific Creationism" may serve to orient the reader to the history of this sectarian movement.

The Creation Research Society was founded in 1963 by a dozen or so disaffected members of the American Scientific Affiliation, an organization founded in 1951 to bring about a union between Evangelical Christianity and science. Professor Laurie Godfrey deals with this aspect of the subject in her essay in this volume. "Scientific Creationists" now have organizations in a number of cities, chief among which is the Institute for Creation Research and the Creation Science Research Center situated in San Diego. The declared purpose of these organizations is to promote the teaching of "Scientific Creationism" or "Creation" on a par with evolution. Bills have been, and are being introduced in fifteen state legislatures, and have been passed into law in Arkansas, Mississippi, and Louisiana, requiring that wherever in public schools evolution is taught "Creation Science" also be taught. Each bill is entitled "Balanced Treatment for Creation Science and Evolution Science Act." The Arkansas act, in a historic decision has been declared unconstitutional. That decision is reprinted in full on pp. 365–97, and it is magnificent. On the day that Judge Overton handed that decision down, January 5, 1982, the State of Mississippi enacted into law "Balanced Treatment for Creation Science and Evolution Science Act." The manner in which such bills are created and carried through the legislature, and signed by the governor, is detailed in Roger Lewin's contribution on pages vii–xvii. It constitutes a shocking commentary on the legislative process in the United States, not to mention the failure of education.

The "Scientific Creationists" are fundamentalists and have repeatedly gone on record to the effect that "Bible-believing students of the biological sciences possess a guide for their interpretation of the available data, the biblical record of divine creation contained in Genesis."[1] This statement, made jointly by two prominent officials of the Creation Science Research Center at San Diego, represents the linchpin of the whole edifice the "Scientific Creationists" have erected. It is the basic dogma from which all their arguments and criticisms originate, and which fuels all their criticisms of evolution. It is an old story, long associated with fundamentalism, and well discussed by Professor George Marsden on p. 99, and in his excellent book on the same subject.[2] It is also ably discussed by Professor Sidney Ratner in his fascinating article, published as long ago as 1936, on pp. 398–415.

It is of some interest to note that when the "Scientific Creationists" speak of divine creation in Genesis, they fail to mention to which creation story they are referring. With respect to the creation of humankind there are three separate creation stories in Genesis: the first is in Genesis I and is completed in Genesis 2:26 and 23; the second 21 is in Genesis 2:7, and the third in Genesis 2:18–22. In the first story God creates man and woman (1:27), whereas in the second he forms man alone of the dust of the earth (2:7), and in the third he creates woman from one of man's ribs (2:21–22).[3]

As a student of biblical literature I see these three different creation stories as probably derived from three distinct traditions, combined by the authors of Genesis into a single chapter, without fusing the stories into a single consistent narrative. To folklorists and ethnologists this is not an unfamiliar turn of events; in many creation myths, and whatever the historical antecedents of Genesis,[4] it represents but one of innumerable creation myths which different peoples at different times have invented in order to account for the manner in which Earth and everything upon it came into being. That supernaturals formed the first men and women out of dust or clay forms a part of the creation myths of many peoples.[5] As Sir James Frazer remarked, to suppose that a world circumscribed in space and time was created by the efforts or the fiat of a being like himself imposed no great strain upon the credulity of our early compeers.[6] It certainly does not place a great strain upon the credulity of millions of our contemporaries.

To make men and women out of the dust or clay of the ground is a natural idea to occur to a reflective mind, for the ground is both elemental and firm, yet its products are evanescent, as is the life of man.

A charming example of this kind of creation myth is that of the Ewe-speaking people of Togoland, in West Africa, for they believe that God still makes humans out of clay. "When a little of the water with which he moistens the clay remains over, he pours it on the ground and out of that he makes the bad and disobedient people. When he wishes to make a good man he makes him out of good clay; but when he wishes to make a bad man, he employs only bad clay for the purpose. In the beginning God fashioned a man and set him on the earth; after that he fashioned a woman. The two looked at each other and began to laugh, whereupon God sent them into the world."[7]

It is not to be wondered at that the Ewe-speaking people are still notable, in spite of serious tribal disruption by whites, for their joviality.

If people choose to believe that Genesis represents an account of the manner in which the natural world was created, they should be free to do so; they should not, however, be free to impose such creation myths upon others.

Fundamentalists may believe what they choose, but they have

no right to insist that their particular views of origins be taught in the public schools. They have no right, first because those views are religious, second because under the guise of "creation science" they clearly desire to smuggle the teaching across the borders guarded by the First Amendment of the Constitution, and third because while there can be not the slightest objection to teaching Genesis in our public schools, as a creation myth among many others of a similar kind in courses on anthropology, it is the sheerest humbug to claim that such stories have anything to do with science.

It should, of course, be understood that a great deal more than simple cosmological explanation is involved in creation myths, for they usually have deep emotional and individually meaningful significances for the lives of the believers. To such believers their views are not myths but eternal truths, and it is the nature of such truths that they are subject neither to verification nor falsification. If that is so, it may well be asked, why the present book? The answer to that question is, that since the creationists have claimed their beliefs to be scientific, and have at times stated that evolution is unscientific, and have gone so far as to term it a religion, it is necessary to set the record straight for readers who may be interested in the truth of the matter. This is rendered all the more necessary in view of the political pressures this small sect has been bringing to bear upon legislators who are often quite ignorant of the facts of evolution, and are only too ready to yield to those who have a vested interest in seeing their views prevail.

The "Scientific Creationists" may call themselves so and refer to their manipulations as "creation science," but they are no more scientific than Christian Scientists or Scientologists. A scientist is characterized neither by a willingness to believe or a willingness to disbelieve, nor yet a desire to prove or disprove anything, but by the desire to discover what *is,* and to do so by observation, experiment, verification, and falsification. So doing, the scientist expects that others will take the trouble to check his findings, for it is only by such independent testing that his findings can be verified.

As a scientist, no scientist can be a fundamentalist in his attitude toward truth. Scientists do not believe in fundamental and absolute certainties. For the scientist, certainty is never an end, but a search; not the ordering of certainty, but its exploration. For the scientist, certainty represents the highest degree of probability which attaches to a particular judgment at a particular time level, a judgment or conclusion that has been arrived at by experiment, inference, or observation, for all good observation and inference, in fact, involves experiment of some sort, by a consensus of independent scientists. Scientists lack a superstitious regard for the catchwords of science, and believe that all knowledge is infinitely perfectible. As Leonardo put it,

> Wisdom is the daughter of Experience,
> Truth is only the daughter of Time.

To quote once more, the scientist finds

> Life's treasure is an endless quest,
> And peace of mind is infinite unrest.

There are people who have taken degrees in one or another science or technology, but that doesn't necessarily make them scientists. They may be scientists in the practice of their particular science, whatever it may be, but the moment they step out of their special field of competence their opinions on any other field or subject are not necessarily of any more value than those of the proverbial man-in-the-street. Their opinion can be of value only when it is based on the application of scientific method to whatever it is they are evaluating. "Scientific Creationists" claim that they have many members who hold degrees in various sciences and technologies, as if that automatically conferred unarguable authority on their judgment of matters scientific and unscientific generally. From the study of one scientific subject to others, or to anything else for that matter, there is no necessary transfer of insight or skill. One may be a chemist, physicist, mathematician, engineer, or whatever, with a Ph.D. or without, and remain nothing but a glorified technician who is unable to see the wood for the trees, that is, unable to think scientifically outside the narrow bounds of his own specialty. Quite possibly there are as many bigots among such "scientists" as there are among other classes of people. It was Oliver Wendell Holmes, Sr., who likened the bigot to the pupil of the human eye: the more light you expose it to the narrower it grows. Bigotry and science can have no communication with each other, for science begins where bigotry and absolute certainty end. The scientist believes in proof without certainty, the bigot in certainty without proof. Let us never forget that tyranny most often springs from a fanatical faith in the absoluteness of one's beliefs.

I am not aware of any observations or experiments that the "Scientific Creationists" have carried out in order to support or reinforce their claims, and therefore there must be the strongest objection to their terming what they are doing "science" and to calling themselves scientists. Some creationists have now admitted that creationism is not really science, yet they persist in calling themselves scientists. What they are in fact engaged in is special pleading for a religious myth which they wish to foist upon the public as a fact, an historic occurrence.

By Orwellian "Doublespeak," the creationists—as I will hereafter call them, for they are in no sense scientists—easily persuaded the more than willing legislators of the State of Louisiana that evolution is really a religion, the evolutionists a venality of narrow-minded dogmatists; that creationism is a science, and fundamentalism, freedom. Senator William Keith, a verbose lay minister from Mooringsport, Louisiana, sponsored Senate Bill 89, which found overwhelming support in both House and Senate and was signed into law on July 21, 1981. All that Senator Keith asked was "that creation-science receive balanced treatment" with "evolution-science," that wherever in the schools "evolution-science" is taught, "creation-science" also be taught.

This is, of course, the equivalent of requiring that wherever chemistry is taught alchemy should be given equal time, or wherever psychology is taught phrenology be given balanced treatment, or astrology be given equal time with astronomy.

Science cannot by any stretch of the imagination be referred to as a religion. Religion is the belief in supernaturals. Science deals with the world of nature, the discovery and ordering of the world of facts and their relations, with concepts that have been tested by the facts. What creationists attempt to do is to measure the facts by their conformity to Genesis, and this is absurd. I gather that creationists are willing to go so far as to grant that our planet may be older by some thousands of years than Bishop Ussher's 4004 B.C. However that may be, generations of scientists, with increasingly refined methods and instruments, have arrived at measurements which yield an age for the Earth, at its lower limits, of 4.5 billion years.[8] Every scientific study confirms that figure.

The creationists go on to claim that even that amount of time is insufficient to account for the evolution of so many different forms of life. They assert that mathematicians have shown that it is impossible. Something must be wrong with such statements, for in the first place evolution represents an outstanding example of the maximization of the improbable, and in the second place, evolution has proceeded at very rapid rates in the development of some forms as compared with others.[9] Quantum evolution, that is, very rapid evolution within a small population, leading to a new plant or animal form with no intermediate transitional forms occurring, has been with us for some time,[10] and in its contemporary form, punctuational evolution, there is now very good reason to believe that speciation has occurred at very rapid rates in many instances.[11]

Another venture into their failure to understand the facts is the claim by the creationists that evolution is contrary to the second law of thermodynamics, that energy tends to become unavailable for use, and that evolution is therefore a completely untenable notion. As Henry M. Morris, a leading creationist puts it, "evolution in the 'vertical' sense (that is, from one degree of order and complexity to a higher degree of order and complexity) is completely impossible."[12] Such ideas are quite unsound.[13] What may be true of physical systems is not true of living ones, for living organisms are continually taking in energy from the environment and using it very efficiently to live and reproduce. The universe may eventually "run down" but the movement of life is in the opposite direction. Furthermore, creationists claim that the order of evolution cannot be derived from disorder, the disorder which is said to have characterized Earth from its early beginnings. Again, the truth is otherwise, for it is from random mutation, in a highly disorderly and unpredictable process, that the environment has selected for survival those genes and genetic systems that possessed adaptive value, and permitted others, under conditions of genetic drift, to become stabilized. Life on earth is the product of inconceivably fortunate accidents—that is, if we consider the variety and richness of life on Earth fortunate.

The work of Ilya Prigogine, the Nobel laureate in chemistry for

1977, thermodynamics constitutes an utter repudiation of the creationist claims.[14]

One of the most frequently reiterated false claims among the collection of canards to which the creationists are addicted, is that the geological record shows few if any transitional fossil forms, and that this constitutes the best evidence for special creation. This is quite untrue. There is a wealth of evidence of such transitional forms from the geological record, and Professor Cuffey in his contribution to the present volume has presented many examples of them. There are many others, perhaps the most spectacular of which is represented by our own genus *Homo* and its immediate forerunners, the fossil manlike forms of East and South Africa and Ethiopia, the australopithecines. There are transitional forms among the australopithecines themselves, as well as between them and humans,[15] and among the latter from *Homo erectus* (formerly known as *Pithecanthropus erectus*), through Neandertal man, and from Neandertal man, to forms very like ourselves such as those from Tabūn and Skhūl in Israel, to contemporary man. To those who erroneously point to the alleged paucity of transitional forms allegedly disproving the reality of evolution, Edwin Drinker Cope, the great American paleontologist, had already returned the perfect answer a century ago.

> It is true [he wrote] that the cases of transition, intermediate forms, or diversity in the brood, observed and cited by naturalists in proof of evolution, are few compared with the number of well-defined, isolated species, genera, etc., known; though far more numerous than the book-student of natural history is apt to discover. But although the origin of most species by descent has not been observed, every one knows the worthlessness of argument based on a negative. Unless these cases exhibit opposing evidence of a positive character, they are absolutely silent witnesses.
>
> He who cites them against evolution commits the error of the native of the Green Isle who testified at a murder trial. "Although the prosecuting attorney brought three witnesses to swear positively that they saw the murder committed, I could produce *thirty* who swore that they did not see it done!"[16]

Cope was too generous in observing that "every one knows the worthlessness of argument based on a negative." Unfortunately, not everyone does, and most especially creationists do not.

In addition to quantum or punctuational evolution, there is another important evolutionary process which serves to explain how it comes about that, in many cases, transitional forms will never be found between one genus or species and another. This developmental process is known as *neoteny* or *paedomorphosis,* the process whereby evolutionary change occurs as a consequence of descendants retaining the larval, fetal, or juvenile traits of their immediate ancestors into adult life. This is in part achieved by the stretching out of each period of development. In neoteny or paedomorphosis there is a displacement of ancestral features to later stages of development. Certain ancestral adult traits are, as it were, "pushed off" the end of individual development and therefore do not appear, thus leaving a "gap" between ancestral and descendant forms. As Julian Huxley put it: "Previous adult characters . . . never appear because their formation is too long delayed: they are lost to the species by being driven off the time-scale of its development."[17] And, again, in another work, "The old characters may be swept off the map and be replaced by characters of a quite novel type."[18] Some years later, the distinguished embryologist Sir Gavin de Beer was even more explicit. De Beer showed that neoteny allowed for the possibility of evolution occurring "clandestinely," in the young stages of the individuals of a phylogenetic line, the changes becoming revealed when the old adult stages are discarded. In other words, new features of adult organization are often derived from characteristics initially evolved in embryonic, fetal, or youthful stages of the ancestors. A consequence of this mode of evolution would be that the preliminary steps in the evolution of successful major groups would take place in young stages, which would be unlikely to be preserved as fossils, and would thus remain forever unknown, and as de Beer concluded: "The sudden appearance of new types, and the apparent gaps in phylogenetic series, receive a logical explanation on these lines."[19]

The neotenic (paedomorphic) mode of evolution enables ancestral specializations to become discarded or lost. As de Beer and others have made clear, this means that new trends of evolution can be superimposed upon the old, so that the old traits are supplanted by the youthful ones.[20]

It is of great interest here to note that Darwin had already made this point in the 6th edition of *The Origin of Species* (1872), even before the term "neoteny," or paedomorphism, had been coined, when he wrote that "if the age for reproduction were retarded, the character of the species, at least in its adult stage, would be modified; nor is it impossible, that the previous and earlier stages of development would in some cases be hurried through and finally lost."[21]

It should furthermore be said that the remains of many animals ancestral to others have over the course of millions of years disintegrated and left no trace behind. For the remains of animals to undergo fossilization rather special conditions are required; when such conditions are wanting, the remains may completely disappear. The inevitable incompleteness of the fossil record, as Stebbins has pointed out, is dramatically illustrated by the fact that while millions of carrier pigeons flourished in North America only a hundred years ago, no one has ever come upon the fossilized remains of a single representative of these extinct birds.[22]

The creationists repeatedly assert that evolution is a theory, and, indeed, not a scientific theory at all because it is unable to predict future developments. Both statements are untrue. Evolution is a fact, not a theory. It once was a theory but today, as a consequence of observation and testing it is probably the best authenticated actuality known to science. There are theories concerning the mechanisms of evolution, but no competent student doubts the reality of evolution.

Scientists in all fields are able to predict a great deal that would otherwise remain unpredictable; they cannot predict everything from their particular specialty, but the coefficient of predictability increases with increase in knowledge and experience. In the field of evolutionary studies that coefficient has grown considerably.

The physical basis of evolution is the gene, the gene and its interaction with the environment. Because this is so, scientists are able to study the mechanisms of evolutionary change at both the individual and the population levels. It is by this means that they are able in many cases to predict not only what will occur by way of physical development, in individuals, families, and populations, but also to predict the exact frequencies with which such changes may be expected. In many instances it is, by this means, possible to forestall and prevent the development of various conditions. The laws of genetics, for example, are not only statements of fact, but they and their effects are dependably predictive.

In conclusion, on the matter of religion and science, it needs to be said that there is no real incompatibility between the two. There is no incompatibility between a belief in God and the belief that evolution is the means by which all living things have come into being. What *is* incompatible with science, religion, and civility is the attempt by a narrow fundamentalist sect to impose its particular brand of a creation myth as a substitute or alternative to the findings of science, and to insist on having that myth taught as a fact in the schools, to the exclusion of all other religious teachings.

That evolution is beyond all else a creative process has never been doubted by anyone acquainted with the facts. One of the earliest books on the subject, written by the distinguished French philosopher and Nobel laureate Henri Bergson (1859–1941) was entitled *Creative Evolution* (1907), a book which William James termed "one of the great turning-points in the history of thought. . . . It tells of reality itself." Bergson wrote of the *élan vital,* the creative force operative in nature, through the process of evolution. There have been many such books on the same themes since. In this sense Darwin remains the most clear-headed and foremost creationist.[23] This, of course, in no sense identifies him with the so-called "Scientific Creationists." Darwin was, as he described himself in his autobiography,[24] an agnostic, who saw evolution as a creative process. As he wrote of the evolutionary process in the concluding passage to the first edition (1859) of *The Origin of Species.* "There is a grandeur in this view of life, with its several powers, having been originally breathed into a few forms or into one; and that, whilst this planet has gone cycling on according to the fixed law of gravity, from so simple a beginning endless forms most beautiful and most wonderful have been, and are being evolved."[25]*

In conclusion, the purpose of this book may be said to affirm the statement made by Chauncey Wright in 1865:

* In later editions, as a concession, probably, to his wife's religious views, Darwin added the words "by the Creator" between the words "breathed" and "into."

If the teachings of natural theology are liable to be refuted or corrected by progress in knowledge, it is legitimate to suppose, not that science is irreligious, but that these teachings are superstitious; and whatever evils result from the discoveries of science are attributable to the rashness of the theologian, and not to the supposed irreligious tendencies of science.[26]

FOR FURTHER READING

P. W. Atkins, *The Creation.* Oxford & San Francisco, W. H. Freeman, 1981.

Eldredge, Niles. *The Monkey Business: A Scientist Looks at Creationism.* New York, Washington Square Press, 1983.

Douglas J. Futuyma, *Science On Trial: The Case For Evolution.* New York, Pantheon, 1983.

Laurie R. Godfrey (ed.), *Scientists Confront Creationism.* New York, W. W. Norton, 1983.

C. Leon Harris, *Evolution: Genesis and Revelations.* Albany, State University of New York Press, 1981.

Philip Kitcher, *Abusing Science: The Case Against Creationism.* MIT Press, Cambridge, 1982.

Nelkin, Dorothy, *The Creation Controversy.* New York, W. W. Norton, 1982.

Normand D. Newell, *Creation and Evolution: Myth or Reality?* New York, Columbia University Press, 1982.

NOTES

1. R. E. Kofahl and K. L. Segraves, *The Creation Explanation* (Wheaton, Ill.: Shaw, 1975), p. 69.
2. George W. Marsden, *Fundamentalism and American Culture: The Shaping of Twentieth-Century Evangelicism 1870–1925* (New York: Oxford University Press, 1981).
3. Sir James Frazer, *Folklore in the Old Testament,* 3 vols. (London & New York: Macmillan, 1919); Jack P. Hailman, "Creation Stories," *BioScience,* vol. 32 (1981), pp. 120–30.
4. F. B. Jevons, *The Idea of God in Early Religions* (Cambridge: at the University Press, 1910); Sir James Frazer, *Folklore in the Old Testament,* vol. 1; J. Strachan, "Creation," in James Hastings (ed.), *Encyclopaedia of Religion and Ethics,* vol. 4 (New York: Scribner's

Sons, 1911), pp. 226–31; W. Robertson Smith, *The Religion of the Semites,* revised and enlarged by S. A. Cook (London & New York: Macmillan, 1927); Paul Radin, *Primitive Man as a Philosopher* (New York: Appleton-Century, 1927); Sir James Frazer, *Creation and Evolution in Primitive Cosmogonies* (London & New York: Macmillan, 1935); Sir James Frazer, *The Dying God,* vol. 4 of *The Golden Bough* (London & New York: Macmillan, 1935); Samuel Noah Kramer (ed.), *Mythologies of the Ancient World* (New York: Anchor Books/Doubleday, 1961); Philip Freund, *Myths of Creation* (London: W. H. Allen, 1964); Barbara C. Sproul, *Primal Myths: Creating the World* (San Francisco: Harper & Row, 1979); Charles Doria and Harris Lenowitz, *Origins: Creation Texts from the Ancient Mediterranean* (New York: Anchor Press/Doubleday, 1976); Paul Radin, *The World of Primitive Man* (New York: Grove Press, 1960).

5. Ibid.
6. Sir James Frazer, *The Dying God,* p. 109.
7. Sir James Frazer, *Creation and Evolution in Primitive Cosmogonies,* London & New York, Macmillan 1935, p. 13. Frazer's account from J. Spieth, *Die Ewe-Stämme, Material zur Kunde des Ewe-Volkes in Deutsch-Togo* (Berlin, 1906), pp. 828, 840.
8. Patrick M. Hurley, *How Old Is the Earth?* (New York: Anchor/Doubleday, 1959); Eric Chaisson, *Cosmic Dawn: The Origins of Matter and Life* (Boston: Little, Brown, 1981); David M. Smith (ed.), *The Cambridge Encyclopedia of Earth Sciences* (New York: Crown Publishers/Cambridge University Press, 1981).
9. Niles Eldridge and Stephen Jay Gould, "Punctuated Equilibria: An Alternative to Phyletic Gradualism," in T. M. Schopf (ed.), *Models in Paleobiology* (San Francisco: Freeman, Cooper, 1972); Niles Eldridge and Joel Cracraft, *Phylogenetic Patterns and the Evolutionary Process: Method and Theory in Comparative Biology* (New York: Columbia University Press, 1981).
10. Steven M. Stanley, *The New Evolutionary Time Table* (New York: Basic Books, 1981); George Gaylord Simpson, *Tempo and Mode in Evolution* (New York: Columbia University Press, 1944).
11. Ibid.
12. Henry M. Morris, "Entropy and Open Systems," *Acts & Facts,* (October 1976), Impact No. 40.
13. William Thwaites and Frank Awbrey, "Biological Evolution and the Second Law," *Creation/Evolution,* vol. 4 (1981), pp. 4–7; Stanley Freske, "Creationist Misunderstanding, Misrepresentation,

and Misuse of the Second Law of Thermodynamics," *Creation/Evolution,* vol. 4 (1981), pp. 8–16.

14. See "Entropy," in *Academic American Encyclopedia,* vol. 7 (Princeton, N.J., 1980), p. 209.
15. Richard E. Leakey and Roger Lewin, *Origins* (New York: Dutton, 1977); Richard E. Leakey, *The Making of Mankind* (New York: Dutton, 1981; C. Loring Brace and Ashley Montagu, *Human Evolution* (New York: Macmillan, 1977); John Reader, *Missing Links: The Hunt for Earliest Man* (Boston: Little, Brown, 1981).
16. Edwin Drinker Cope, *The Origin of the Fittest* (New York: Appleton & Co., 1887), pp. 5–6.
17. Julian Huxley, *Problems of Relative Growth* (New York: Lincoln McVeagh, 1932), p. 239.
18. ——— *Evolution: The Modern Snythesis* (New York & London: Harper & Brothers, 1942), p. 532.
19. Gavin R. de Beer, "Embryology and the Evolution of Man," in Alex L. Du Toit (ed.), *Robert Broom Commemorative Volume* (Cape Town: Royal Society of South Africa, 1948), p. 184.
20. ——— *Embryos and Ancestors,* 3rd ed. (New York: Oxford University Press, 1958; see also Stephen Jay Gould, *Ontogeny and Phylogeny* (Cambridge: Harvard University Press, 1977), and Ashley Montagu, *Growing Young* (New York: McGraw-Hill Book Co., 1981).
21. Charles Darwin, *The Origin of Species,* 6th ed., London, John Murray, 1872, p. 160.
22. C. Ledyard Stebbins, *Darwin to DNA, Molecules to Humanity* (San Francisco: W. H. Freeman, 1982).
23. C. S. Gillispie, *Charles Darwin and the Problem of Creation* (Chicago: University of Chicago Press, 1981).
24. Norah Barlow (ed.), *The Autobiography of Charles Darwin 1809–1882* (London: Collins, 1958), p. 94.
25. Charles Darwin, *The Origin of Species* (London: John Murray, 1859), p. 490.
26. Chauncey Wright, "Natural Theology as a Positive Science," *North American Review* (1865), vol. 100, p. 184.

KENNETH R. MILLER

SCIENTIFIC CREATIONISM VERSUS EVOLUTION: THE MISLABELED DEBATE

The history of science is a history of conflict. No new idea has ever been developed without challenging an old idea. In the struggle that ensues, scientific ideas are born and discarded, built up, and torn apart. From time to time, the old order is overthrown (sometimes with great drama), and the body of ideas and statements that we call science changes, in some cases actually from day to day. Everyone who works in science is familiar with that kind of conflict. The stuff of scientific research is essentially the work of those who seek facts—by experiment or observation—which will either support or undermine an existing scientific concept about the nature of things. Remarkably, it is the latter, the experiment which overthrows the established order, that ambitious young scientists burn the lights of their laboratories searching for. Nothing establishes the personal reputation of a scientist quite as quickly as producing a new set of results which upsets an existing theory, or producing a radically new theory which displaces an older one because it fits the facts much better. In science, as in perhaps no other profession, the ambitions of nearly all the practitioners of the craft are to be revolutionaries!

One popular characterization of the great scientists of the past—honed by classics such as Paul deKruif's *Microbe Hunters* (1935) and Jacob Bronowski's remarkable television series *The Ascent of Man* (1974)—is that of a lonely rebel struggling against the scientific establishment of his day. Tireless, determined, and persistent, he endures rejection, ridicule, and exile from the scientific community in order to break the new ground which allows science to advance. We have so many examples of the final triumph of such lonely pioneers that the scientific community has developed a kind of automatic fondness for the intellectual rebel and the eccentric spirit. We take great care that scientific meetings are open to all who wish to present their theories; we provide careful levels of appeal and review for those who object when a paper is not accepted or a grant request declined; and we are mindful, above all, that today's well-established theory may be only years away from a sudden demise at the hands of the eager group of youngsters who are ever forcing their way into the scientific establishment. For this reason, as many writers on the nature of the scientific endeavor have allowed, there are certain positive political and social values which the scientific enterprise demands, even if that enterprise takes place in an otherwise totalitarian society. The foremost of these is the principle of free expression. Whatever our political attachment to this ideal, the process of science depends on freedom of expression almost as much as it depends on the scientific method itself. Censorship and control of scientific expression are objectionable in science not merely because we believe that freedom of expression is a basic human right, but also because such controls stifle the advance of science itself. Science depends on the freedom to criticize the currently accepted theory, to test it by experimentation, and to propose a new theory, not without criticism, but without hindrance.

In recent years in the United States we have been confronted by a well-organized group calling themselves "scientific creationists" who propose to challenge one of the most basic ideas in the biological sciences: Darwin's theory of evolution. These people are well read, well funded, and fully appreciative that the open character of science all but demands them a hearing in the scientific community. Indeed, in a debate with Prof. Russell Doolittle of the University of California, Dr. Duane Gish of San Diego's Institute for Creation Research (ICR) made heroic comparisons with great scientists of the past:

> Ladies and gentlemen, three or four centuries ago, the notion that the sun and other planets revolved around the earth was the

> dogma of the scientific establishment. Galileo faced determined opposition from fellow astronomers when he suggested otherwise. Louis Pasteur and others, about a century ago, overturned the established dogma of centuries when they showed that living things never arose spontaneously from dead matter. Today, even though thousands of scientists are creationists, and the number is growing rapidly, the notion of evolution remains a stifling dogma.[1]

Even overlooking the gratuitous misreading of history concerning Galileo (whose most vicious opposition came not from the scientific but from the religious establishment), Gish seems to complain that science is closed to his theories. He should be aware, as other scientists are, that censorship is not even *possible* in scientific societies where every member is entitled to present a public paper at the annual meeting. Is a fair hearing in the scientific community the principal demand of the creationists? Apparently not. Federal Judge William Overton made the following observation in his written opinion throwing out the "creation-science" law in Arkansas:

> The journals for publication are both numerous and varied. There is, however, not one recognized scientific journal which has published an article espousing the creation-science theory described in section 4 [of the Arkansas "creation-science" law].
>
> Some of the state's witnesses suggested that the scientific community was "close-minded" on the subject of creationism and that explained the lack of acceptance of the creation-science arguments. *Yet no witness produced an article for which publication had been refused.*[2]

Obviously, Dr. Gish and his associates have not availed themselves of the open platforms of scientific meetings and societies (for reasons which will become clear) but have taken their case to the general public. In the political forum, they have asked that state legislatures write their ideas into school curricula in the name of fairness, and at this writing two states, Arkansas and Louisiana, had done exactly that. (The law in Arkansas, of course, was recently found unconstitutional as the result of a lawsuit brought against the State of Arkansas by the American Civil Liberties Union and several residents (including many clergy) of the State of Arkansas.

A similar lawsuit is currently pending in Louisiana.) In this chapter I will examine some of the ideas of the "scientific creationists," will discuss whether or not these ideas meet the standards of scientific inquiry, and will consider whether any of the points raised by these groups should teach a lesson to those of us in science.

The Scientific Creationists

We begin with a dilemma. Who are the creationists? Simply stated, a creationist should be anyone who believes in creation, in a universe formed by a supreme being. In other words, a creationist is someone who believes in God. By that standard of ordinary usage, I am a creationist (I'm a Roman Catholic), and so is any other scientist who professes a religious belief. However, in the context in which I must write this article, ordinary usage will not do. We will be forced to use another definition for the word *creationist,* a definition which has been forced on us by the current of the political debate in the United States. In this sense, a creationist is someone who believes that each and every kind of living organism was directly created by a supreme being, and that no organisms have arisen by the process of descent with modification advanced by Charles Darwin more than a century ago. In short, a creationist is an antievolutionist.

There are many groups who wear the creationist label with pride, and these groups often find themselves in agreement about little else other than the need to oppose evolution and the teaching of evolution. Some of these groups have moved so far beyond the limits of scientific inquiry as to make meaningful discussion absolutely impossible. I recently received a book from a creationist by the name of Partee Fleming entitled *Is God's Bible the Greatest Murder Mystery Ever Written?*[3] in which he argued (with little fact but much flamboyance) that dinosaurs and other extinct organisms in fact never existed, but were created as fossils in the earth in order to demonstrate the "wit of Jesus." I will not spend time here discussing the other grand pranks that Mr. Fleming believes the Creator may have played on us, but rather will turn instead to the scheme of natural history advanced by the Institute for Creation Research (ICR) of El Cajon, California. I do this in part because it is necessary to identify a set of creationist ideas in order to discuss modern creationism coherently, and in part because this institute has been the moving intellectual force behind the drafting of the "creation-science" laws in Arkansas and Louisiana and other states.

Henry Morris, Ph.D., is the director of the ICR, and Duane Gish, Ph.D., is its associate director. The ICR lists a professional staff of several scientists, all with advanced degrees, and claims a membership list of several hundred active "creationist scientists." Although the organization claims to be a purely scientific one, the religious character of its work is clearly evident in its literature. Monthly letters to "Friends" of the ICR are signed "Sincerely yours, in Christ," by Morris, and scientists applying for membership in the ICR must sign a statement attesting to the literal inerrancy of the Christian Bible. Although the ICR often emphasizes that it is the scientific nature of creationist theory which brings scientists to a belief in a supreme being, it is curious that they include a requirement for membership (the inerrancy of the Christian Bible) which effectively excludes Jews, Muslims, Hindus, Buddhists, and the majority of Christian sects (who do not accept a literal reading of all parts of the Bible) from membership. It is clear that the ICR, which is the most respected of creationist groups in its attempts to appear scientifically legitimate, is essentially an organization composed solely of Christian Fundamentalists.

Many scientists, once they recognize this fact, dismiss the creationists as an unimportant religious fringe group. An idea advanced by such a group is surely being promoted mainly for religious and political reasons, and most of us in science realize that such an idea is not likely to have any scientific foundation. The majority of American biologists have recognized the creationists for what they are—a religiously motivated group—and dismissed them. However, the creationists, realizing that the enormous weight of scientific evidence is stacked in favor of evolution, have taken a different route, one which demands the kind of response that scientists are not accustomed to giving.

The American creationist movement has entirely bypassed the scientific forum and has concentrated instead on political lobbying and on taking its case to a fair-minded electorate. In so doing, they have presented the interested layperson with a convincing scientific case (which looks at least as good as the case that evolutionists seem to be able to make), and asked, in the spirit of open-mindedness, for "fairness" or equal time in the presentation of what they call "creation-science" alongside of "evolution-science." When objections are raised on the grounds that "creation-science" has no general credence within the scientific community, the creationists are quick to raise cries of censorship, academic freedom, and even Lysenkoism. Pointing out that every available scientific forum is open to them is useless, because they are not interested in convincing their "peers" in the scientific community. They are aiming directly for the textbook and the school classroom. The reason for this strategy is overwhelmingly apparent: no scientific case can be made for the theories they advance. Therefore, rather than present these ideas to an audience of specialists (geologists, geneticists, molecular biologists, biochemists, and paleontologists) who would at once point out the factual contradictions of their ideas, they have chosen instead the general audience of interested public. In the public forum they can be confident that a specific, point-by-point refutation of their theories will not be made, and even if it is made, the scientists carrying it out may not be clear enough in their explanations to get the message across. How can such an approach be countered? By doing precisely what the creationist does not expect. By taking the scientific case to the public.

In the public forum it is important to accomplish several things at once. First of all, each of the creationist arguments against evolution should be answered in a clear and precise way. Second, the creationist scheme of natural history must be clearly exposed (something creationists avoid doing at all costs), so that the remarkable contradictions between it and scientific fact are easily seen by anyone caring to look. And finally, the attempt by creationists to pretend that evolution is an inherently atheistic theory (thereby calling all Christian citizens automatically to their camp) must be exposed and refuted. A critical fact which is often lost in the debate, namely the lack of conflict between modern science (including evolution) and belief in God, must be brought out. This final point must be made in order to expose the creationists for what they are—not scientists trying to leave a place for religion per se in the teaching of human origins, but rather people trying to inject a specific religion, Christian Fundamentalism, into the schools in the guise of science and to the exclusion of all other religions.

Creationist Arguments Against Evolution

My space in this volume is too limited to scramble about like a lone fielder at batting practice, chasing after every stray ball the creationists have hit in our direction. I will, instead, choose some of the major arguments they have made and see whether or not they have made valid scientific points. My sources for creationist arguments are some of the many writings and publications of the Institute for Creation Research. I will consider at length some of their arguments relating to the fossil record and the age of the earth, and will direct the reader to other sources for well-considered answers to other creationist criticisms.

The Age of the Earth

One of the great moving forces behind the development of modern geology was the gradual recognition, in the 18th and 19th centuries, that the earth was old, much older than the 6,000 or 7,000 years suggested by a literal reading of Biblical histories. This realization developed basically as a consequence of the industrial revolution and not because of the demands of any theory of evolution. The need to extract large quantities of ore and coal for the developing industrial age led directly to a rapid expansion of mining activities, resulting in a dramatic increase in earth-moving to make way for new roads, canals, and railways. The pioneers of the young science of geology were now able to examine large cross sections of rock and soil cut away by these activities, and began to realize that these enormous formations of great depth were put together in a way which suggested that they had been first formed by gradual sedimentary deposition and then uplifted by massive earth movements to form large parts of the land masses of the continents. The slow rate at which these sedimentary rocks seemed to have been formed (rapid sedimentation from great floods can be distinguished from gradual sedimentation by the sizes of sedimentary particles) suggested to Lyell and many other early geologists that it must have taken millions of years for these formations to develop prior to their uplifting to form modern land masses. They concluded that the earth must be as much as 10,000 times older than suggested by Biblical chronologies.

A knowledge of this enlarged time scale was already available to Charles Darwin when he took a copy of Lyell's *Principles of Geology* with him on board the *Beagle.* It is, of course, fair to say that an awareness of the earth's antiquity influenced him as he began to develop what we now call his theory of evolution. An immensity of time would surely have been required for the process of natural selection to cause the sort of descent with modification which was the very center of his work. Creationists are fond of pointing this out, and allow themselves, in contrast, to appear reasonable and open-minded:

> As a matter of fact, the creation model does not, in its basic form, *require* a short time scale. It merely assumes a period of special creation sometime in the past, without necessarily stating when that was. On the other hand, the evolution model does *require* a *long* time scale. The creation model is thus free to consider the evidence on its own merits, whereas the evolution model is forced to reject all evidence that favors a short time scale.[4]

This sort of open-mindedness, however, is essentially a rhetorical device to make the evolutionist seem to be bound by an intellectual straightjacket. The creationists do indeed have a real answer to the age of this planet, and elsewhere in their writings both the answer and its ultimate source are made quite clear:

> The only way we can determine the true age of the earth is for God to tell us what it is. And since He *has* told us, very plainly, in the Holy Scriptures that it is several thousand years in age, and no more, that ought to settle all basic questions of terrestrial chronology.[5]

Such statements are remarkably straightforward, because they illustrate quite clearly the motivation behind scientific creationism: the fervent desire to construct a pseudoscientific natural history which will not contradict a literal reading of Biblical history. Nonetheless, the age of the earth is a legitimate question of great scientific importance. We inhabit this beautiful planet, and we would like to know something about its history. We would like to know when it was formed. We would like to know if all the features of its surface were formed at the same time, and we would like to know the dates at which specific rocks holding the fossils of living organisms were laid down.

Is there a scientific way by which we can begin to answer these questions? There most certainly is. Since the discovery of radioactivity by A. Henri Becquerel in 1896, it has been obvious that the radioactive decay process provides an excellent opportunity for estimating the age of this planet and the dates of formation of specific rocks. For example, let us consider the process of radioactive decay starting with the isotope of uranium known as uranium 235 (^{235}U). This atom decays through a series of intermediates to an isotope of lead known as lead 207 (^{207}Pb). The half-life of this complete series is 713 million years. From this one fact, coupled with the observation that there are significant amounts of ^{235}U available in the crust of the earth, we can make a bold and unequivocal statement: The earth could not have existed forever!

Because the rate of decay of ^{235}U into ^{207}Pb is constant, if this planet had existed essentially "forever," then by now all of the ^{235}U would have decayed and we would not be able to find it in any significant amount. Our problems with the machinery and weaponry of the atomic age, however, show quite clearly that there is more than enough ^{235}U around. Therefore, the beginning of this planet is not in the infinite past. Can we go a little further and make some estimates as to just how long ago the planet might have been formed?

In order to do this, we must put together a method for estimating the amount of ^{207}Pb which was present when the planet was formed, and then using that amount to determine how much "new" ^{207}Pb has been added by radioactive decay over the ages. The initial ^{207}Pb is known as primordial lead, and the "new" ^{207}Pb is termed radiogenic lead. There is no way to distinguish chemically an atom of one from an atom of another, so a direct determination is not possible. Fortunately, there are several ways to make reasonable estimates of the amounts of primordial lead in the crust or in a specific rock. For example, one of the isotopes of lead, ^{204}Pb, is nonradiogenic, meaning that it is not formed by any known decay process. Therefore, all of ^{204}Pb is primordial. When a mineral is formed, it will incorporate Pb and U in a specific ratio, and among the lead molecules there will be a ratio of ^{204}Pb to ^{207}Pb determined by the abundance of these isotopes in the crust at that time. Once the mineral is formed, ^{235}U will decay to ^{207}Pb over time. Many, many years later a geologist may analyze the mineral, and if billions of years have passed, he will discover that (1) the mineral seems to be deficient in uranium, and (2) the ratio of ^{204}Pb to ^{207}Pb is very low. If the planet were recently formed, then all minerals should show lead isotope ratios and uranium/lead ratios which are reflective of the current abundances of these materials on the surface of the earth. Instead, the ratios indicate clearly that the most reasonable date for the formation of the planet is 4 to 5 billion years ago.

There are problems with the uranium methods, and one is the need for certain assumptions (no matter how reasonable) concerning the amount of primordial lead. However, there are other methods which do not require any assumptions and from which extremely reliable dates for the formation of rocks can be determined. One of these is the potassium-argon method, which measures time by the production of argon 40 (^{40}Ar) from potassium 40 (^{40}K). For the determination a crystalline mineral can be chosen whose chemistry demands a specific number of potassium atoms at fixed positions in the lattice of the crystal. One can then determine the actual amount of ^{40}K still present and the amount of ^{40}Ar trapped in the crystal. Because the passage of time in such a mineral results in the conversion of potassium 40 to argon 40, there should be a loss of one exactly balanced by a gain of the other. Furthermore, because a molecular crystal with a constant ratio of potassium to other materials can be used for the determination, the starting amount of potassium in the rock need not be estimated but can be determined directly. Potassium/argon ratios from various rocks can be used for age determination in many situations. The result? An age for the earth of 4 to 5 billion years, entirely consistent with other methods.

At this point I will ask you to suffer through a third example of the power of radiometric methods in the dating of rocks. This example I will discuss in some detail because it illustrates quite nicely the power of the technique and the lengths to which geologists have gone to make these determinations as accurate as possible. This method takes advantage of the decay of rubidium 87 (^{87}Rb) to strontium 87 (^{87}Sr), a process of direct decay (no intermediates) with a half-life of 47 billion years. There are three other isotopes of strontium (84, 86, and 88) which are not produced by any radioactive decay process but, of course, are chemically indistinguishable from ^{87}Sr. Once a rock containing these five components is formed, what will happen to each of them? As seen in Figure 1, while the amounts of the nonradiogenic isotopes of strontium remain constant (only ^{86}Sr and ^{88}Sr are shown, but ^{84}Sr is also present), ^{87}Sr will increase by the exact amount that ^{87}Rb decreases.

Therefore, in a very old rock, the ratio of ^{87}Sr to a nonradiogenic isotope of strontium (like ^{86}Sr) will have increased, and will have increased in proportion to the decrease in ^{87}Rb.

As can be seen from Figure 1, if we were aware of the starting conditions of the mineral at time-zero, then at any time in the future we would be able to tell the exact age of the rock by measuring the increase in the level of ^{87}Sr. However, it would seem that we have the same problem we encountered earlier in the uranium-lead method: the need to estimate the starting conditions. Here is where the rubidium-strontium method sets itself apart: it provides a method to do just that!

Consider a rock which is composed of several different minerals (which is not at all an uncommon situation). Some of those minerals will have very small amounts of strontium but plenty of rubidium, some will have large amounts of strontium with very little rubidium, and many will have intermediate quantities of each. At the date the rock is formed, the rubidium-strontium composition of three representative minerals might look something like Figure 2. Notice that although each mineral has a different strontium content, the ratios of each of the various strontium isotopes in each (for example, the ratio of ^{86}Sr to ^{87}Sr) are constant. This occurs because each of the isotopes is chemically identical, and when a rock is formed there is no chemical way to include one of the isotopes preferentially over the others. Therefore, they are in-

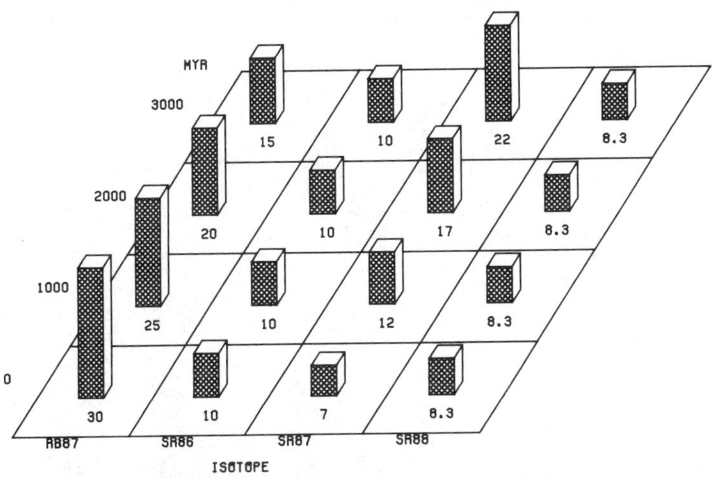

Fig. 1. Highly diagrammatic illustration of the change through time in the amounts of rubidium 87 and the various strontium isotopes. At the time of formation, the three strontium isotopes (for the sake of simplicity, 84-Sr is not shown, and 88-Sr is shown at approximately 1/10 of its actual value) are present in a fixed ratio. As the rock ages, the amounts of the non-radiogenic isotopes (Sr-86 and Sr-88) do not change, but the amount of Sr-87 *increases* in direct proportion to the amount of Rb-87 which decays over any period of time. As millions of years pass, the *increase* in Sr-87 is exactly balanced by the decrease in Rb-87.

cluded in a set of constant proportions determined merely by their relative abundance at the earth's crust as the mineral is formed. How can we represent the striating conditions in each of these minerals in a graphic way? Well, although each mineral has a different Rb/Sr ratio, the ^{86}Sr/^{87}Sr ratio in each is the same. We can represent these starting conditions by plotting each mineral as a point on a graph of ratios like that in Figure 3. So, at the time of its formation, the points of each mineral will lie on a straight line. What will happen as time passes? In each of the two minerals, of course, the amount of ^{87}Sr will increase. But it will increase at a different rate in each, because each mineral starts with a different amount of ^{87}Rb. This situation is also shown in Figure 3.

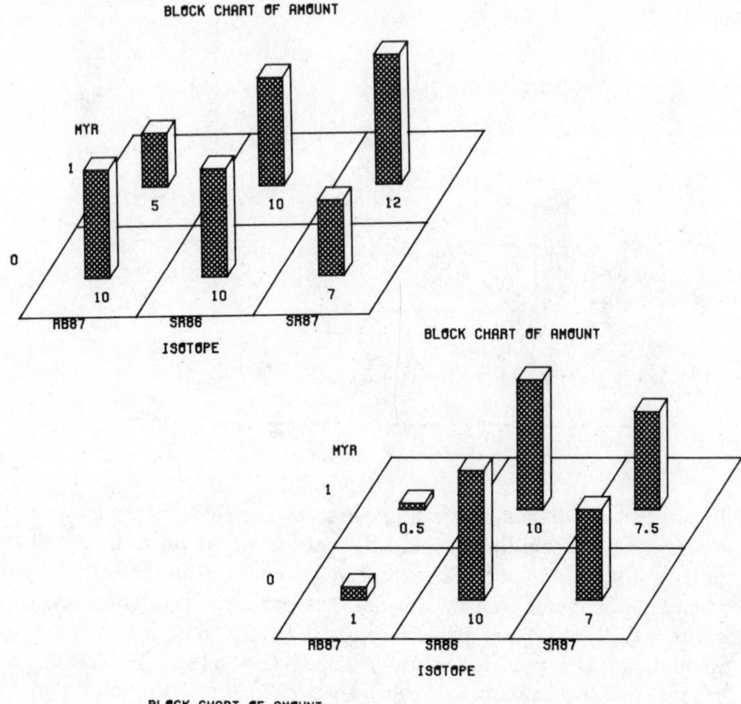

BLOCK CHART OF AMOUNT

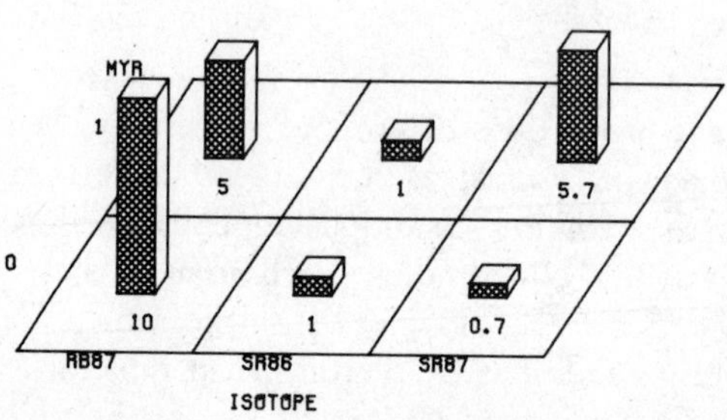

Fig. 2. Representations of strontium and rubidium concentrations in three idealized minerals in a single rock. By combining measurements of isotope ratios in several minerals, the starting conditions of the rock (and hence its actual age) can be determined with great accuracy. Idealized minerals shown include a mineral with strontium and rubidium

Now, let us consider what will happen to our graphic representation of the four minerals. The $^{87}Sr/^{86}Sr$ ratio in each will have changed, but because in each case it will have changed in direct proportion to the rubidium/strontium ratio, something very interesting happens to our graph, as can also be seen in Figure 3.

The ratios of the minerals will, as you see, still form a straight line, but the line will now have a slope. The slope of the line gives us the amount of time which has passed since the formation of the rock! We do not need to make an estimate of the starting conditions, because the starting conditions can actually be determined directly. The intersection of the line with the ordinate of the graph tells us the $^{87}Sr/^{86}Sr$ ratio at the time of the rock's formation. An experimental graph of this kind from an actual rock is shown in Figure 4.

The power of the method is remarkable. Every single mineral in the rock lies on the line (which is known as an isochron), and therefore every single mineral in the rock "agrees" on the age which is given by the slope of the line! Now, as someone whose training is in cell biology and biochemistry, I feel obliged to step in here and make a comment about the kind of experimental result you see above. The agreement of various minerals in a single rock (or many rocks from the same site) is nothing short of spectacular. Very seldom have I obtained data on biological systems which even approaches the consistency and precision of this method. The rubidium-strontium method therefore gives self-calibrating and self-checking experimental results. If geological processes have removed or added either strontium or rubidium, the method will show it at once, because the points will fail to lie on a straight line. If a rock has been homogenized by melting and recrystallization,

---

present in large amounts (2a), one with a large excess of strontium (2b), and one with a large excess of rubidium (2c). For diagrammatic purposes, we have shown 50 percent of the strontium-87 decaying in the time period. Because the half-life is much longer than the age of the earth, the actual percentage decaying is much less. Nevertheless, the extreme accuracy of present-day isotopic measurement techniques permit such determinations to be made with very great confidence.

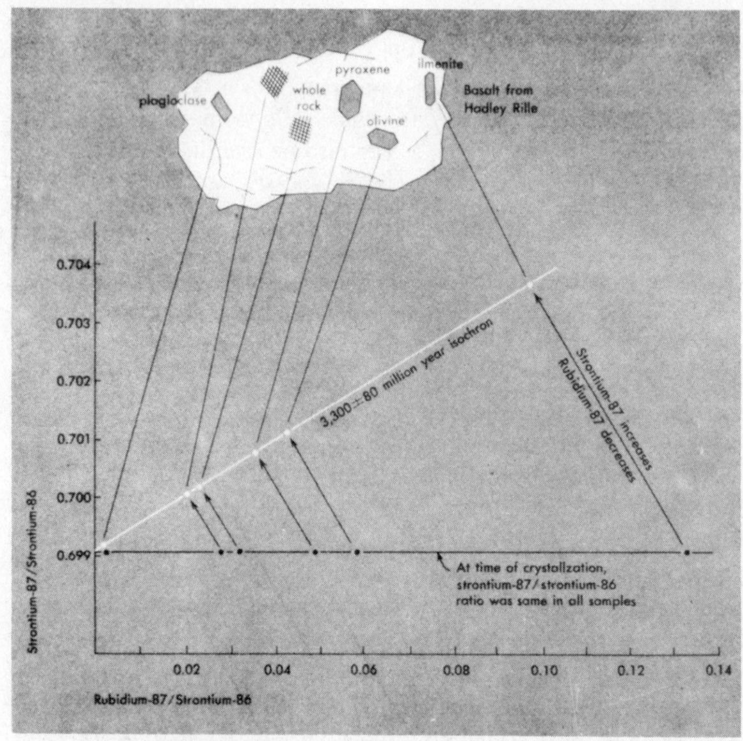

Fig. 3. Diagram illustrating the isochron method for age determination. At the date of formation of a rock, the Sr-87/Sr-86 ratios of each mineral will be exactly the same (see the idealized minerals in Figure 2). However, as time passes, each mineral in the rock will gain Sr-87 at a rate proportional to the amount of rubidium in that mineral, and the time since formation. If the mineral is not disturbed by geologic processes, then the rubidium-strontium ratios will form a straight line, whose slope directly indicates the age of the rock. [Fig. 3 courtesy of Dr. Donal Eicher. From *Geologic Time,* Prentice-Hall. Englewood Cliffs, N.J. 1976. Page 128.]

the isochron line will be reset to zero, and the age may be an underestimate. But no process has ever been suggested which could cause this technique to give an artificially old age for this planet.

What does the rubidium-strontium method say about the oldest rocks on this planet? It says that they are about 4.2 billion years old. The agreement between this method and those based on other isotopes with entirely different chemistries is remarkable. The earth is a very old place.

There is even more to the story. We can use these methods to determine the dates of formation for rocks from the moon and rocks which have fallen from space (meteorites). If we were to discover that meteorites, lunar rocks, and terrestrial rocks were formed at very different times, we might have to make serious revisions in our estimates for the age of the solar system. What is the actual result? The ages for rocks from each of these sources are dazzlingly consistent. Some examples are seen in Figure 5.

With all of this in place, let us consider what the problems associated with this method might be. To begin with, is there any chance that we might get an age which is incorrectly old because some of the parent isotope in the decay series (rubidium) has been removed by geological or chemical processes, or because an extra amount of the daughter isotope (^{87}Sr) has been added? If we were dealing with a system like the uranium-lead series in which only a single pair of measurements are made for a single rock, coupled with estimates (reasonable though they may be) of starting conditions, then such considerations would be a real problem. They would not be insurmountable, however, because there is no known way for one isotope of strontium (^{87}Sr) to be added to a rock without also adding the other three isotopes, nor, for that matter, for just the right amount of ^{84}Sr, ^{86}Sr, and ^{88}Sr (but not ^{87}Sr) to be removed along with rubidium, which would be necessary to give an incorrectly large value for the age of the rock. However, in the case of a rock for which a rubidium-strontium isochron can be constructed, the possibility of determining an age which is off by a factor of say, 500,000 (or 50 million percent, which is the difference between 10,000 and 5 billion years) is zero. In every single mineral in the rock, the hypothetical geochemical process would have had to remove exactly the same percentage of rubidium,

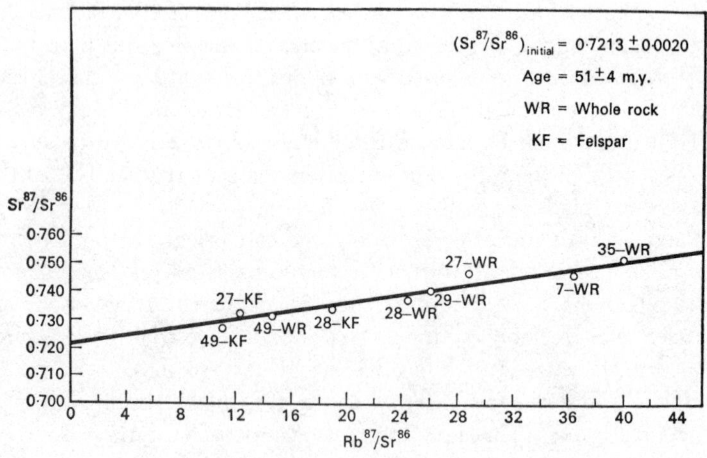
$(Sr^{87}/Sr^{86})_{initial}$ = 0·7213 ± 0·0020
Age = 51 ± 4 m.y.
WR = Whole rock
KF = Felspar
Sr^{87}/Sr^{86}
0·760
0·750
0·740
0·730
0·720
0·710
0·700
27–KF
49–KF
49–WR
28–KF
28–WR
29–WR
27–WR
7–WR
35–WR
0 4 8 12 16 20 24 28 32 36 40 44
Rb^{87}/Sr^{86}

Glauconite

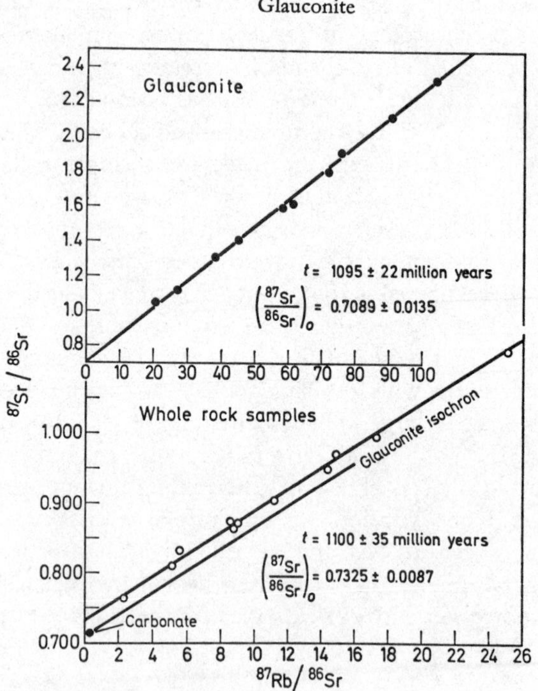
Glauconite
2.4
2.2
2.0
1.8
1.6
1.4
1.2
1.0
0.8
t = 1095 ± 22 million years
$\left(\frac{^{87}Sr}{^{86}Sr}\right)_o$ = 0.7089 ± 0.0135
0 10 20 30 40 50 60 70 80 90 100
$^{87}Sr/^{86}Sr$
Whole rock samples
Glauconite isochron
1.000
0.900
0.800
0.700
t = 1100 ± 35 million years
$\left(\frac{^{87}Sr}{^{86}Sr}\right)_o$ = 0.7325 ± 0.0087
Carbonate
0 2 4 6 8 10 12 14 16 18 20 22 24 26
$^{87}Rb/^{86}Sr$

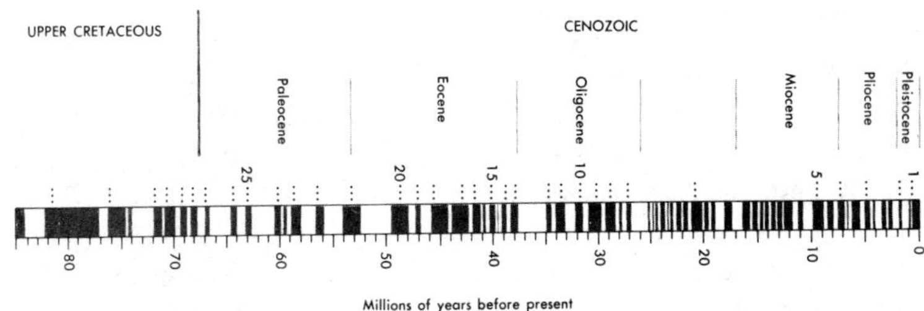

Fig. 5. A chart of known magnetic reversals occurring during the last 85 million years. Creationist arguments that the age of the earth can be estimated by measuring small fluctuations in the field strength are contradicted by the geologic record in a most spectacular way: the field has actually reversed itself countless times in the past history of the planet. [Fig. 5 courtesy of Dr. Donald Eicher. From *Geologic Time,* Prentice-Hall. Englewood Cliffs, N.J. Page 81.]

and then have added ^{87}Sr, not in proportion to existing strontium but in proportion to the previous rubidium content of each and every mineral, so that each one would give a false age in total agreement in every mineral in the rock. Not even the ICR creationists are bold enough to suggest that such a process could ever have occurred in a single rock, much less the tons and tons of samples from every geological age and from every single sample brought back from the moon as well.

Finally, there is another estimate of the age of the planet, beautiful for its simplicity and appealing for those who may have gotten lost in the forest of graphs and charts on the preceding pages. This analysis was first discussed by Stanley Freske.[6]

Fig. 4. Actual rubidium-strontium isochrons used for age determinations. A formation dating to 51 million years is shown in the upper panel, and rocks dating to 1.1 billion years are shown in the lower panel. [*upper panel* courtesy of Dr. Derek York. From *The Earth's Age and Geochronology.* Pergamon Press. New York (1972). Page 58. *lower panel* courtesy of Dr. Gunter Faure. From *Strontium Isotope Geology.* Springer-Verlag. New York (1972). Page 89.]

Collectively, the isotopes of the various elements are known as nuclides. Sixty-four of the nuclides show radioactive half-lives in excess of 1,000 years. Of these, 47 range between 1,000 and 50 million years. There are 7 which we must exclude from the analysis because they are constantly formed by other natural processes such as cosmic ray interactions or other decay series. Consider for a moment the suggestion that the earth was a recent creation. If it were formed, say, 10,000 years ago, then significant amounts of each of these nuclides should still be around in the material of the earth's crust for us to measure. If it were billions and billions of years old, then by now each of these 40 nuclides should have decayed away leaving no trace, and we should only be able to find radioactivity in the nuclides with long half-lives. Now, how many of the 40 short-lived isotopes are still around to bear evidence of a recent creation? None. Zero for 40. An interesting result. There are 17 nuclides with half-lives greater than 50 million years. How many of these long-lived nuclides are still present in detectable amounts? Seventeen—17 out of 17. The age of this planet is written into the very material upon which we stand. This is a very old place.

How do scientific creationists deal with this wealth of information which demolishes one of their basic tenets? They employ a variety of strategic devices. One involves suggesting that the radiometric methods are filled with errors and are useless. A second (and completely contradictory) argument is to say that the methods are fine and the creator, when he made the planet 10,000 years ago, made it look 5 billion years old. And a third marvelous strategy is to claim that, yes, the methods we have discussed yield a large value for the age of the earth, but other equally reliable methods yield much lower values. Therefore, we cannot determine the age of the earth and must rely on another method to estimate its age. And what is the other scientific method? Divine revelation.

As I have suggested, although all the radiometric methods are powerful ones, the rubidium-strontium method is particularly useful because of the sensitivity of the isochron line to any disturbances which would affect the age determination. The manner in which one of the leading texts on scientific creationism discusses the method is therefore quite instructive:

THE RUBIDIUM METHOD

In addition to all the difficulties encountered in these methods [the other radiometric methods, including uranium-lead and potassium-argon], they have been of limited usefulness because of the extreme rarity of uranium and thorium minerals, especially in fossiliferous rocks. Consequently, much attention has been given in the past decade to the development of methods involving the radioactive isotopes of the alkali metals, rubidium and potassium. These are much more common, and the potassium minerals especially are commonly found in sedimentary rocks.

One of the main workers in the development of the rubidium-strontium method has been Dr. Otto Hahn. The main question about the method has been the lack of agreement concerning the disintegration rate of rubidium. Hahn says:

"For this method, however, a knowledge of the transformation rate of rubidium into strontium is necessary. The final decision regarding the half-life has yet to be made."

Ahrens, another leading worker in the field, gives a list of different determinations of the half-life of rubidium as made by various scientists showing a variation all the way from 48 to 120 billion years. A further limitation is the very small amount of strontium present and the fact that much of this may be non-radiogenic.[7]

As we can see, there is absolutely no attempt to explain the method in detail to the readers of the text (because this would only serve to enhance its credibility), and instead one is left with the impression that only a single measurement of each isotope is made, coupled with unsupported assumptions about starting conditions and unreliable data on the decay half-life. In truth, the decay half-life has now been determined with very great accuracy. There is, in fact, only one way to explain the rubidium-strontium method and its consistency with other methods, coupled with the complete absence of the 40 short-lived nuclides: to claim that the rates of radioactive decay have varied in the past to give an incorrect appearance of age. When Henry Morris debated me at Brown in April 1981, he referred to my arguments on radiometric dating and said, "We now know that the decay rates do vary," implying that a substantial known variability made the determinations unreliable. Here you can see one of the reasons why the scientists who are willing to confront creationists publicly often have a problem controlling their tempers. How much do these rates actually vary? Enough to be consistent with a much younger age for the earth? Not at all. The maximum reported variability (at very great extremes of temperature and pressure) is less than 4 percent.[8] Stacked up against the 50 million percent discrepancy which all of this demands, this objection looks a little pale, to say the least! Yet, just when the scientific creationist is finished trying to undermine our confidence in these techniques, something remarkable then happens. He spins 180 degrees in a flash of illogic and says that the 5-billion-year-old earth is just what he expected!

> Furthermore, it will be maintained that even though any given age measurement may be completely erroneous due to leaching or emanation or some other effect, there are many cases now known where the age estimate has been checked by two or more different methods, independently. It would seem improbable that the elements concerned would have each been altered in such a way as to continue to give equal ages; therefore, such agreement between independent measurements would seem to be strong evidence that alteration has not occurred and that the indicated age is therefore valid.
>
> We reply, however, that the Biblical outline of earth history, with the geologic framework provided thereby, would lead us to postulate exactly this state of the radioactivity evidence. We would expect radiogenic minerals to indicate very large ages and we would expect different elements in the same mineral, or different minerals in the same formation, to agree with each other![9]

Just exactly *why* does one expect the rocks to have *"very large ages?"* What is there about the character of a creator which would lead one to infer that he would, as Morris says, have put this 50 million percent error in the material of the earth? The only refuge of the creationist is a device known as the "appearance of age." He must postulate that a creator, for reasons of his own, made the universe *look* much older than it actually *is,* thereby denying us any opportunity to gather actual facts about the age of the planet. This argument allows one to dismiss *any* scientific evidence as simply an act of deception on the part of the creator. The argument can be used for results in other areas of science as well. The stars in the sky look the way they do not because there are distant galaxies and stars, but because the creator created streams of photons emanating from nonexistent objects. He created an "apparent" universe. This line of reasoning is, of course, nondisprovable, because a deceptive creator would by definition be capable of nearly anything! But the argument, ultimately, denies science itself. It suggests that the physical universe is a collossal joke played on us by a prankster who will remain forever beyond our reach. The "appearance of age" argument is a complete retreat from scientific argument into the realm of nontestable mysticism. It is the ultimate admission of scientific defeat.

Finally, after objecting to radiometric methods, then turning around and saying the methods are fine but we expect great ages, the creationists have a third and final strategy: to claim that other methods exist which are equally as valid as the radiometric ones. The evolutionists, the creationist implies, ignore these excellent methods because evolution demands a very great age for the earth, and therefore the methods which indicate smaller ages are ignored. Henry Morris summarizes this line of reasoning:

> Other processes do give much younger ages, however. For example, the present rate of sedimentary erosion would have reduced the continents to sea level in six million years and would have accumulated the entire mass of ocean-bottom sediments in 25 million years. Present rates of volcanic emissions would have produced all the water of the oceans in 340 million years and the entire crust of the earth in 45 million years.
>
> There is no measurable accumulation of meteoritic dust on the earth's surface, but present rates of influx of such dust from space would produce a layer 1/8 inch thick all over the earth in a million years and a layer 54 feet thick in 5 billion years. The comets of the solar system are disintegrating so rapidly that they could only have come into existence less than a few million years ago at most. The earth's magnetic field is decaying so rapidly that its origin cannot have been more than about 10,000 years ago.
>
> The rate of uranium influx into the ocean indicates a maximum age for the oceans of about one million years. Sodium influx indicates perhaps 100 million years, but chlorine, sulfates and other materials give much less.[10]

Is their any truth to this? The obvious internal contradiction (of claiming that the continents are *both* being eroded too quickly and being built up too quickly to be very old) we may overlook, but what of the other methods which Dr. Morris seems to claim would convince us of a much younger planet? One of the alternate dating methods often suggested by creationists involves measurements of the strength of the earth's magnetic field:

> This evidence is found in a remarkable study by Dr. Thomas G. Barnes, Professor of Physics at the University of Texas in El Paso. Dr. Barnes is author of many papers in the fields of atmospheric physics and a widely used college textbook of electricity and magnetism. He has pointed out that the strength of the magnetic field (that is, its magnetic moment) has been measured carefully for 135 years, and also has shown, through analytical and statistical studies, that it has been decaying exponentially during that period with a most probable half-life of 1400 years.
>
> This would mean that the magnetic field was twice as strong 1400 years ago than it is now, four times as strong 2800 years ago, and so on. Only 7000 years ago it must have been 32 times as strong. It is almost inconceivable that it ever could have been much stronger than this. Thus, 10,000 years ago, the earth would have had a magnetic field as strong as that of a magnetic star! This is highly improbable, to say the least.
>
> Magnetic stars have thermonuclear processes with which to establish and maintain magnetic fields of such strength, but the earth has no such source. Dr. Barnes shows beyond reasonable question that the only possible source for the earth's magnet must be free circulating electric currents in the earth's iron core. Electric currents, however, must flow against resistance, and such resistance generates heat, which is then dissipated through the surrounding medium and lost. Such currents must gradually decay because of this heat loss and this, in turn, accounts for the decay of its induced magnetic field.
>
> Thus, 10,000 years seems to be an outside limit for the age of the earth, based on the present decay of its magnetic field. Any objections to this conclusion must be based on rejection of the same uniformitarian assumption which evolutionists wish to retain and employ on any process from which they can thereby derive a great age for the earth.[11]

Now that sounds like a fine argument. Except for one troublesome little fact which is never even hinted at in creationist texts.

The rocks of the earth's crust contain a record of the direction and the strength of the earth's magnetic field at the time of their formation. This phenomenon is known as paleomagnetism. From analysis of sea-floor spreading in the vicinity of the mid-Atlantic (and other areas) and paleomagnetism in core samples of sedimentary rock in other places, an interesting fact emerges: the earth's magnetic field has gone through innumerable fluctuations and reversals throughout geological history! An example is seen in Figure 6. If a scientific paper were to be published containing the magnetic field "method" for estimating the age of the earth and ignoring the thousands of published studies showing the hundreds of reversals of the magnetic field over geologic time, a kindly editor might suggest that the creationist authors of such a paper would do well to read an elementary textbook on geology. Another editor might have something less diplomatic to suggest.

Why is the magnetic field "method" not used to determine the age? Because pretending that the strength of the field has decreased smoothly throughout all the history of the earth would involve ignoring our knowledge that the field has reversed hundreds of times! And that would be outright deception, nothing less. A factual evaluation of the magnetic field data (when one considers the very large number of field fluctuations and reversals which have been documented) actually supports a very great age for the earth. As the creationists themselves indicate, the present rate of flux indicates that a reversal may take thousands of years. Because the geologic history of the crust indicates that hundreds of such reversals have taken place, we are immediately forced away from our 10,000-year dogma for the creation of this planet. Otherwise, one might imagine complete reversals taking place every few years, and that is completely contradicted by the geologic record and the historical record.

The other methods offered by scientific creationists are, if anything, even less credible than the magnetic field method. One argument concerns the rate of accumulation of meteoric dust from outer space:

> 2. INFLUX OF METEORIC MATERIAL FROM SPACE
>
> It is known that there is essentially a constant rate of cosmic dust particles entering the earth's atmosphere from space and then

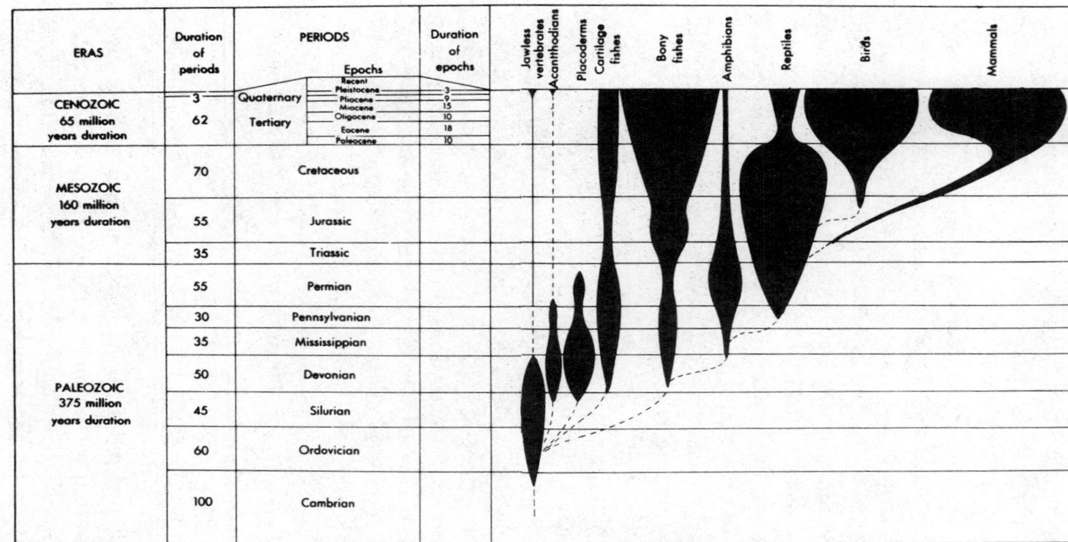

Fig. 6. The successive character of the fossil record. The creationist assertion that living forms appeared "suddenly" in the fossil record is not accurate. The various vertebrate categories appear in a successive fashion which is completely supportive of evolution. Creationist theories which demand a *simultaneous* creation of all living forms are completely contradicted by the facts of the fossil record. [Fig. 6 courtesy of Dr. Edwin Colbert. From *Evolution of the Vertebrates.* John Wiley & Sons. New York. 1980. Pages 10–11.]

> gradually settling to the earth's surface. The best measurements of this influx have been made by Hans Pettersson who obtained the figure of 14 million tons per year. This amounts to 14×10^{19} pounds in 5 billion years. If we assume the density of compacted dust is, say, 140 pounds per cubic foot, this corresponds to a volume of 10^{18} cubic feet. Since the earth has a surface area of approximately 5.5×10^{15} square feet this seems to mean that there should have accumulated during the 5-billion-year age of the earth, a layer of meteoritic dust approximately 182 feet thick all over the world!
>
> There is not the slightest sign of such a dust layer anywhere of course. On the moon's surface it should be at least as thick, but the astronauts found no sign of it (before the moon landings, there was considerable fear that the men would sink into the dust when they arrived on the moon, but no comment has apparently ever been made by the authorities as to why it wasn't there as anticipated).
>
> Even if the earth is only 5,000,000 years old, a dust layer of over 2 inches should have accumulated.
>
> Lest anyone say that erosional and mixing processes account for the absence of the 182-foot meteoritic dust layer, it should be noted that the composition of such material is quite distinctive, especially in its content of nickel and iron. Nickel for example is a very rare element in the earth's crust and especially in the ocean. Pettersson estimated the average nickel content of meteoritic dust to be 2.5 per cent, approximately 300 times as great as in the earth's crust. Thus, if all the meteoritic dust layer had been dispersed by uniform mixing through the earth's crust, the thickness of crust involved (assuming no original nickel in the crust at all) would be 182 × 300 feet, or about 10 miles. Since the earth's crust (down to the mantle) averages only about 12 miles thick, this tells us that practically all the nickel in the crust of the earth would have been derived from meteoritic dust influx in the supposed 5 × 109 year age of the earth![12]

I found this argument to be particularly interesting because the source they provided for the rate of dust accumulation is an entirely legitimate one, a *Scientific American* article published in 1959 by Hans Pettersson.[13] The measurements made in this article were subject to a severe problem, of which the author was painfully aware. Pettersson made his measurements on some high mountains in Hawaii where he hoped they would be less affected by earthly sources of nickel dust such as metal smelting and refining operations. When I went to the library and read this article, the solution was immediately apparent to me. Surely other astrophysicists interested in the same question would have investigated the rate of dust accumulation from platforms less susceptible to terrestrial contamination: space satellites. The satellite exploration programs of our country and the Soviet Union were just in their infancy when this article was written, and I realized at once that the proper place to go for more recent data would be NASA. Surely, they must have been interested in the rate at which particulate matter was banging into their spacecraft, and I was sure that more recent (and more accurate) figures would be available just for the asking.

I sat down and wrote a letter to a rather well-known astrophysicist who had been connected with the space program, and a letter came back within two weeks with the answer.[14] NASA had indeed been interested in the rate of dust accumulation (as had the Soviets), and measurements made by many satellites and many types of instruments indicated quite clearly that the rate of dust accumulation in space around our neighborhood in the solar system is from 10^{-16} to 10^{-17} grams of dust per square centimeter per second. That's actually a rather large rate! Now that we have the correct figure, let's do a little calculation and see how much dust we would expect to find on the surface of a 5-billion-year-old moon:

> Five billion years = 10^{17} seconds. The maximum rate of dust accumulation times 5 billion years—$10^{17} \times 10^{-16}$ g/cm^2 = g/cm^2.
>
> So, an upper limit on dust accumulation would be about 10 g/cm^2, assuming the surface had been absolutely undisturbed for 5 billion years. That would not happen on earth, of course (sprinkle a few pounds of dust on your driveway and then try to find it after a few weeks of rainy weather), but something like that might happen on the moon.
>
> How much dust is actually there on the surface of the moon? Now, the density of dust varies, but let's take a minimum estimate (1 g/cm^3—the density of water) to make the largest reasonable prediction of dust volume. We would now expect that 10 g of dust per cm^2 to form a layer 10 cm high. The summaries of geological information about the moon show that the surface of the moon is covered with a debris blanket (known as the regolith) which varies in thickness from 5 to 10 meters. The regolith has been formed by constant bombardment from meteorites (which

have pulverized the rock at the lunar surface), and from its elemental composition the amount of meteoric dust in the regolith can be determined. How much is there? The regolith is about 1.5% meteoric dust.[15] How much is 1.5% of 10 meters? 15 cm. Bingo.

Incredibly, when one actually looks into the details of the "methods" the scientific creationists suggest as a reasonable gauge for the age of the earth, the result that emerges is a consistent confirmation of prevailing scientific conclusions. Now, once again, one has to ask some difficult questions about those who are willing to advance such arguments to the general public. Did the "scientists" of the Institute for Creation Research really fail to look into any work done on the subject of meteoric dust accumulation done during the last 20 years? And now that the error has been pointed out, may we expect that these arguments will disappear from their popular writings? I doubt it. The least that can be asked of a scientist is that he never use evidence which he knows to be in error to support a scientific argument. A scientist may often argue for an idea which does not seem well supported by the evidence and occasionally will eventually be proven right as more work is done and older observations are challenged. But he will never knowingly use an incorrect "fact" to support an idea, even if that idea is something he fervently believes in. Why? Because there is no greater scientific crime than a knowing deception. And a scientist who is discovered to have falsified a result will find his career to be over in a remarkably short time. In this case, one cannot say that the scientific creationist has falsified a result . . . in fact, he has produced no results at all (which is characteristic of "creation-science") but has argued his point of view based on the work of others. The problem here is that the creationist has used data which he knows to be in error merely to support his ideas. He has allowed himself to present his readers and listeners with "false facts" because he believes that they will advance his arguments, his cause. And one is left to wonder at the motives, the thoughts, the consciences of those who are willing to do all of this in what they consider to be the service of the Christian God.

There are other arguments advanced by the scientific creationists relating to the age of the earth, of course. The creationists often claim that there are scores of such methods and enjoy impressing lay audiences with long lists of "natural processes" yielding a "young age" for the earth. They imply that narrow-minded evolutionists have ignored these methods because of their requirement for a great age for the earth. In fact, of course, these alternate natural processes tell us nothing about the real age of the planet. Most of the methods, for example, are based on the concentrations of several minerals in the ocean. The scientific creationist assumes that the oceans of the world are a closed system (that is, that no salts or minerals can escape from them), measures the content of salt or potassium or iron or magnesium in seawater, and then makes an estimate of how much of that particular mineral is washed into the oceans each year by the rivers of the world. If one millionth of a mineral is washed in each year, then he reasons, the world cannot be older than one million years. Here are the maximum values for several such materials:

CHEMICAL ELEMENT	YEARS TO ACCUMULATE IN OCEAN FROM RIVER INFLOW
Sodium	260,000,000
Magnesium	45,000,000
Silicon	8,000
Potassium	11,000,000
Copper	50,000
Gold	560,000
Silver	2,100,000
Mercury	42,000
Lead	2,000
Tin	100,000
Nickel	18,000
Uranium	500,000

To their credit, the scientific creationists admit that some of the minerals in the ocean give ages which are a little strange (although they don't mention it in all their publications)—for example, aluminum gives a maximum age for the oceans of 100 years! Now, I think even Henry Morris would be willing to admit the earth is older than 100 years. What these values actually measure, of course, are the "residence times" of various materials in seawater.[16] Aluminum is a very reactive element, and very readily precipitates with other compounds to form kaolinite, one of the main chemical constituents of clay. The other materials have similar stories. Many of them are taken up constantly into microscopic plants and animals in the oceans, where they enter the food chain as these organisms are eaten by larger marine animals. Eventually, plants and animals in large numbers perish in the seas, and the materials of which they are composed (including sodium, potassium, magnesium, etc.) are added to the floor of the ocean. The floors of various oceans show only moderate piles of such accumulated material, because at several places in the world, such as the deep trenches near the Mariana Islands, the oceon floor is subducted into the mantle of the earth. Should we really use what we know to be the "residence time" of a mineral to measure the age of the oceans? Of course not, unless we are willing to consider the 100-year-old age for aluminum right up there along with everything else.

What this array of "methods" really indicates is something that will come up again and again as we examine "creation-science." Confronted with a well-established scientific theory with very strong experimental support extending over many decades, they will not attempt to establish anything that can be recognized as an alternative, testable, theoretical system. Instead, they will blow off a shotgun full of mutually contradictory arguments (the age is 10,000 years . . . the radiometric methods are inaccurate . . . the radiometric methods do indicate a great age but we "expected" that . . . we cannot determine the age of the earth by any scientific evidence, but, by the way, we've got some evidence which says it's young! . . . the age of the earth doesn't matter because there's still not enough time for evolution . . . and so it goes) designed essentially to confuse and mislead, and even to misinform.

There is a very interesting point worth making here. If the world really had been created 10,000 years ago, the radiometric methods used by modern geologists could prove it. The rubidium-strontium isochrons would be nearly flat (the slope of an isochron for a 10,000 year old rock is very slight), potassium-containing rocks would show only minuscule amounts of trapped argon, uranium-based minerals would show very little lead accumulation, the lunar regolith would show only 10,000 years of bombardment, and the most distant star in the night sky would be 10,000 light years away. Such observations are *exactly* the kinds of empirical evidence a scientist would expect to find in a recently created universe. Needless to say, we don't find that kind of evidence. By any fair and reasonable standard, the most basic prediction of scientific creationism, a recent creation, is disproven.

Creationism and the Fossil Record

While creationists often like to claim that because the events of the past were unique and nonrepeatable, no theory of natural history can ever be truly scientific, they are indeed very much aware that a factual record of the past does exist in the form of the fossil record. How do they treat this record? In a most remarkable way:

> If millions of species have gradually evolved through hundreds of millions of years, the fossil record must contain an immense number of transitional forms—museums should be overflowing with them. The fossil record shows, however, an explosive appearance of a great variety of highly complex creatures for which no ancestors can be found and systematic gaps between all higher categories of plants and animals. The fossil record is thus highly contradictory to evolution but remarkably in accord with creation.[17]

The bold denial that the fossil record provides any proof for evolution is found in virtually all creationist literature. A publication of the ICR known as the Impact series states flatly:

> All present living kinds of animals and plants have remained fixed since creation, other than extinctions, and genetic variation in originally created kinds has occurred within narrow limits.[18]

The insistence that the fossil record does not document evolution is a bold one and is calculated to win public attention if the scientific community is too uninterested to respond in kind. The fact of the matter is that the fossil record not only documents evolution, but that it was the fossil record itself which forced natural scientists to abandon their idea of the fixity of species and look instead for a plausible mechanism of change, a mechanism of evolution.

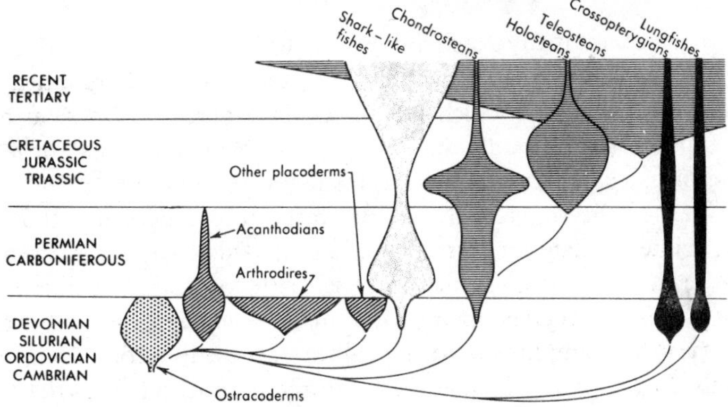

Fig. 7. The sequential character of the fossil record is continued even within the various vertebrate subdivisions. The various classes of fishes also show a pattern of appearance through geologic time which completely supports evolution, and completely contradicts any suggestion of simultaneous creation. [Figure courtesy of Dr. Edwin Colbert. From *Evolution of the Vertebrates.* John Wiley and Sons. N.Y. 1980. P. 60.]

The fossil record not only demonstrates evolution in extravagant detail, but it dashes all claims of the scientific creationists concerning the origin of living organisms.

For example, the supposed "explosive appearance" of all living phyla at one point in the fossil record (the Cambrian period, from 450 to 600 million years ago) is in fact a hideous distortion of the actual nature of that record. Although there is agreement that the first examples of vertebrates, crustaceans, molluscs, and echinoderms are found in the Cambrian, their actual appearances in that geological period cover some 100 million years. I find very little justification for claiming that one event follows *suddenly* upon the heels of another if they are separated by 100 million years! Moreover, the major groups of living organisms (comprising only certain subdivisions of the major phyla) did not arise in the Cambrian. Jawed fishes did not appear until 400 million years before the present, land plants did not appear until 375 million years ago, land animals 350 million years ago, insects 350 million years, mammals 150 million years, flowering plants 135 million years, primate mammals 25 million years; and our own species, *Homo sapiens,* did not appear until 2 to 4 million years ago (depending upon which of many fossil intermediates one first begins to classify as human). This sequential character can be diagrammed in many ways, some of which are shown in Figure 7. This sequence of appearances is hardly suggestive of a lack of change through time, or of a single creation event. It suggests evolution.

What of the suggestion that the fossil record does not show a single intermediate, or transitional, form? This charge is nonsense. The criticism of the creationists essentially consists of complaints that the intermediates are not intermediate enough. For example, the very first amphibian to appear in the fossil record is of the genus *Ichthyostega.* Duane Gish of the ICR complains that he would like to see some intermediates between *Ichthyostega* and the Devonian crossopterygian fishes from which it supposedly developed:

> Not a single transitional form has ever been found showing an intermediate stage between the fin of the crossopterygian and the foot of the ichthyostegid. The limb and the limb-girdle of *Ichthyostega is already of the basic amphibian type, showing no vestige of a fin ancestry.*[19]

I have the devilish suspicion that were such an intermediate to be found, Dr. Gish would inform us that now our problem had increased: we would have two gaps to account for instead of one! Instead, the factual record of fossils from the Devonian period shows that the very first amphibian to appear looked, in simplified terms, more like a fish and less like a typical amphibian than any that were to follow. As Edwin Colbert points out, *Ichthyostega* looks precisely like a fish that had begun to make its way upon the land:

> In the postcranial skeleton *Ichthyostega* showed a strange mixture of fish and amphibian characteristics. The vertebrae had progressed but little beyond the crossopterygian condition, whereas in the caudal region the fin rays of the fish tail were retained. In contrast to the primitive vertebrae and persistent fish tail, there were strong pectoral and pelvic girdles, with which were articulated completely developed limbs and feet, quite capable of carrying the animal around on the ground.[20]

While one may demand yet another intermediate for every pair of fossils discovered, it is worth noting how the various vertebrate classes actually appear in the fossil record. As noted, the first amphibian is the most fishlike of all amphibians. The first true reptiles? These can be seen in a group where amphibian and reptilian characteristics are mixed, known as *Seymouria.* Here again, the first known reptiles are remarkably amphibian in character. The pattern is repeated in the development of birds, where the first known complete skeletons, including the remarkable fossils of *Archaeopteryx,* are completely intermediate in character, and some specimens would in fact actually have been classified as reptiles but for the lucky fact that certain fossils were found with preserved feather impressions.

> Indeed, if feather impressions had not been preserved, all *Archaeopteryx* specimens would have been identified as coelurosaurian dinosaurs. The only reasonable conclusion is that *Archaeopteryx* must have been derived from an early or mid-Jurassic theropod.[21]

Finally, there is a remarkable evolutionary series of fossils connecting reptiles and mammals. Is this final series represented by a single intermediate which the creationist may insist is not quite intermediate enough? Not at all:

> . . . the mammals almost universally offer sharp and obvious contrasts to the reptiles. But in the Permo-Triassic sequence of fossils the contrasts were established only slowly and gradually through groups of intermediates, and became indefinite in the earlier members of the theromorphs so that "what is a mammal and what is not a mammal is pretty much an academic question . . ."[22]

The transition between reptiles and mammals is filled by nearly a dozen intermediates, and the case for evolution is stronger with every new fossil discovered in the rocks of this period.[23] Here we have the first and most direct response to the creationist criticism of the fossil record. Intermediate forms do exist, there are thousands of them, and we can tell the public what they are, why they are "intermediate," and show that however bold the creationist attacks are, they are totally wrong at the level of the fossil record itself. However, there is a more telling point to be made here, and this point has to do with the alternative theory the creationists are attempting to establish.

If we were to accept creationist criticism of the fossil record at its face value, what scheme could we then propose for the natural history of this planet? Accepting the premise of divine creation and the impossibility of evolution, we would have to suggest that a creator formed the first jawless vertebrates some 600 million years ago by an act of special creation, so that these animals appeared suddenly and without ancestors. Nearly all these jawless forms died out shortly after being created, and those that do survive are quite different. Some in fact survive by parasitizing species of fish that did not appear until some 200 million years after the jawless fishes were specially created, a curious fact indeed. Then, nearly 100 million years later the creator made bony fishes, somewhat like the kinds which now occupy the oceans. Later he specially created primitive amphibians, stepping in again and again over the next 50 million years to create the many amphibian groups which appear and then disappear in the fossil record. Still later, he formed primitive birds and primitive mammals, intervening again and again to carry out a series of special creation events so closely graded that the scientists of the present would misinterpret these progressive appearances and disappearances as the result of evolutionary change and extinction. Is this scheme of a progressive creator acting over millions of years the alternative explanation of natural history which the creationists would like to open our minds to? Not at all.

The actual natural history which the creationists proposed is, in a sense, their most closely guarded secret. They avoid mentioning it at all costs, and yet, when pressed closely for a genuinely scientific alternative to the natural history which so closely supports evolution, bits and pieces of the scheme emerge:

> Consequently, the vast fossil record, comprising as it does, a worldwide cemetery preserved in stone for men everywhere to see, is not at all a record of the gradual evolution of life, but rather of the sudden destruction of life.[24]
>
> The earth's geologic features were fashioned largely by rapid, catastrophic processes that affected the earth on a global and regional scale (catastrophism).[25]

And, in less veiled language, Henry Morris speaks of his "theory" in the creationist classic, *Genesis Flood:*

> Within this basic framework we have attempted to reinterpret the basic data of historical geology and other pertinent sciences, which at present are popularly interpreted in a context of uniformitarianism and evolutionism. We have tentatively suggested a categorization of the various geological strata and formations in terms of the Biblical periods of earth history, although retaining as far as possible the terminology of the presently accepted geological periods.
>
> Thus, it seems reasonable to attribute the formations of the crystalline basement rocks, and perhaps some of the Pre-Cambrian non-fossiliferous sedimentaries, to the Creation period, though later substantially modified by the tectonic upheavals of the Deluge period. The fossil-bearing strata were apparently laid down in large measure during the Flood, with apparent sequences attributed not to evolution but rather to hydrodynamic selectivity, ecologic habitats, and differential mobility and strength of the various creatures.[26]

More simply put, the fossil record is an illusion. It was formed in 300 days by the Flood of Noah. All animals and plants were not merely created, but were created side by side. The flood, an unlucky accident, sorted their corpses into strata which gave the appearance of evolution. The American creationist movement does not merely suggest creation as a plausible mechanism for the changes among living organisms observable in the fossils of the past. It claims a *simultaneous* creation event for all living "kinds," followed by a flood which serendipitously sorted them into layers which look like an evolutionary sequence.

The "flood geology" demanded by this model has indeed been considered by science and was dismissed by even "creationist" geologists and naturalists such as Cuvier more than a century ago. The world's geological features are not explicable in terms of a single catastrophic flood, and neither is the sequence of organisms found in the fossil record. The contradictions and fallacies and weaknesses of flood geology are almost too numerous to mention, but one point is worth mentioning. Mammals occupy virtually every corner of this planet. Some are very large, some are extremely small, some are quick, some slow, some burrow into the ground, some swim in the ocean, some climb into the highest trees. They differ enormously, as Henry Morris might say, in terms of their "hydrodynamic" properties (shape and weight), "ecologic habitats," "differential mobility and strength." Yet, not a single mammalian fossil appears until the very last few strata from the creationist "flood" were laid down. And when they do appear, with incredibly bad luck, the fossils arrive in just the right sequence to piece together imaginary evolutionary sequences in a dozen different families. Why is it that the first mammal to appear happens to be the most reptilelike of all subsequent mammals and just happens to appear just after the most mammallike of all reptiles? Shouldn't a single family of moles near the shore have been trapped in the rampaging waters and fossilized in the Cambrian period? Shouldn't swimming mammals have been fossilized alongside the jawless and jawed fishes in the early stages of the flood? And why doesn't a single human fossil appear anywhere in the hundreds of millions of years of life represented by the fossil record, until the last 3 or 4 million years?

The flood geology model is so hopelessly contradicted by the fact of the fossil record that creationists will avoid revealing it at all costs. Added to the impossible difficulties of the fossil record, the creationist must also develop a method which would precisely sort radioactive isotopes so that the "apparent" ages of the various geologic periods could be accounted for. Flood geology represents the desperate attempts of the scientific creationists to reconcile the doctrine of a single, week-long, creation event with the facts of the geological record. Against such a backdrop, Duane Gish's boast that the fossil record is "remarkably in accord with creation" rings hollow indeed.

Other Creationist Arguments

Regrettably, we will not have time here to discuss each of the creationist arguments against evolution, but I do hope that we have covered two of them in detail, which is at least sufficient to convey the nature of creationist objections. An excellent source of well-documented replies to specific creationist arguments is found in a small journal recently organized by Fred Edwords, Philip Osmon, and Christopher Weber, and known as *Creation/Evolution.* Within its pages are more complete discussions of flood geology[27] and the age of the earth[28] than I have been able to present here.[29] Creationists are also fond of arguing that the second law of thermodynamics forbids evolution. This represents either a deliberate attempt at deception or an incredible misunderstanding of chemical thermodynamics, and is fully discussed in articles by William Thwaites and Frank Awbrey,[30] and Stanley Freske.[31] Creationist arguments regarding gaps in the fossil record are covered by Niles Eldredge[32] and creationist distortions, misrepresentations, and hoaxes such as the "human" footprints in the Paluxy River bed are dealt with in many other articles by leading scientists. I should mention as a matter of interest to some readers that the Paluxy footprints (which supposedly show that humans and dinosaurs existed at the same time) are even an embarrassment to reputable creationists, and have been thoroughly exposed by such workers as the creationist geologist B. Neufeld.[33] The only manlike tracks ever to have come from this formation show clear traces of carving, and those that are represented as manlike are generally poor photographs of authentic dinosaur tracks painted with oil to simulate the outline of a human foot.

As a general summary, the creationists have failed in their attack on evolution in two principal ways. First, they have failed to deal with the enormous amount of evidence which supports the general notion of evolution: the fact that living things on the earth were different in the past, and that as time led to the present, the kinds of animals and plants on this planet changed (evolved) into those we see today. Second, and most important, they have failed to provide anything which even resembles an alternative theory of natural history. Proclaiming "creation" every time we are puzzled by the mechanism of evolution is not the same thing as providing an alternative theory. "Creation" merely implies the existence of a creator (God) and says nothing about the real issue: the natural history of this planet. When creationists do provide a scheme for natural history, we discover that it postulates a 6,000-year age for the earth (without any scientific support) and the simultaneous creation of all living things, both fossil and contemporary (again, without any scientific support). Where do these postulates come from? Are they derived from careful and unbiased observation?

Are they derived from the courageous ability to shake off the intellectual yoke of Darwinism? Henry Morris of the Institute for Creation Research shows us very clearly the source of each of these "scientific" postulates:

> Thus it is obvious that one can logically reject the historicity of Genesis 1–11 only if he likewise rejects the rest of the Bible as well, and even the infallibility of Christ himself. Many modern-day religious liberals and even some supposedly conservative Christians have done exactly that. Most Christians, however, are unwilling to go this far. Some try to avoid the issue altogether, but this tactic almost inevitably is a prelude to compromise.
>
> The only Bible-honoring conclusion is, of course, that Genesis 1–11 is the actual historical truth, *regardless of any scientific or chronologic problems thereby entailed* [my italics].[34]

The practitioners of scientific creationism have not, as they contend, discovered flood geology and simultaneous creation and a young earth as the result of scientific evidence. Rather, as Henry Morris so clearly indicates, they have found the Bible first, come to these conclusions as a matter of faith, and then tried to force these ideas back into a secular framework, as Morris says, *"regardless of any scientific or chronological problems."* The religious character, indeed the character of a particular religious sect, is part and parcel of scientific creationism.

"Creation-Science" and the Authentic Question of Creation

Earlier in this article I confessed my own religious beliefs to the reader and began to redefine creationism for the purposes of discussing the American "creation-science" movement. Using the word *creation* in this way does some violence to the language but is clearly necessary for reasons of convenience. I now beg the indulgence of the reader for a reversal of form. What do we know of the real question of creation? Can we say, as scientists, whether or not there is a supreme being who brought the universe into existence? A majority of the American people hold to this view, and one might suggest that a majority of scientists do as well. Are there valid scientific reasons for believing in creation? Some respected scientists clearly seem to think so.

Chandra Wickramasinghe and Fred Hoyle of University College in Cardiff, Wales, have argued as much in recent years. They have concluded that the 10 or 15 billion-year age suggested for the universe does not allow enough time for evolution of the genetic codes found in living cells. Wickramasinghe has colorfully compared the probability of life arising from inanimate matter to the probability that "a tornado sweeping through a junkyard might assemble a Boeing 747."[35] These authors have therefore argued that such improbabilities point toward an intentional and intelligent creation of living organisms.

Jacques Monod and Francis Crick, both Nobel laureates, have also argued that the enormous complexity of living cells makes prospects for their development from nonliving matter simply too slim to be worth considering. In each case, the distinguished scientist finds himself at one with the idea that life was either created by a supreme being or transported to this world from another, more hospitable place in the universe. These are grand and interesting ideas, and although they are in each case greatly flawed they point out some areas where scientists must be careful to distinguish between factual observation and hypothesis.

Our knowledge of the evolution of plants and animals is based on a *factual* set of observations. We have a large number of facts (fossils) on which to draw, and these fossils paint an enormous, elaborate, and consistent picture of change and orderly succession among living things up to the present time. However, we have no such set of observations concerning the development of the first living cells on this planet. All that we can hope to do is to reproduce something resembling the initial conditions on the earth and hope to get a feeling as to whether or not the evolution of the first living organisms seems possible or impossible. Many such experiments have been done, and the results of those experiments have led most scientists to conclude that the development of living cells, given millions of years and the whole planet as a laboratory, is certainly possible and even probable. Most scientists, therefore, would disagree with Wickramasinghe, Hoyle, Crick, and Monod. Simple chemical mixtures of methane, ammonia, water, and carbon dioxide can give rise to several complex chemicals when exposed to raw discharges of electrical energy or ultraviolet light. Hydrogen cyanide can polymerize, without a catalyst, into adenine, one of the most important molecules in living organisms and an essential component of both DNA and RNA. And recent work has shown that even simple strands of RNA are capable of nonenzymatic self-replication[36] and elongation of the type which might allow the organization of much larger molecules necessary for the development of structures resembling the genetic apparatus. It is a fair statement that when the steps on the pathway to a process of chemical evolution from non-living to living matter are examined, no step seems to be forbidden, and some take place with surprising speed.

Nevertheless, we do not have anything resembling the fossil record which would show us that the steps of chemical evolution have in fact occurred, and the suggestion that the first living cell may have been created is entirely proper. However, I do not agree with the suggestion that such a process is so improbable that it could not have happened. I am reminded of Lord Kelvin's calculations demonstrating the impossibility of powered flight and would suggest that perhaps we would do well to learn a little more about the genetic apparatus and chemical evolution before we go on to apply a final label of impossibility on the evolution of the first living cell. However, speculation such as this is entirely reasonable (however much one may disagree with the specifics), and science must make clear to the general public that the *first* cause of life on this planet is neither factually established, nor is it wedded to the process of biological evolution in any way.

A similar statement can be made about the origin of the universe. Our understanding of astrophysics is too primitive to permit us to put a meaningful answer to the question of why there is a universe to begin with. Some scientists even suggest that the fortunate "choice" of physical constants for atomic and subatomic particles makes life itself possible and therefore implies conscious choice of these constants by a clever and powerful creator. Fair enough. What we know about the origin of life and the origin of the universe, therefore, is certainly consistent with the existence of a creator. We would be very foolish to maintain that our advancing understanding of the cosmos and the biological world in any way argues against the existence of God. I, like many other scientists, therefore see no conflict between my religious beliefs and the work of science. This is a point worth making to all who teach and write about science. Consciously or not, many of us have conveyed the impression that the advance of science has meant a retreat for religion. Religious people have felt threatened by this perceived hostility and sought refuge in such barren shelters as scientific creationism. We owe it to ourselves and to the public who trust us to investigate, to teach, and to explain, to make clear that the question of God is a nonscientific one and therefore is entirely beyond our reach as scientists.

I am somewhat saddened by the proofs advanced by the likes of Wickramasinghe and Hoyle for the existence of a creator, because their argument is based on something so tenuous in an empirical sense. Because we cannot explain how life might have originated (it looks too improbable), there must be a creator, or so they argue. One could have made the same argument 50 years ago and argued that because we cannot find (or even imagine) a chemical basis for heredity, this proves that there is a vital principle in living things traceable only to a creator and beyond our understanding in chemistry and physics. Today, thanks in large measure to the likes of Monod and Crick, we know that there is no such vital principle, and we understand a great deal about the chemical nature of heredity. If 30 more years of scientific work show that chemical evolution from nonliving matter is in fact possible, must we now conclude that God does not exist? I think not. We must therefore be very careful if we claim that our inability to understand or imagine a natural process proves creation (God's existence). Why? Because in the not too distant future we may understand that natural process all too completely and wind up as hopelessly confused as, say, Lord Kelvin being offered a ride in a Boeing 747.

As we look back on the last two centuries of science, we find that the authentic question of creation is very much an open one. The American scientific creationists have sought to play a semantic trick by pretending that the support of individual scientists for creation per se is the same as support for their outmoded, Bible-derived theories. This strategy (of pretending that the genuine question of creation is the same as the kind of "creation-science" they advocate) backfired on the scientific creationists when they invited Wickramasinghe to testify on behalf of an Arkansas law requiring the side-by-side teaching of "creation-science" and evolution:

> The most accomplished scientist called by the state to defend the Arkansas creation law yesterday said he thought most of the law was "claptrap" and that much of the so-called "creation-science" in the law was scientifically unsupportable.[37]
>
> Cross-examination closed with the witness (Wickramasinghe) undermining the case of the team who paid his air fare to Little Rock.
>
> Q) Could any rational scientist believe that the earth's geology can be explained by a single catastrophe?
>
> A) [Wickramasinghe] No.
>
> Q) Could any rational scientist believe that the earth is less than one million years old?
>
> A) [Wickramasinghe] No.[38]

Judge William Overton, writing his opinion which overturned the Arkansas "creation-science" law, confessed that he was "at a loss to comprehend" why Wickramasinghe had been called by the State, since his testimony tended to undermine the State's contention that theirs was the only scientifically valid "creation-science" theory.

> The Court is at a loss to understand why Dr. Wickramasinghe was called on behalf of the defendants [The State of Arkansas, defending the "creation-science" law.] Perhaps it was because he was generally critical of the theory of evolution and the scientific community, a tactic consistent with the strategy of the defense. Unfortunately for the defense, he demonstrated that the simplistic approach of the two-model analysis of the origins of life is false. Furthermore, he corroborated the plaintiff's witnesses by concluding that "no rational scientist" would believe the earth's geology could be explained by reference to a worldwide flood or that the earth was less than one million years old.[39]

"Creation-Science" in the Schools

The final stage of any argument between scientific creationism and evolution has to do with the teaching of evolution in the public schools. What creationists propose to ask for, after all, is merely a fair treatment of what they suggest are two opposing points of view: either the universe was created and there is a God ("creation-science"), or it was not created ("evolution-science") and there is no God. They mistakenly argue that since the latter view is already offered in the schools, why not the former?

As I hope to have made clear in the last few pages, the latter view is *not* taught in the schools, because it is *not* an authentic representation of evolution. The scientific nature of evolution, like all scientific theories, *does not* exclude a creator. Indeed, the "fairness" argument for the inclusion of an admittedly religious viewpoint requires intentional misrepresentation of the nature of the theory of biological evolution! It contrives to establish a false duality between the idea of a creator and biological evolution, and constantly encourages students to choose between these intentionally misrepresented extremes. Such a process misrepresents both creation and evolution. Further, it is a great mistake to suggest that there are only two points of view on the ultimate question of "origins." Which theistic view must be considered? Henry Morris's? Mine? Fred Hoyle's? For this reason, all religious aspects of the question of origins are quite properly left out of the scientific curriculum. What must remain as science are the factual conclusions of geology, chemistry, biology, and astronomy: The universe is a vast and ancient place. We know little of its ultimate origins, but we can clearly state that our species was not always here. We developed from a grand and magnificent process known as evolution which unites us in a most remarkable way with every other living thing. We are one with all of the universe, or if you prefer, with all of creation.

NOTES

1. From a transcript of a taped debate between Dr. Duane Gish and Dr. Russell Doolittle (Univ. of Calif. at San Diego) held at Liberty Baptist College, Lynchburg, Va., October 13, 1981.
2. From the opinion of Federal Judge William Overton, setting aside the Arkansas "creation-science" law. As reported in the *New York Times,* January 6, 1982, p. B8.
3. *Is God's Bible* . . . (Memphis, Tenn.: A-M Press, 1980.)
4. Henry M. Morris, *Scientific Creationism* (San Diego: Creation-Life Publishers, 1974), p. 136.

5. Henry M. Morris, *The Remarkable Birth of Planet Earth* (San Diego: Creation-Life Publishers, 1972), p. 94.
6. *Creation/Evolution,* 2(1980):34–39.
7. John C. Whitcomb and Henry M. Morris, *The Genesis Flood* (Grand Rapids: Baker Book House, 1961), pp. 341–42.
8. G. T. Emery, *Ann. Rev. Nuc. Sci.,* 22(1972):165–202.
9. Whitcomb and Morris, *Genesis Flood,* pp. 343–44.
10. Morris, *Remarkable Birth,* p. 92.
11. Morris, *Scientific Creationism,* pp. 157–58.
12. Morris, *Scientific Creationism,* pp. 151–52.
13. Hans Pettersson, "Cosmic Spherules and Meteoritic Dust," *Scientific American,* 202:132.
14. Carl Sagan to K. R. Miller, March 15, 1981.
15. S. R. Taylor, *Lunar Science: A Post-Apollo View* (New York: Pergamon Press, 1975), p. 92. For a complete discussion of regolith and composition, see pp. 55–119.
16. A complete discussion of the phenomenon of residence times can be found in F. MacIntyre's article "Why the Sea Is Salt," *Scientific American,* 223[5](1970):104–15.
17. Duane Gish, *Science Digest,* October 1981, p. 84.
18. Impact Series No. 95 (Institute for Creation Research), May 1981.
19. Duane Gish, *The American Biology Teacher,* March 1973, p. 136.
20. Edwin Colbert, *Evolution of the Vertebrates* (New York: J. Wiley, 1980), p. 75.
21. J. H. Ostrom, *Nature,* 242(1973):136.
22. This quotation is from an article by T. N. George, *Sci. Prog. Ox.,* 189(1960):13. The internal quotation is from Olson, *Evolution,* 13(1959):344.
23. Excellent discussions of the reptile-to-mammal transition can be found in several recent accounts: Crompton and Parker, *Am. Sci.,* 66(1978):192–201; and Crompton and Jenkins, *Ann. Rev. Earth and Plan. Sci.,* 1(1973):131–53.
24. Morris, *Remarkable Birth,* p. 77.
25. Impact Series No. 95 (IRC), May 1981.
26. Whitcomb and Morris, *Genesis Flood,* p. 327.
27. Weber, 1(1980):24–37 and 2(1980):10–25.
28. Freske, 2(1980):34–39.
29. An excellent account is also found in Stephen G. Brush's "Finding the Age of the Earth: By Physics or by Faith?" *J. Geol. Ed.,* in press, 1982.
30. *Creation/Evolution,* 4(1981):5–7.

31. *Ibid.*, 8–16.
32. *Ibid.*, 17–20.
33. *Origins,* 2(1975):64/76.
34. Morris, *Remarkable Birth,* p. 82.
35. *New York Times,* December 27, 1981, p. 48.
36. Eigen et al., *Scientific American,* April 1981, p. 88.
37. *Washington Post,* December 23(?), 1981.
38. From testimony in the Arkansas "creation-science" trial, as reported in *Science* magazine, 215:146.
39. U.S. District Judge William Overton, from the text of his opinion overturning the Arkansas "creation-science" law, pp. 27–28.

ROBERT ROOT-BERNSTEIN

ON DEFINING A SCIENTIFIC THEORY: CREATIONISM CONSIDERED[1]

Controversies rarely concern what they purport to concern. If they did, they would be readily resolved. Instead, controversies usually result from the conflict of hidden assumptions. The evolutionist-creationist controversy is a case in point. So far, evolutionists and creationists alike have argued as if the issues were scientific ones, resolvable on the basis of which theory best explains the most data. Were this assumption correct, then the controversy would have been decided long ago by scientific research. Anyone familiar with the history of this controversy[2] knows, however, that scientific discourse has only exacerbated the conflict. Its cause must lie elsewhere.

The thesis of this essay is that the evolutionist-creationist controversy is not a scientific one. The assumption that evolutionism and creationism are comparable and competitive scientific theories is false. The controversy is, in fact, due to the promulgation of a religious belief—creationism—as a scientific idea. Thus, the controversy results not from scientific issues, but from a confusion between science and religion. Only when this confusion is generally recognized will the controversy be resolved.

To dispel the confusion and to reveal the hidden assumptions underlying this controversy, three questions will be considered: what is a scientific theory? what is a religious belief? and how do they differ? These are not new questions. Historians and philosophers of science and of religion have addressed these questions often and at length. I have integrated their diverse ideas and conclusions in an attempt to place the evolutionist-creationist controversy in perspective. Then, the underlying assumptions can be recognized and the controversy can be seen to be what it is: a conflict between science and religion. Towards this end, this essay is divided into three parts. In the first, a scientific theory is characterized according to well-established criteria. In the second, these criteria are applied to "scientific" or "special creationism." In the third, differences between scientific theories and religious beliefs are discussed, and it is demonstrated that creationism is in fact a religious belief rather than a scientific theory.

Some Criteria for Theory Evaluation

Before one can evaluate the comparative scientific merits of evolution and of creationism, one must first answer the question, what is a scientific theory? A survey of the literature on the history, philosophy and sociology of science[3] reveals that there are at least four fundamental categories of criteria by which theories are judged: (1) logical criteria; (2) empirical criteria; (3) sociological criteria; and (4) historical criteria. Selected samples of each will be discussed.

There are four primary logical criteria for a theory. It must be (1.a) a simple, unifying idea that postulates nothing unnecessary ("Occam's Razor"); (1.b) logically consistent internally; (1.c) logically falsifiable (i.e., cases must exist in which the theory could be imagined to be invalid); (1.d) clearly limited by explicitly stated boundary conditions so that it is clear whether or not any particular data are or are not relevant to the verification or falsification of the theory.

The need for these four criteria should be obvious upon reflection. An idea that is too complex or deals with observations piecemeal can have no practical explanatory value for a scientist. Theories must make clear patterns of things and relationships between things. These patterns and relationships must be internally logical and consistent since these are required attributes of all sound explanations. The explanation must be falsifiable, at least logically if not by actual experiment, or else tautologies or other logically sterile constructions might be admissible as theories. Finally, a theory must be limited by boundary conditions or else there will be no criteria for determining whether or not any particular observations or experiments should or should not be explainable by the theory. In fact, if a theory is totally unbounded, then it is not possible to imagine any observation that is irrelevant to verifying the theory. Thus, an unbounded theory would not be falsifiable. And, if a theory cannot be falsified, it cannot be self-corrected. Yet self-correctability is precisely the characteristic that gives scientific theories their epistemological power: a theory that is incorrect or incomplete can, by attempts to falsify it, reveal its faults or limitations and so be corrected or extended.

Three empirical criteria are of primary importance as well. A theory must (2.a) be empirically testable itself or lead to predictions or retrodictions that are testable; (2.b) actually make *verified* predictions and/or retrodictions; (2.c) concern reproducible results; (2.d) provide criteria for the interpretation of data as facts, artifacts, anomalies, or as irrelevant. The basic point is that not all data are valid for testing any particular theory.[4] Some data may be interpreted as factual (that is, they fall within the boundary conditions specified by the theory and verify its predictions or retrodictions); some may be artifactual (that is, the result of secondary or accidental influences lying outside the boundaries set for the validity of the theory); some are anomalous (that is, demonstrably valid within the bounds of the theory, but also at odds with predictions or retrodictions made by the theory); some are irreproducible and so, invalid; and some are irrelevant since they address the theory not at all.

Once again, the necessity of these criteria should be relatively self-evident. It is possible to imagine theories that are logically falsifiable but experimentally untestable (as when the technology does not exist to convert the logical test into an empirical test). Such a theory cannot be either verified or falsified and so is useless *in practice* to an experimental scientist. One can, at best, only leave such empirically untestable ideas in the scientific purgatory of doubt. A theory must also limit what may possibly be observed in the universe. Otherwise, one would not know what to look for, where, how to look for it, under what conditions, or know whether what one saw was relevant or irrelevant to the theory. If everything is possible under an explanation, then no tests for it can exist.

A theory must also make *verified* predictions and retrodictions to validate itself. It is possible otherwise to imagine theories that make predictions and retrodictions all of which are falsified. A theory whose predictions and retrodictions are falsified, either by the prior existence of relevant data or the subsequent discovery of relevant data, cannot be considered a *valid* theory. It is in need of correction or extension. And, of course, a theory based upon irreproducible results is, in effect, invalidated by the very fact that the results cannot be reproduced; for either the boundary conditions governing the collection of the data have not been properly set, or the original data may have been due simply to coincidence rather than any mechanism proposed by the theory.

Sociological criteria also exist for determining the validity of a theory. A theory must (3.a) resolve recognized problems, paradoxes, and/or anomalies irresolvable on the basis of preexisting scientific theories; (3.b) pose a new set of scientific problems upon which scientists may work; (3.c) posit a "paradigm" or problem-solving model by which these new problems may be expected to be resolved; (3.d) provide definitions of concepts or operations beneficial to the problem-solving abilities of other scientists.

Once again, the need for these criteria is manifest. An idea that does not resolve any recognized scientific problems cannot be called a scientific theory. It can have no effect upon the research activity of scientists. Similarly, an explanation that does not pose new problems does not allow scientists to learn anything they were unable to learn before. A theory therefore has no sociological value unless it provides a model for new or more efficient sorts of scientific activity. Thus, a theory must be stated in terms that are operationally useful to the community of scientists who might use the theory. If the definitions or concepts upon which the theory is based are not operationally useful,[5] then the explanation says nothing experimentally verifiable about nature. Such an explanation cannot, therefore, be scientific.

Finally, there is a fourth set of theory criteria as well: historical ones. A theory must (4.a) meet or surpass all of the criteria set by its predecessors or demonstrate that any abandoned criteria are artifactual; (4.b) be able to accrue the epistemological status acquired by previous theories through their history of testing—or, put another way, be able to explain *all* of the data gathered under previous relevant theories in terms either of fact or artifact (no anomalies allowed); (4.c) be consistent with all preexisting ancillary theories that already have established scientific validity.

These criteria are necessary to ensure that theories are *correctable.* Correctability, in turn, ensures the increase in explanatory power of scientific theories with time, and promotes the consistency and integration of all scientific explanations. Without such criteria, scientists would be free to pick and choose data that favor their explanations while ignoring previously recorded evidence and theories that falsify their own ideas. The historical criteria make such unscientific procedures nearly impossible. The historical records of each science stand as a monument to the series of tests any new theory must pass to prove its mettle against the accumulated knowledge of nature. The longer a science has existed, the greater the accumulated knowledge concerning any particular aspect of nature. Consequently, the more difficult it becomes to reformulate all of that knowledge into a new and consistent framework.

"Scientific" or "special" creationists themselves recognize as valid the four categories of criteria listed above.[6] Part of the present controversy stems from disagreements over the interpretation of these criteria and their application to evolutionism or creationism.[7] As far as I have been able to determine, neither creationists nor evolutionists have attempted to apply these criteria in a formal manner to creationism. Thus, it has never been established that "scientific" or "special" creation is, in fact, a scientific alternative to evolutionism as the creationists claim. On the other hand, the creationists, by selectively quoting from the philosopher of science, Sir Karl Popper, have attempted to make the case that evolution is not a valid scientific theory.[8] Since Popper himself has changed his mind on this issue,[9] it is worth a few moments of time to discuss how to apply these criteria properly in order to yield a valid testing of evolution and creation.

Theories meeting the foregoing criteria obviously do not spring fully formed from the brains of scientists as did Athena from the head of Zeus. Theories evolve. While the process of theory evolution has been little studied, it appears to have the following characteristics:[10] a problem is propounded; numerous ideas for resolving the problem are formulated; each possible resolution is evaluated against logical criteria and preexisting data concerning it; this process of idea generation and evaluation continues until a possible solution having the requisite historical and logical characteristics is found; then the implications of the idea are stated and systematically tested. At this point, one has an hypothesis. If the hypothesis passes empirical testing and has viable sociological characteristics, then it will become an acceptable theory.

Clearly the process of theory invention as a whole is an historical process taking *time* and involving people. The history and sociology of an idea or a theory is therefore a measure of its epistemological status: the greater the number of tests performed and passed by an idea and the longer an idea has performed its sociological functions, the greater its epistemological value as a scientific explanation. It follows that an hypothesis that successfully incorporates or unifies several established theories thereby acquires the combined epistemological value and validity accrued during the historical development of the incorporated theories. Therefore, if a new theory is proposed to replace an old one, the new must not only explain everything that previous, relevant theories explained, but explain even more data (i.e., be tested and verified), more economically ("Occam's Razor") and more completely. Otherwise, there is no logical or sociological reason to prefer the new theory to the old.[11]

How, then, does one determine when a new theory is needed? Since theory building begins with the definition of a set of problems, clearly the problem set defines the need.[12] There are, however, many sorts of problems that do not and cannot logically require resolution by means of a new theory. As this is a point of confusion in the evolutionist-creationist debates, it is one worth examining in detail here. One must ask whether the problems that creationists see with evolutionary theory are the sort that require a new theory (such as creationism) or are resolvable in other ways.

Every properly stated problem defines the method in which it is to be resolved. Because I have written about this subject in detail elsewhere,[13] I will only illustrate this assertion without trying to justify it here. For example, problems concerning the means of collecting data are problems of technique and can only be resolved by the application or invention of data-collecting techniques. Whether or not any particular technique is or is not adequate to the resolution of the data-collecting problem is a problem of criteria and can only be resolved by the development of criteria for technique evaluation. The falsification or verification of theoretical predictions and retrodictions is a problem of data evaluation that can only be resolved by the collection and interpretation of data. Anomalies, paradoxes, and contradictions, on the other hand, create theory problems that have historically required new theory building.[14] Other types of problems also exist, but only one more will be mentioned here: problems that result from interpreting one type of problem to be another or from utilizing artifactual data. These I call artifactual problems. They are irresolvable until recognized to be other sorts of problems. The evolution-creation controversy as a whole is, in my opinion, an example of an artifactual problem, because it is based upon false assumptions.

Clearly, only certain types of problems—the ones I call theory problems—necessitate theory building for their resolution. The vast majority of problems do not. This conclusion should be obvious upon reflection, since there would be no need of empirical research if science were comprised solely of theory problems. In fact, as was stated above, a theory *must* pose a variety of new empirical problems if it is to make predictions, be tested, and be sociologically viable. One must therefore employ great care in evaluating problems to see them for what they are—otherwise one risks raising artifactual problems instead of valid and resolvable ones.

The "scientific creationists" have, unfortunately, fallen prey to exactly this folly. They have interpreted *all* problems of *all* classes to be theory problems requiring the abandonment of evolutionism in favor of creationism. Harold Slusher, Duane Gish, Thomas Barnes, M. E. Clark, and A. E. Wilder-Smith,[15] for example, have written technical critiques of radiometric dating, the age of the universe, evolutionary mechanisms of adaptation, and experiments relating to the origin of life. They raise many problems for evolution, but I have failed to find a single instance in *any* of the cited works of a *theory* problem. Thus, even if one accepts all of the problems creationists raise as valid ones—though many are demonstrably artifactual—the existence of these problems does not invalidate evolution as a scientific theory. On the contrary, it demonstrates just how vibrant the tradition of research in evolutionary sciences is. For if one examines the creationist literature closely, one finds that in no instance have they raised a valid problem that has not already been stated in the evolutionary literature itself. Attention to footnotes, where the creationists use them, clearly demonstrates that to attack evolution, they have had to resort to the evolutionists' own critiques of themselves. Far from denying that evolution has problems, evolutionists have been even more critical than creationists of the theory.[16] Indeed, two evolutionists, Stebbins and Ayala, have just completed their own review of the problems faced by evolutionary theory and have concluded that while many problems do exist, none are of the sort that require abandonment of the theory or invention of a new one to replace it. What is needed, they have concluded, is more research. Even the briefest glance at almost any scientific journal will demonstrate that the evolutionists are carrying out that research.

In short, there is no *scientific* need for a new theory to replace evolution. Neither is there a philosophical, sociological, or historical need. Many respected philosophers of science—notably Barbour, Beckner, Haldane, Hempel, and Hull[17]—have examined evolution in the light of logical and empirical criteria and have concluded that evolution is a valid theory. Popper himself recently stated as much,[18] though the creationists are either ignorant of his statement or choose to ignore it. Reference to any good history of biology or geology will demonstrate that evolution certainly qualifies as a theory according to all of the historical criteria.[19] The same sources, in addition to contemporary scientific journals, will demonstrate that evolution has performed its sociological functions of problem raising and problem solving admirably well.[20] In fact, it is almost impossible to find any major post-Darwinian biological or geological discovery that was not motivated by or has not added to our understanding of evolution. Nor can there be any doubt that evolution has provided the fundamental framework for the planning and evaluation of most biological, geological, anthropological, and ethological research during the past century. Thus, creationist counterclaims to the contrary, scientists and historians, philosophers, and sociologists of science—even many religious scholars[21]—agree that evolution *is* a valid scientific theory.

The problem to be addressed here reduces to this: if creationism is to be considered a valid *scientific* alternative to evolution, then it must also demonstrate itself to be free of theory problems and equivalently or better qualified as a theory according to the same logical, empirical, sociological, and historical criteria by which evolution has been judged. Creationists have simply *assumed* that creationism meets these criteria. They have never demonstrated that it does. In the next section I shall demonstrate that it does *not.*

Is "Special Creation" a Scientific Theory?

Consider creationism in the light of the criteria for theory evaluation. (1.a) Is "special" or "scientific" creationism a simple unifying idea? Certainly. What could be simpler than an omniscient, omnipresent, omnipotent Creator? (1.b) Is creationism stated in a consistent, logical manner? One's answer to this question depends largely upon one's interpretation of the Bible as literally true, true but incomplete, allegorical, or mythical. There is, for example, no consensus as to whether or not the Genesis account of creation is meant literally or is even internally consistent. One's view depends upon one's particular religious belief. Thus, this part of the question cannot be answered in scientific terms and must be deferred. Some aspects of the "special creation" explanation—for example, the dependence upon the "argument from design"—have, however, been refuted on purely logical grounds by a number of eminent philosophers.[22] At least part of the creationist explanation is not, therefore, logically consistent and needs to be revised. (1.c) Is, then, creationism logically falsifiable? Can one imagine any observation that could not be explained by the mechanism of a Creator? No. In Christian belief, the Creator is everything imaginable and more. If, as creationists argue, every natural object is evidence of design, and every evidence of design argues for the existence of a Designer, then everything that can be observed argues for the Designer.[23] Thus the argument from design is tautological: one argues both from design to Designer and from Designer to design. But tautologies cannot be valid scientific explanations because they cannot be falsified. Furthermore, the mechanism by which "special creation" took place—i.e., the Creator—is unbounded. There is nothing that the Creator cannot do. Therefore, creationism fails to postulate explicitly stated boundary conditions within which the validity of the idea of the Creator is limited (1.d). In short, the very dependence of creationism upon the existence of a Creator makes the entire explanation unscientific.

The empirical criteria confirm the conclusion that creationism is not scientific. According to the creationists, creationism does not lead to experimentally testable predictions or retrodictions (2.a). The Bible states Truth, not hypothesis or theory. Truth need not be tested. At least, that is the argument that H. M. Morris, Director of the Institute for Creation Research, has made with regard to the catastrophic Flood geology by which creationists explain the history of the earth:

> The main trouble with catastrophist theories is that there is no way of subjecting them to empirical test . . . There seems to be no restraint on imagination or speculation when catastrophism is espoused, and this is one reason why it has been in such poor repute for over one hundred years. And yet catastrophism, as we have seen, is necessary [because uniformitarian theories do not explain everything]. It is *not* necessary to speculate, however, since the Biblical record has provided a clear description of the causes, nature, and results of *true* catastrophism: The Noahic Flood . . . We cannot verify it experimentally, of course, any more than any of the various other theories of catastrophism [e.g., Velikovsky], but we do not need experimental verification; God has recorded it in His Word, and that should be sufficient.[24]

Duane Gish, another leading proponent of creationism, has gone even further in taking creationism out of the realm of science. Gish maintains that:

> We do not know how the Creator created, [or] what processes He used, *for He used processes which are not now operating anywhere in the natural universe.* This is why we refer to creation as special creation. We cannot discover by scientific investigation anything about the creative processes used by the Creator.[25]

Science, then, is irrelevant to creationism since the mechanism being proposed by the creationists to explain the history of life on earth is neither imaginable nor observable.

The scientific-historical view of creationism is, however, rather different. One might, for example, consider Genesis 1–11 to be not a revelation, but a statement of retrodictions concerning the origin and development of life on earth. It would then be reasonable to ask: (2.b) have these retrodictions been verified by scientific research? Once again, the answer is no. Morris has listed 23 major geological retrodictions made in Genesis 1–11 that are, by his own admission, contradicted by empirical research.[26] Creationism, in other words, has been falsified. Morris, however, contends that it is the Bible that falsifies the geological observations, rather than the other way around, for he explicitly states that "no geological difficulties, real or imagined, can be allowed to take precedence over the clear statements and necessary inferences of Scripture."[27] Clearly, then, there is no limit on the data applicable to testing creationism (2.c), since no data are applicable. It follows that creationism obviously has no need to, and does not, provide criteria for evaluating data as fact, artifact, or anomaly, since *all* data are irrelevant. In consequence, statements by "scientific" creationists that creationism is a "better theory" than evolution because it fits the data better are utter nonsense. On the one hand, there are no imaginable data that could not be explained by a Creator; while on the other hand, creationists state that data are irrelevant to testing creationism in any case. Thus, creationism fails the empirical tests for a scientific theory.

Creationism is no more successful in meeting sociological criteria. It *can* be argued that "special creation" resolves recognized problems currently left unanswered by evolutionary theory (3.a), but not by means of scientific investigation (2.a, above), and only at the cost of rejecting much of the rest of science (see 4.c, below). Creationism does not pose any new *scientific* problems for *scientists* to work on (3.b), as was discussed in the last section. Instead, reference to creationist essays such as those in Lammerts's *Why Not Creation?* or *Scientific Studies in Special Creation*[28] demonstrates that what creationists call "science" in fact resembles the scholasticism practiced by medieval faculties of theology. Their problems are exegetical: how can the data of science best be squared with the statements of the Bible? While these problems are certainly valid ones, they are not scientific, since the Bible, not nature, forms the centerpiece of their research.

Indeed, I have not been able to find any evidence whatsoever that creationists are engaged in *experimental research concerning creationism.* While they are more than happy to point out problems associated with evolutionary theory, they do not contribute to the scientific resolution of those problems. Thus, although some creationists are indeed scientists,[29] their *scientific* research seems invariably to focus upon noncreationist problems. In fact, if one examines carefully all of the data employed by the creationists in arguing for creationism and against evolution, one quickly discovers that, except in the rarest instances, all of it is derived from the research of evolutionists. Sociologically, then, while evolution has been incredibly fruitful of new scientific knowledge, creationism since Darwin has been almost totally fruitless. The difficulty is that, as Gish stated above, the Creator's powers are unknown and unknowable and so cannot be modeled, duplicated, or explained. Thus, creationism cannot and does not provide scientists with a problem-solving model or "paradigm" by which to understand nature (3.c). In fact, Kofahl and Segraves have written that scientific research into the history of life and of the earth is unnecessary: "if one's philosophy is Biblical Theism, the Genesis record [already] provides the answers."[30] In instances like this, creationism is not only unproductive of new scientific knowledge; it is actually antiscientific in its rejection of the need for research.

Creationism also fails sociological evaluation because it does not provide a set of scientifically useful definitions or concepts (3.d). The Creator, Himself, is undefinable and His existence subject only to proof by revelation, but not by reason. In consequence, the mechanism by which the Creator created is also undefinable, as Gish has stated. Thus, the concept of "creation" is itself undefined and has absolutely no explanatory value to scientists. To quote T. H. Huxley, "the hypothesis of special creation is, in my judgment, a mere specious mask for our ignorance."[31]

Creationists also fail to define many of the other terms they employ in their creation explanation. A case in point concerns the term "kind." Creationists claim, on the basis of Biblical authority, that the diversity of forms within any "kind" is so limited that evolution is impossible. Unless "kind" can be defined in an operationally testable manner, however, no research can be conducted either to verify or falsify the creationist claim. "Kind" has not, in fact, been operationally defined. Gish, for example, gives examples of "kinds" being species, genera, families, and classes, and finally concludes that "we cannot always be sure, however, what constitutes a separate kind."[32] Other creationists, e.g., Morris and Zimmerman,[33] have reached the same conclusion. In consequence, when creationists argue that variation in living organisms is limited to "kinds," they are neither providing a testable statement, nor are they providing a valid argument against evolution. They are simply using a scientifically meaningless word to attempt to refute science.

Not surprisingly, creationism also fails to meet the historical criteria required by a scientific theory. In the first place, since "scientific" creationism is being espoused as an alternative to evolutionary theory, one must ask whether it meets or surpasses all of the historical criteria set forth in previous theories of the origin and development of life (4.a). Clearly not: any good history of biology or geology explains the many ways in which creationism has failed to account for the data discovered by these sciences during the past century.[34] Creationism is not a new explanation—it is older than evolution—and therefore it has had more than ample time to prove its utility as a scientific explanation. It has failed. To bring it forth again at this late date, without modification and in the absence of demonstrated scientific need, is unwarranted.

The second historical criterium (4.b) is that theories must be able to explain all data gathered under previous, relevant theories in terms either of fact or artifact. Evolutionary theory succeeded in doing this when displacing creationism during the nineteenth century. It is reasonable to expect creationism to be able to do the same if it is to be an alternative to evolutionary theory today. But creationism cannot explain all of the data collected by evolutionists. Morris, as we have seen, admits that the major retrodictions of creationism have already been falsified. G. M. Price, a widely cited proponent of geological catastrophism has written:

> That many (but minor) scientific objections can be raised against this interpretation [Biblical catastrophism], cannot be regarded as

> entirely precluding this as the true explanation of the facts of the rocks. It would be quite unreasonable to ask us to explain in detail just *how* everything took place, both in the way of the formation of the strata and in the recovery of the world by the subsequent redistribution of mankind and animals over the earth.[35]

Yet creationists insist that evolution be able to explain, in detail, just how these events occurred, and evolutionists are attempting to do so. It seems that creationists employ a double standard with regard to this criteri only to draw attention away from their own failures.

Finally, creationism also fails the historical criteria by being inconsistent with preexisting and well-established ancillary theories (4.c). As Walter Lammerts, one of the organizers of modern "scientific creationism," once stated, "Our aim is a rather audacious one, namely the complete reevaluation of science from the theistic viewpoint."[36] Such a reevaluation is necessary for the creationists because they deny, among other long-established theories, explanations of radioactive decay, the constancy of the speed of light, and equilibrium thermodynamics[37] (which can be used to date the age of the universe, the earth or the fossils in it[38]); explanations of sedimentation, fossilization, and geological change[39] (which are used to understand the geological record[40]); chemical affinity and reaction rates[41] (which are used to demonstrate that chemicals do combine into ordered aggregates spontaneously[42]), etc. The list is almost infinite. In short, the creationists have set themselves up, not only against evolution, but against accepted interpretations of major aspects of every science. To establish creationism, then, it is not sufficient for creationists simply to establish the plausibility of their explanation of the fossil record—they must also be able to establish the plausibility of an entirely new "science from the theistic viewpoint" upon which the creationist explanation rests. Anyone familiar with creationist literature is surely aware that creationists have not as yet come close to establishing a "theistic physics" or a "theistic chemistry." Nor should we expect them to do so: for to do so would be to demonstrate that every major scientist from Copernicus to Newton to Einstein and Lavoisier to Pauling has misunderstood nature. It seems more reasonable at present to assume that the creationists have misunderstood the subtle thought of these great minds.

Creationism: A Religious Belief

If creationism is not a scientific theory, then what is it? As previous passages have hinted, creationism is a religious faith. For faith "has been well defined as a belief which is not altered or shaken by evidence to the contrary."[43] Creationists do indeed ignore evidence falsifying the retrodictions of Genesis 1–11. They reject aspects of almost every established science in their attempts to validate the creation explanation. In consequence, their belief in creationism is based on faith. Indeed, some believers in Creation are honest enough to admit this point. A. R. Peacocke certainly maintains this view;[44] and he states that

> . . . for [Karl] Barth and his followers, the doctrine of creation was an article of *faith;* it was not knowledge which a man might procure for himself or which would ever be accessible to observation or logic.[45]

G. M. Price, one of the founders of modern creationism, has written similarly that "we must not expect even a true view of natural science to do much more than this: it removes difficulties, and thus makes it possible for us to believe [in the Bible]. To do more than this would be to coerce our minds, leaving no room for faith."[46]

Religoius beliefs are characterized by more than just faith, however. Philosophers of science and of religion have documented a wide range of differences between scientific and religious explanations.[47] Briefly, the two sorts of explanations are characterized as follows. Scientific theories are (1) comprised of contingent or tentative knowledge which is (2) organized to be operationally useful for (3) solving problems concerning *particular* aspects of nature that (4) exist in the here and now. Scientific explanations may not have recourse to final causes and may only be stated in terms of secondary ones.[48] Because scientific knowledge is contingent, and because the causes invoked can never be final ones, science must promote (5) skeptical consideration of (6) alternative explanations that (7) are evaluated against one another on the basis of empirical and logical tests. Religious beliefs are usually characterized very differently. Religious beliefs are (1) comprised of absolute knowledge ("Truth") (2) concerning values and morals that

(3) direct universal aspects of human existence and (4) emphasize the supernatural, either in time (e.g., afterlife) or in space (e.g., Heaven). Religious explanations are stated in terms of a final cause (i.e., some sort of god). Because religious beliefs are absolute, and because they are based upon supernatural (and thus unobservable) causes, religion promotes (5) faith in (6) an orthodox doctrine that is (7) established by reliance upon authority (e.g., a holy man, a sacred text, or a revelation).

Careful scrutiny of creationist passages already cited in this essay leaves little doubt that creationism falls into the category of a religious belief. Creationism cannot exist separate from a belief in a universal, omnipotent, supernatural Creator. Otherwise, there is no mechanism by which to explain the Creation. The Creator is, in addition, a final cause as Gish, Morris, A. E. Wilder-Smith, and Kofahl and Segraves make abundantly clear in their creationist texts.[49] Employing a final cause in an explanation automatically disqualifies it as a scientific explanation and qualifies it as a religious one.

Further, careful reading of creationist texts reveals that creationism has its basis not just in *any* creator, but in the particular Creator described in the Christian Bible. The Bible represents the creationists' religious authority. Creationist research, as we have already seen, is carried out not in the skeptical pursuit of alternative explanations, but solely to promote the authority of the Bible and to verify the absolute truths contained in it. Thus, one finds that even *"scientific"* creationists place the authority of the Bible *above* science itself. Harold W. Clark, for example, has written:

> Some object to [the "New Creationism"] saying that the Bible is being placed on a scientific basis and being used as a textbook of science. To this objection it may be pointed out that the Bible is not a textbook, but that it does contain fundamental truths by which all scientific study must be oriented . . . The Bible must not be judged by men's ideas of science, for scientific theories come and go; on the other hand the World of God abides forever, and human theories must be brought to the unerring standard of the Word. The Bible may not give details, they are left for man to discover; it does, however, lay down basic philosophical principles on which science is to be interpreted.[50]

Clark's statement, taken in conjunction with the passages from other "scientific" creationists cited in the last section, establishes beyond doubt that their criteria for choosing explanations are not scientific ones. The creationists judge explanations according to religious criteria: consistency with orthodox doctrine as it is stated in the Bible.

It can further be demonstrated that the creationists' reliance upon Biblical authority stems from a preoccupation with an orthodox, doctrinaire set of values and morals, rather than with operational knowledge. Creationists do not, as was pointed out above, perform empirical research that results in operational knowledge concerning creation. They cannot, because they deny that it is possible to know *how* the Creator operates. Instead, they are primarily concerned with upholding the values and morals contained in the Bible. "Scientific" creationists do so by insisting upon the scientific and historical truthfulness of their sacred text. They are afraid that if one questions the scientific and historical basis of the Bible, then one will then reject all Biblical doctrines and thus Christianity itself.

The preoccupation with theistic values and morality is clear in most creationist texts, if one looks carefully. A. E. Wilder-Smith, for example, concludes a basically scientific critique of evolution not by claiming that creationism is a better theory, but by calling for a revolution in biology that will restore Christian values to their rightful place in society:

> If [scientists and politicians] become convinced that Darwin was mistaken, then the standards of evolution will no longer be applicable to morals, politics, and religion. *Thus a revolution within the sphere of biology would be followed by an even greater revolution in religion, morals, and politics.* The theoretical scientific considerations of the previous pages are pregnant with even more significant consequences for our morals, religion, and politics.[51]

Morris is also motivated by concern for Christianity. In his book *Biblical Cosmology and Modern Science,* he attacks evolution not on scientific grounds, but on the religious grounds that to believe in evolution, one must deny a literal interpretation of Genesis 1–11:

> One hesitates even to consider the unfortunate type of exegesis which treats Genesis 1–11 as allegorical or mythical, rather than

historical. Nevertheless there seem to be an increasing number of evangelical scholars today who are advocating the notion that this section is only a great hymn, or liturgy, or poem, or saga—anything except real history! They seem unaware or unconcerned that this type of interpretation inevitably undermines all the rest of Scripture. If the first Adam is not real, as Paul taught, and if therefore the Fall did not really take place, then neither is the second Adam real and there is no need of a Savior . . . If we are permitted to interpret Genesis in this fashion, what is to prevent our interpreting any other part of Scripture in the same way? Thus, the Virgin Birth may, after all, be only an allegory, the Resurrection could be only a myth of supra-history, the Ten Commandments only a liturgy, the Crucifixion only a dream. Every man may interpret Scripture as suits his own convenience, and thus every man becomes his own God![52]

Morris goes on to say that he fears the same breakdown of morals that participants in the famous Scopes Trial feared:

If I lose faith with Genesis, I'm afraid I'll lose faith in the rest of the Bible; and if I want to commit larceny, I'll say I don't believe in the part of the Bible that says "Thou shalt not steal." Then I'll go out and steal. The same thing applies to murder.[53]

No Adam, no fall; no fall, no atonement; no atonement, no Saviour. Accepting Evolution, how can we believe in a fall?[54]

Clearly, the creationists' persistent attacks upon evolution as being "purely atheistic, materialistic, and mechanistic" and excluding "an explanation based on theism"[55] can only be motivated by religious, rather than scientific, concerns. They are invoking historical and scientific arguments not in the interest of furthering historical or scientific knowledge in its operational sense, but rather to prevent the authoritative basis of their religion from being undermined. In consequence, creationists are concerned foremost with a religious issue. Scientific issues are only a peripheral means to their end.

In summary, the evolutionist-creationist controversy is not a result of problems posed by two competing scientific theories. It is the result of problems posed by a religious belief competing with a scientific theory. Indeed, Richard Bliss, Director of Curriculum Development for the Institute for Creation Research, has recently admitted this himself. Bliss said:

> An individual once reported to the news media that the two models [evolution and creationism] were "apples and oranges"; I agree. We are dealing with two entirely different "world views" based upon an interpretation of the scientific data in some cases and based upon religious philosophies in others.[56]

Conclusion: What Relationship Between Science and Religion?

The evolution-creationism controversy is not what it first appears to be. Creationism is not a scientific alternative to evolution, no matter how often the creationists insist upon placing "scientific" in front of "creationism." Creationism is a religious belief. The present controversy is not, therefore, born of scientific problems; it is born of the misrepresentation of a religion as a science. In consequence, the real issue to be discussed is not whether creationism is a valid scientific alternative to evolution, but what relationship should exist between scientific ideas and religious beliefs.

This is no small problem. Alfred North Whitehead wrote in 1925 that "when we consider what religion is for mankind and what science is, it is no exaggeration to say that the future course of history depends upon the decision of this generation as to the relations between them."[57] Whitehead's generation has passed, however, and still the problem of the relationship between science and religion remains. Indeed, it may remain for every generation to wrestle with, for perhaps there is no single solution to it. And if there is no single solution, then attempts by creationists to legislate their own particular answer into law must be seen as misdirected. For to paraphrase Kant, creationism is *a* religion, but it is not Religion.

The creationists' idea of a proper relationship between science and religion is not the only one possible. Scientists and theologians alike have suggested many alternatives.[58] Indeed, there is reason to believe that the creationists' idea is unacceptable not only scientifically, but on religious grounds as well. The Vatican II Council, for example, concluded that "research performed in a truly scientific manner can never be in contrast with faith because both profane and religious realities have their origin in the same God."[59] Evolution and Catholicism have been deemed by the Roman Catholic Church to be compatible.[60] The result is a mixture of science and religion called "evolutionary theism" in which God created the matter and laws of the universe; evolution is the manner in which these laws have unfolded. Most Jews and many non-Fundamentalist Protestants also believe in some form of "evolutionary theism." Yet the "special creationists" have gone to great lengths in their books to deny "evolutionary theism" any validity.[61] Indeed, for some, research into evolutionary problems such as the origin of life is not just invalid; it is the work of the devil:

> Finally, even though the artificial production of a living being (should it come about) need be no threat to our faith, yet the attempt to do such a thing may be something in which no Christian (or Jew, or, for that matter Moslem) should be engaged. The building of the tower of Babel, so far as we know, threatened no one's faith, but for all that it was presumption. Incidentally, if it be said that the magicians of Egypt had help from dark powers [in spontaneously creating life (Exodus 8:7)] can we be sure (I say this in all seriousness) that the same might not be true of similar attempts today [Revelations 13:15]. The devil can think of many ways of working with man to man's harm, nor is he limited to appearing complete with horns and tail and stirring the cauldron with his pitchfork.[62]

Believe it or not, this passage may be found printed in a book entitled *Scientific Studies in Special Creation,* yet the message is clear enough: there are some sorts of scientific research that, to some people, are not compatible with belief in God.

What is a devout individual to believe? Is scientific research compatible with faith, or is it often the work of the devil? Clearly, at least in this case, one's answer will differ dramatically depending upon one's particular religion. Most Catholics, Jews, and Protestants see little or no incompatibility between science and religion. Most Fundamentalists see a great deal of incompatibility. Thus, it is adherence to a particular religious belief, not science, that is the cause of this controversy. Scientific knowledge is acceptable to men and women all over the world, regardless of race, religion, creed, or country; but only Biblical Fundamentalists can and do believe in "special" or "scientific" creationism.

The resolution that I offer the evolution-creationism controversy is neither new nor in any way novel. In 1930, twelve eminent English scholars, including five distinguished representatives of various churches and a number of highly respected scientists, agreed unanimously "that conflicts between Science and Religion are avoidable when each of them avoids encroaching upon the domain which naturally belongs to the other."[63] Controversy results only when either science invades the domain of religion, or when, as in the present case, a religion invades the domain of science. In the latter case, concluded the Bishop of Birmingham, invading "fundamentalists and magic-mongers alike merely do harm to true religion . . . [for] in every such direct battle since the Renaissance, science has been the victor."[64] More recently Harvey Cox, Professor of Divinity at the Harvard Divinity School, has reached the same conclusion:

> The notorious 19th century "Warfare Between Science and Religion" arose from mistaken notions of what religion and science are. Although there are still occasional border skirmishes, most theologians and scientists now recognize that religion overstepped its boundaries when—at least in the West—it tried to make geological and biological history into matters of revelation.[65]

One can only rue the fact that "scientific creationists" feel the necessity once again to overstep those boundaries, provoking yet another controversy on the same old theme. If one understands what science is and what religion is, and what are the limitations of each, then these two most important of man's activities need not be at odds.[66]

NOTES

1. I must acknowledge that my ideas have been shaped by very fruitful discussions with Ruth Doan, Don McEachron, Jonas Salk, and Fred Westall. My wife, Michèle Root-Bernstein, has, as always, acted as both sounding board and editor. The opinions I have expressed are, however, mine alone. Many thanks to Barbara Robinson for typing the manuscript.
2. See: Andrew D. White, *A History of the Warfare of Science with Theology in Christendom,* 2 vols. (New York: Dover Publications,

1960). James R. Moore, *The Post-Darwinian Controversies* (Cambridge and New York: Cambridge University Press, 1979).

3. It is impossible to list all of the many sources integrated in producing the list of criteria presented here. Some of the most influential and relevant works were:

Barbour, Ian G., *Issues in Science and Religion* (Englewood Cliffs, N.J.: Prentice-Hall, 1966).

Barnes, Barry, *Scientific Knowledge and Social Theory* (London: Routledge, 1974).

Beckner, Morton, *The Biological Way of Thought* (Berkeley and Los Angeles: University of California Press, 1968).

Bernard, Claude, *An Introduction to the Study of Experimental Medicine,* H. C. Greene, ed. (New York: Dover Publications, 1957).

Carmichael, R. D., *The Logic of Discovery* (Chicago and London: The Open Court Publishing Co., 1930).

Elsasser, W. M., ed., *The Chief Abstractions of Biology* (New York and Amsterdam: North Holland-Elsevier, 1975).

Gillispie, C. C., *The Edge of Objectivity* (Princeton: Princeton University Press, 1960).

Haldane, J. S., *The Philosophical Basis of Biology* (Garden City, N.Y.: Doubleday, Doran and Co., 1931).

Hempel, Carl, *Philosophy of Natural Science* (Englewood Cliffs, N.J.: Prentice-Hall, 1966).

Hull, David, *Philosophy of Biological Science* (Englewood Cliffs, N.J.: Prentice-Hall, 1974).

Kuhn, Thomas, *The Essential Tension* (Chicago and London: University of Chicago Press, 1977).

Kuhn, Thomas, *The Structure of Scientific Revolutions,* 2nd ed. (Chicago and London: University of Chicago Press, 1970).

Merton, Robert K., *Sociology of Science,* N. W. Storer, ed. (Chicago and London: University of Chicago Press, 1973).

Pierce, Charles S., *Essays in the Philosophy of Science,* Vincent Tomas, ed. (New York: The Liberal Arts Press, 1957).

Polanyi, Michael, *Personal Knowledge* (Chicago and London: University of Chicago Press, 1958).

Popper, Karl, *The Logic of Scientific Discovery* (New York: Basic Books, 1959).

4. Robert Root-Bernstein (letter to the editor), *Science,* 212 (1981), pp. 1445–48.

5. Examples of non-operationally defined concepts abound in the history of science. Two of the best-known are Driesch's embryological notion of "entelechy" and Mesmer's hypnotic "magnetic fluid." No evidence was found for the existence of either.
6. See:

   Gish, Duane (letter to the editor), *Discover,* July 1981, p. 6.
   Klotz, John W., "Philosophical and Theological Background [to Creationism]" in Walter E. Lammerts, ed., *Why Not Creation?* (Presbyterian and Reformed Publishing Co., 1970), pp. 5–24.
   Kofahl, R. E. and Segraves, K. L., *The Creation Explanation* (Wheaton, Ill.: Shaw, 1975), pp. 107–14.
   Morris, Henry M., *The Scientific Case for Creation* (San Diego: Creation-Life Publishers, 1977).

7. Creationists often alter the criteria to suit their needs. E.g., Klotz, (1970, p. 8), Gish (1981, p. 6), and R. E. Kofahl (letter to the editor), *Science,* 212 (1981), p. 873, each asserts that "theories must be testable *by direct observation.*" Since no one can actually see a new species evolve, they argue, then evolution cannot be a scientific theory (see also H. M. Morris, ed., *Scientific Creationism* (San Diego: Creation-Life Publishers, 1974), pp. 4–13). No major theory has ever been tested *by direct observation.* One cannot directly observe gravity or energy or entropy or electricity or magnetism. Yet creationist scientists and evolutionist scientists alike use these theories. They are valid scientific theories because they make predictions and retrodictions of *effects* that are directly observable. So does evolution. Creationism does not.
8. Gish, 1981, p. 6; Kofahl, 1981, p. 873; Duane Gish, "The Genesis War," *Science Digest,* October 1981, p. 82. Gish and Kofahl refer to Sir Karl Popper, *Unended Quest* (1976). The same passage is in P. A. Schilpp, ed., *The Philosophy of Karl Popper* (La Salle, Ill.: Open Court Press, 1974), pp. 134–43.
9. Karl Popper (letter to the editor), *New Scientist,* 87 (1980), p. 611.
10. Linus Pauling has summarized theory invention as follows: ". . . you just have lots of ideas and throw away the bad ones. And this, I think is part of it, that you aren't going to have good ideas unless you have lots of ideas and some sort of principle of selection." (Transcript of "Linus Pauling: Crusading Scientist," produced by WGBH-TV for NOVA, 1977, p. 24.)
11. Kuhn (1970) makes this point particularly strongly.

12. Robert Root-Bernstein, "The Problem of Problems," *Journal of Theoretical Biology* 99 (1982), pp. 193–201.
13. *Ibid.*
14. Kuhn (1970) discusses this at length. See also Werner Heisenberg, *Physics and Philosophy* (New York: Harper and Brothers, 1958), pp. 30–44.
15. Thomas G. Barnes, *Origin and Destiny of the Earth's Magnetic Field* (San Diego: Creation-Life Publishers, 1973).

    Clark, M. E., *Our Amazing Circulatory System . . . By Chance or Creation?* (San Diego: Creation-Life Publishers, 1975).

    Gish, Duane, *Speculations and Experiments Related to Theories on the Origin of Life: A Critique* (San Diego: Institute for Creation Research, 1972).

    Slusher, Harold, *Critique of Radiometric Dating* (San Diego: Creation-Life Publishers, 1981).

    Wilder-Smith, A. E., *The Natural Sciences Know Nothing of Evolution* (San Diego: Creation-Life Publishers, 1981).

16. G. Ledyard Stebbins and Francisco J. Ayala, "Is a New Evolutionary Synthesis Necessary?" *Science,* 213 (1981), pp. 967–71. For other evolutionist critiques of evolution, see the "Bibliography of Creationism," part II, in Morris's *Scientific Creationism,* p. 205. What Morris does not tell his readers is that each of these critiques was written to *correct* some aspect of evolutionary theory, not to argue against evolution itself as the creationists maintain (e.g., Duane Gish, *Evolution? The Fossils Say No!* 3rd ed. (San Diego: Creation-Life Publishers, 1979), pp. 1–29).
17. See note 3 for sources.
18. Popper in *New Scientist.*
19. See:

    Allen, Garland, *Life Science in the Twentieth Century* (New York: John Wiley and Sons, 1975).

    Coleman, William, *Biology in the Nineteenth Century* (New York: John Wiley and Sons, 1971).

    Daniel, Glyn, *The Idea of Prehistory* (Harmondsworth: Penguin Books, 1971).

    Gillispie, C. C., *Genesis and Geology* (New York: Harper & Row, 1959).

    Nordenskiold, Eric, *The History of Biology,* L. B. Eyre, trans. (New York: Tudor Publishing Co., 1928).

Rudwick, Martin J. S., *The Meaning of Fossils* (London: MacDonald; New York: American Elsevier, 1972).

20. A particularly apropos example of how evolutionary problems are raised and resolved—and one which proves, incidentally, that evolutionary theory makes retrodictions that can be falsified—is Niles Eldredge's "The Elusive Eureka," *Natural History,* August 1981, pp. 24–26. See also S. J. Gould, "On Paleontology and Prediction," *Discover,* July 1982, pp. 56–57.
21. In 1978, theologian John A. T. Robinson debated Duane Gish (*Christianity Today,* April 1980, p. 50). The theological tradition supporting evolution goes back much further, however. One of the first clergymen to support Darwin was Charles Kingsley (1819–1875), who wrote to Darwin in 1860: "I have gradually learnt to see that it is just as noble a conception of Deity to believe that He created animal forms capable of self-development into all forms needful . . . as to believe that He required a fresh act of intervention to supply the lacunae which He Himself made. I question whether the former be not the loftier thought" (Brenda Colloms, *Charles Kingsley* (London: Constable; New York: Barnes and Noble, 1975), pp. 243–44). See notes 64 and 65 for other, more modern theologian adherents of evolution.
22. See:

Darwin, Charles, *On the Origin of Species* (Cambridge, Mass. and London: Harvard University Press, 1964), reprint of 1859 edition. See particularly pp. 62–79 and 201–6.

Haldane, J. S., "Biology in Religion," *The Modern Churchman,* 14 (1924), pp. 269–82.

Hume, David, *Dialogues Concerning Natural Religion* (Indianapolis and New York: Bobbs-Merrill Co., 1947), N. K. Smith, ed., pp. 45 ff.

Root-Bernstein, Michèle, and Root-Bernstein, Robert (letter to the editor), *New York Times Magazine,* July 19, 1981, p. 54.

23. The most blatant example of this sort of reasoning by "scientific creationists" is to be found in R. E. Kofahl and K. L. Segraves, *The Creation Explanation* (Wheaton, Ill.: Shaw, 1975). The entire book is no more than one extended argument from design. Almost all other creationist texts also include at least a chapter on the argument from design.
24. H. M. Morris, *Biblical Cosmology and Modern Science* (Nutley, N.J.: Craig Press, 1970), p. 30.
25. Duane Gish, *Evolution? The Fossils Say No!,* p. 42. Italics in the

original. It should be noted that the Public School Edition uses the word *Creator,* whereas the unedited edition uses the word *God.*

26. Morris, *Biblical Cosmology,* pp. 59–62. E.g., "(d) Geology teaches that fish and other marine organisms developed long before fruit trees; Genesis 1:11, 20, 21 directly contradicts this order. (e) Geology teaches that the sun and moon are at least as old as the earth, whereas Genesis 1:14–19 says they were made . . . on the fourth day [as were the stars (f)] . . . (i) The Bible states that birds and fishes were created at the same time (Gen. 1:21); but geology says fishes evolved hundreds of millions of years before birds developed . . ." etc.
27. Morris, *ibid.,* p. 33.
28. Walter E. Lammerts, ed., *Why Not Creation?* (Presbyterian and Reformed Publishing Co., 1970) and Walter E. Lammerts, ed., *Scientific Studies in Special Creation* (Presbyterian and Reformed Publishing Co., 1971).
29. Unfortunately, some creationists seem to be claiming better credentials than other individuals find reasonable. Bette Chambers of the American Humanist Association writes that "*The Atlanta Journal* reported on April 27, 1979, the embarrassment of school officials in DeKalb County, Georgia, upon discovering that 'Dr.' Richard Bliss, billed as curriculum development director of the Institute for Creation Research, of San Diego, and author of a creationist text just approved in DeKalb County, was a very recent graduate of the University of Sarasota, Florida. That university, the *Journal* stated, '. . . is not accredited, has no campus, and specializes in graduate degrees.' In addition, Clifford Burdick, who claims Ph.D. credentials from the University of Physical Sciences of Phoenix, Arizona, and whose name and title appear on Creation Science Research Center letterhead, may want to explain a letter from the Arizona State Board of Education to a colleague of mine, dated January 31, 1975, that states, 'We know of no University of Physical Sciences in Phoenix. The only accredited universities in Arizona are the three governed by the Board of Regents: The University of Arizona, Arizona State University, and Northern Arizona University.' Yet both these men are billed as 'scientists' " (from *The Sciences,* December 1981, p. 2). Similarly, "Dr." Harold Slusher's D.Sc. degree is in fact an honorary one awarded by Indiana Christian University (see the biographical sketch in his *Critique of Radiometric Dating* [Creation-Life Publishers, 1973]). Careful attention to the biographies given in the Institute for Creation Research's pamphlet "21 Scientists Who Believe in Creation" (Creation-Life Publishers, 1977) reveals

that, even if one counts the six engineers as scientists, only 16 of the 21 have any kind of advanced degree in *science.* And Joel Gurin writes: "It would be interesting to know more about the members of the Creation Research Society, which claims more than six hundred and fifty members with graduate degrees in the sciences. Although I wrote to them for information about their membership, my letter was never answered" (*The Sciences,* December 1981, p. 3). Is "scientific creationism" in fact being practiced by scientists? If not, can it be considered a science? These are important questions in need of answers.

30. Kofahl and Segraves, *The Creation Explanation,* p. 90.
31. Leonard Huxley, *Life and Letters of Thomas H. Huxley,* vol. II (New York: D. Appleton and Co., 1900), p. 320.
32. Gish, *Evolution? The Fossils Say No!,* pp. 34–37.
33. H. M. Morris, *The Scientific Case for Creation* (San Diego: Creation-Life Publishers, 1977), p. 29 note. P. A. Zimmerman, ed., Rock Strata and the Bible Record (St. Louis: Concordia Publishing House, 1970).
34. See sources note 19—particularly Gillispie (1959) and Daniel (1971). Gillispie's *Edge of Objectivity* (1960) also discusses the many ways in which creationism failed to meet the challenge of Darwinian evolution.
35. G. M. Price, *Evolutionary Geology and the New Catastrophism* (Mountain View, Calif.: Pacific Press, 1926), p. 341.
36. Lammerts, ed., *Why Not Creation?,* p. 2.
37. See:

    Morris, *The Scientific Case for Creation.*
    Morris, ed., *Scientific Creationism.*
    Slusher, Harold, *Age of the Cosmos* (San Diego: Creation-Life Publishers, 1976).
    Slusher, *Critique of Radiometric Dating.*
    Whitcomb, John C., *Origin of the Solar System* (Presbyterian and Reformed Publishing Co., 1964).
    Zimmerman, ed., *Rock Strata* . . . (1970).

38. See:

    Blum, Harold F., *Time's Arrow and Evolution,* 3rd ed. (Princeton: Princeton University Press, 1968).
    Oster, G. F., Silver, I. L., and Tobias, C. A., *Irreversible Thermodynamics and the Origin of Life* (New York, London, Paris: Gordon and Breach Science Publs., 1974).

Prigogine, Ilya, *From Being to Becoming* (San Francisco: W. H. Freeman, 1980).
Rutten, M. G., *The Origin of Life by Natural Causes* (Amsterdam, London, New York: Elsevier, 1971).
Whitcomb, *Origin of the Solar System.*

39. See:

Clark, Harold W., *Fossils, Flood and Fire* (Escondido, Calif.: Outdoor Picture, 1968).
Cook, Melvin, *Prehistory and Earth Models* (London: Max Parrish, 1966).
Gish, *Evolution? The Fossils Say No!*
Lammerts, ed., *Scientific Studies in Creationism.*
Marsh, Frank Lewis, *Life, Man and Time* (Escondido, Calif.: Outdoor Pictures, 1967).
Price, *Evolutionary Geology and the New Catastrophism.*
Whitcomb, John C., and Morris, Henry M., *The Genesis Flood* (Grand Rapids: Baker Book House, 1961).

40. Again, Rutten's *Origin of Life by Natural Causes* provides an excellent summary response. So does Stephen Jay Gould's *Ontogeny and Phylogeny* (Cambridge, Mass.: Harvard University Press, 1977).
41. See: Gish, *Speculations and Experiments;* Smith, *The Creation of Life;* Morris, *The Scientific Case for Creationism;* and Kofahl and Segraves, *The Creation Explanation.*
42. A good summary of work to date can be found in S. W. Fox and Klause Dose's *Molecular Evolution and the Origin of Life* (New York: M. Dekker, 1977).
43. E. R. Goodenough, *The Psychology of Religious Experiences* (New York: Basic Books, 1965), p. 168.
44. A. R. Peacocke, *Creation and the World of Science* (Oxford: The Clarendon Press, 1979).
45. *Ibid.*, p. 14.
46. Price, *Evolutionary Geology*, p. 342.
47. I cannot list every book and article that I have used in synthesizing this list of differences between science and religion. The most important were:

Barbour, *Issues in Science and Religion.*
Barbour, Ian G., ed., *Science and Religion* (New York: Harper & Row, 1968).

Dillenberger, John, *Protestant Thought and Natural Science* (Garden City, N.Y.: Doubleday and Co., 1960).

Hesse, Mary, "Criteria of Truth in Science and Theology," *Religious Studies,* 11 (1976), pp. 385–400.

Peacocke, *Creation and the World of Science.*

Pupin, Michael, ed., *Science and Religion* (Freeport, N.Y.: Books for Libraries Press, 1969).

Robin, Horton, "African Traditional Thought and Western Science," in Bryan R. Wilson, ed., *Rationality* (Oxford: Blackwell, 1970), pp. 131–71.

Russell, Bertrand, *Religion and Science* (London: Oxford University Press, 1935).

Schilling, Harold K., *Science and Religion* (London: George Allen and Unwin, 1963).

Schlesinger, George, *Religion and Scientific Method* (Dordrecht [Holland] and Boston: D. Reidel Publishing Co., 1977).

Shideler, Emerson W., *Believing and Knowing* (Ames, Iowa: The Iowa State University Press, 1966).

48. As several of the authors listed in note 47 state, this excludes the use of God, gods, or any other supernatural agent as a scientific cause.
49. Gish, *Evolution? The Fossils Say No!,* p. 42; Morris, *Scientific Creationism,* p. 5; Wilder-Smith, *The Creation of Life* (1970), p. 254; Kofahl and Segraves, *The Creation Explanation,* preface. The use of final causes (i.e., an Intelligence, Creator, or God) can also be found in almost any other creationist text as well.
50. Clark, *Fossils, Flood, and Fire,* pp. 18–19.
51. Wilder-Smith, *Natural Sciences,* p. 148. See also Smith's Foreword in his *Creation of Life* for expressions of similar religious and moral concerns.
52. Morris, *Biblical Cosmology,* p. 57. Compare this statement, which was made to a religious audience, to Morris's public "scientific" stance: "The scientific creationists themselves are men and women who have acquired all of the standard credentials of the scientist, but who maintain that creation explains the facts of science better than evolution does. To them it is not primarily a question of religion (after all, people can be religious and moral while still believing in evolution), but of science" (Morris, *The Scientific Case for Creation,* p. 2). It seems, then, that there are *two* cases for creationism, but when presenting the *scientific* face, creationists hide the *religious* one. See the Bibliography, Part III of Morris's *Biblical*

Cosmology and Modern Science for references to other religiously motivated arguments against evolution. Note also that many creationists texts must be *revised* for public consumption (e.g., Gish, 1979, or Morris, 1970).

53. Ray Ginger, *Six Days or Forever* (Oxford: The University Press, 1958), p. 111.
54. *Ibid.*, p. 63.
55. Gish, *Evolution? The Fossils Say No!*, p. 186. See also these essays: Bolton Davidheiser, "Social Darwinism" in Lammerts, ed., *Scientific Studies in Special Creation,* pp. 338–43; John N. Moore, "Neo-Darwinism and Society," in Lammerts, ed., *Why Not Creation?,* pp. 367–88; Kofahl and Segraves, *The Creation Explanation,* preface.
56. Letter from Richard Bliss to Ben Patrusky of the Council for the Advancement of Science Writing, dated August 25, 1981, and distributed to the participants in the CASW New Horizons in Science Conference held in Columbus, Ohio, November 8–12, 1981, at which Bliss spoke.
57. Alfred North Whitehead, *Science and the Modern World* (New York: The Macmillan Co.; Cambridge: Cambridge University Press, 1925), p. 180.
58. See:

    Huxley, Julian, *Religion Without Revelation* (New York: Harper and Bros., 1927).

    Kaufmann, Walter, *Faith of a Heretic* (Garden City, N.Y.: Doubleday, 1963).

    Millikan, Robert Andrews, *Evolution in Science and Religion* (New Haven: Yale University Press, 1927).

    Montague, William Pepperell, *Belief Unbound: A Promethean Religion* (New Haven: Yale University Press, 1930).

    Peacocke, A. R., *Creation and the World of Science.*

    Teilhard de Chardin, Pierre, *The Phenomenon of Man,* B. Wall, trans. (London: Collins, 1965 [1959]).

59. Quoted from James Hansen, "The Crime of Galileo," *Science,* 81, March 1981, p. 14.
60. See: Garret Hardin, *Nature and Man's Fate* (New York: Holt, Rinehart and Winston, 1959) and John C. Greene, *Darwin and the Modern World View* (Baton Rouge, La.: Louisiana State University Press, 1961).
61. Kofahl and Segraves, in *The Creation Explanation* (pp. 159 ff.),

argue against all non-Christian religions. They relegate their arguments against non-Fundamentalist Christian religions to an appendix so as not to offend readers. Morris, in his *Biblical Cosmology and Modern Science* (p. 16), is more blatant. He states that "paganism, humanism, and pantheism are merely variant forms of evolutionary uniformitarianism, as ultimately are *all* religions and philosophies except Biblical Christianity." Other authors are more subtle, simply asserting a literalist, Fundamentalist viewpoint by adopting a 6 to 10 thousand-year age for the earth, the existence of a real Adam, the actual occurrence of the Genesis Flood, etc.

62. Harold L. Armstrong, "The Possibility of the Artificial Creation of Life," in Lammerts, ed., *Scientific Studies in Special Creation,* pp. 328–29.
63. Pupin, *Science and Religion,* pp. ix–x.
64. The Rt. Rev. E. W. Barnes, F. R. S., Bishop of Birmingham, in Pupin, *ibid.,* p. 57.
65. Harvey Cox, "Religion," in Alberto Villoldo and Ken Dychtwald, eds., *Millennium: Glimpses into the 21st Century* (Los Angeles: J. P. Tarcher, 1981), p. 255.
66. Don McEachron and I have recast the argument made here in terms suitable for high school students and laymen in our paper "Teaching Theories: The Evolution–Creation Controversy, *American Biology Teacher,* 44 (1982), pp. 413–20.

GEORGE M. MARSDEN

UNDERSTANDING FUNDAMENTALIST VIEWS OF SCIENCE

Fundamentalism has long been identified by its liberal critics with obscurantism. "For the first time in our history," declared Maynard Shipley in 1927 in his *War on Modern Science,* "organized knowledge has come into open conflict with organized ignorance." "If the 'self-styled fundamentalists' gain their objective of a political take-over," Shipley warned, "much of the best that has been gained in American culture will be suppressed or banned, and we shall be headed backward toward the pall of the Dark Age."[1]

Shipley wrote in the wake of the Scopes trial of 1925, which had marked the furthest advance in the fundamentalist assaults on liberal culture. Fundamentalists indeed had successfully promoted laws in several states banning the teaching of Darwinism in public schools. John T. Scopes, a young biology teacher in Dayton, Tennessee, precipitated the trial by defying the Tennessee version of the law. The trial at Dayton became a worldwide sensation as the American Civil Liberties Union sent famed trial lawyer Clarence Darrow for the defense and three-time Democratic Presidential candidate William Jennings Bryan volunteered to aid the prosecution. When Bryan, with no better training than long years as a Sunday School teacher, offered himself to testify as an expert on the Bible, Darrow ran circles of skepticism around his simple literal interpretations. The bizarre episode, together with its rural setting, lent credibility to the liberal views of fundamentalists' essential backwardness. It opened the door also for what in the short run proved a successful counterattack—the attempt to laugh fundamentalism to death. "Dullness has got into the White House," wrote H. L. Mencken in thanksgiving that Coolidge and not Bryan ruled the nation, "and the smell of cabbage boiling, but there is at least nothing to compare to the intolerable buffoonery that went on in Tennessee. The President of the United States may be an ass, but he at least doesn't believe that the earth is square, and that witches should be put to death, and that Jonah swallowed the whale."[2]

Such attacks were premised on the widespread assumption that modern scientific and cultural thought was advancing progressively and irreversibly, leaving behind it primitive prescientific outlooks to die from intellectual undernourishment in isolated backwaters. Fundamentalism, especially with the rural and southern image it acquired at Dayton (although in fact it had large urban and northern components), seemed to fit this view. Indeed, fundamentalists sometimes were frankly anti-intellectualistic, themselves resorting to ridicule, as for instance by pleasing crowds with recitations of learned degrees, "D.D., Ph.D., L.L.D., Litt.D.," and ending with "A.S.S." "When the word of God says one thing," fulminated evangelist Billy Sunday, "and scholarship says another, scholarship can go to hell." "All the ills from which America suffers," declared Bryan, "can be traced to the teaching of evolution. It would be better to destroy every other book ever written, and save just the first three verses of Genesis."[3]

Fundamentalism was in fact struggling intellectually. At the Scopes trial Bryan could name only two scientists who held his views. One was George Frederick Wright of Oberlin who had died five years earlier and the other was a Seventh-Day Adventist geologist named George McCready Price, whom Darrow could characterize as a "montebank and a pretender" without fear of contradiction from the scientific community.[4] Little wonder that judicious observers might conclude, as John Dewey did in 1934, that

> the fundamentalist in religion is one whose beliefs in intellectual content have hardly been touched by scientific developments. His notions about heaven and earth and man, as far as their bearing on religion is concerned, are hardly more affected by the work of Copernicus, Newton, and Darwin than they are by that of Einstein.[5]

Such plausible evaluations, however, were wrong on several counts. First, they implicitly and often explicitly assumed that fundamentalism would soon die away as culturally deprived peoples became better acquainted with the teaching of modern science. In fact, however, fundamentalism survived and adapted quite well to the twentieth-century cultural setting. Their recent resurgence, accompanied by a dedicated contingent of degree-holding scientists and an arsenal of arguments defending the fully "scientific" character of their cosmology, has caught some of the guardians of liberal culture off guard. The obituaries were written too soon.[6] Fundamentalism, which we may briefly define as militantly antimodernist evangelical Protestantism, has proven a resilient force in modern America. Fundamentalist institutions are steadily growing in strength, and their intellectual position is probably more viable relative to the alternatives than it was half a century ago.

One factor in the underestimation of the suitability of fundamentalism to modern culture was a major misapprehension in views such as Shipley's or Dewey's of the relation of fundamentalism to modern science. True, fundamentalists adamantly opposed Darwinism and may have known little of Einstein or Heisenberg; yet they were certainly as modern in their scientific views as Newton and indeed were bona fide champions of a venerable version of the scientific method.

In fact, rather than being indiscriminately antiscientific, fundamentalism when examined as a belief system proves to reflect a striking commitment to the assumptions and procedures of the first scientific revolution. The epistemology that prevailed in Western culture in the seventeenth and eighteenth centuries dominates much of their thinking. Fundamentalist thinkers are in a broad sense Baconians—followers of roughly the principles for attaining objective certainty enunciated by the early seventeenth-century philosopher of science Francis Bacon. True science in this view is a matter of induction. It involves careful observation and experimentation to discover the facts, classify these facts, and generalize from them. While some judiciously conceived hypotheses might be entertained to guide generalization and experimentation, speculative hypotheses incapable of verification by observation are beyond the realm of true science.[7] So Henry Morris, the most prolific of the current champions of "scientific creationism" writes:

> Science is *knowledge* and the essence of the scientific method is experimentation and observation. Since it is impossible to make observations or experiments on the origin of the universe . . . the very definition of science ought to preclude use of the term when talking about evolution.[8]

Contemporary fundamentalists did not invent such Baconian-inductionist definitions of science just for the sake of recent arguments. Rather Baconianism is the hallmark of their intellectual heritage, antedating the Darwinist debates. Fundamentalists are the heirs to informal establishment of evangelical Protestantism in nineteenth-century America. Much of their militancy is ascribable to their sense that Christian culture in America has been lost, due to the rise of sinister liberal forces.[9] Furthermore, if one goes back to the era of evangelical prominence, before the Civil War and before Darwin, one finds the American scientific community dominated by Bible-believing evangelicals. These scientists were almost to a man Baconians, as was most of the intellectual community.[10] Romanticism and transcendentalism did eventually challenge this fusion of Christianity and Enlightenment inductionism; yet these currents encountered stiff resistance in America. "It is taken for granted," complained James Marsh, an early champion of American romanticism in 1829, ". . . that our whole system of philosophy of mind, as derived from Lord Bacon especially, is the only one which has any claims to common sense. . . ." So closely was Baconism identified with Protestantism "that by most persons they are considered as necessary parts of the same system. . . ."[11] "Protestant christianity and Baconian philosophy," agreed Presbyterian philosopher Samuel Tyler, "originate in the same foundation."[12]

Far from seeing science and religion at war, nineteenth-century American evangelical thinkers had so thoroughly incorporated Enlightenment inductionism into their thought that natural science had become a keystone in their defenses of the faith. Responding to eighteenth-century challenges, notably from David Hume, that Christianity should be abandoned for lack of evidence, evangelicals replied that on the contrary there was overwhelming evidence in support of Christianity. Particularly, they saw the scientific revolution as presenting irrefutable and awesome evidence in support of the ancient argument for the existence of God on the basis of the design of the universe. The classic statements of this argument were found in William Paley's popular *Natural Theology,* published in 1802, which was widely used as a college text in England and America through much of the nineteenth century. Basing his arguments on empirical evidence accessible to all right-thinking persons, Paley reasoned that the existence of an intelligent designer of the universe could be demonstrated as certainly as could many of our other most firmly held beliefs. If one found a watch on the beach, one could be sure there was a watchmaker. So if one contemplated the intricacies of the human eye, one knew that a very skilled creator had been at work. The many designs of this creator indicated a unified plan and purpose that on balance was beneficial to mankind. As apologist Bishop Joseph Butler had emphasized earlier in another perennially popular work, *The Analogy of Religion, Natural and Revealed* (1736), the God of creation evidenced by a scientific look at nature had the same characteristics as the God of the Bible.

With such considerations establishing a presumption in favor of a revelation such as the Bible, nineteenth-century evangelical apologists wound together arguments for the truth of Christianity. These arguments they considered to yield virtual certainty, much as many strands could form an unbreakable rope. Many historical evidences, they argued, supported the Biblical accounts. Moreover, the moral teachings of Scripture were marvelously well suited to the known moral character and needs of mankind. It was unthinkable, furthermore, that authors who expressed such unsurpassed moral ideals should lie about matters of fact in history or science. "The Christian religion," declared the renowned college teacher and textbook writer Mark Hopkins, "admits of certain proof." We need only to free our minds of anti-Christian prejudices and approach the evidence "in a position of an impartial tone." "This course alone," said Hopkins in a characteristic affirmation of Baconianism, "decides nothing on the grounds of previous hypothesis, but yields itself entirely to the guidance of acts properly authenticated. . . ."[13]

Such confidence in objective scientific certainty was based on a philosophical foundation of trust in the "common sense" of mankind. Following the Scottish Common Sense philosophy best developed by Thomas Reid (1710–1796), ninetenth-century American evangelical intellectuals typically held that the reliability of the commonsense perceptions of mankind provide a firm basis for inductive argumentation. All normal humans are, by the very constitution of their nature, forced to believe certain basic truths, such as the existence of the external world, the existence of other persons, the continuity of one's self and of others, and the reliability (under certain circumstances) of one's sense perceptions, memory, and the testimony of others. Normal people can no more escape such beliefs than they can avoid breathing. Since such beliefs are virtually universal, arguments from evidence carefully built on this foundation can attain as much objective certainty as the human race can ever hope for.[14]

A feature of such views important for understanding them today is that they emphasize the stability and uniformity of human knowledge across time and cultures. By contrast, much Western thought since the later nineteenth century has regarded truth itself as a matter of evolutionary development. What is regarded as "truth" develops in various ways in various cultures and, moreover, differs according to the psychic makeup of persons within cultures. Accordingly, in many areas of human inquiry "truth" has become largely a matter of convention or, as Richard Rorty has recently argued, whatever our peers will let us get away with.[15] For early nineteenth-century American evangelicals as for most Westerners before evolutionary-developmental categories dominated our thought, our perceptions could be (with the proper care) mirrors of nature.

Moreover, scientific truths known with such certainty and truths revealed in the Bible could not conflict. Confident that truth was stable and in principle discoverable, nineteenth-century evangelicals were sure that a truth firmly established in one area of knowledge would harmonize with those known in another area. "Truth is one," insisted Mark Hopkins. "If God has made a revelation in one mode, it must coincide with what he has revealed in another."[16] An important strand in the rope of arguments for the truth of Christianity, then, was to demonstrate some of these harmonies between the teachings of natural science and the teachings of the Bible.

With the advent of geology the harmonies of the Bible and modern science were not always immediately obvious, but the evangelical American scientific community proved resourceful in demonstrating that they were indeed there. The main problem was that geology seemed to demand vast amounts of times for the development of the earth while Genesis spoke of the entire creation taking place in six days. The most popular solution for the Bible-believing American scientists was to posit that the "days" of Genesis represented indefinitely long eras, not twenty-four hour days.[17] The problem still remained, however, to reconcile the order of the days of Genesis with geological theory. Genesis, for instance, had light created on day one, but the sun not until day four. Such problems, nonetheless, could be resolved in keeping with the latest scientific theories. The nebular hypothesis of Pierre Simon Laplace, for instance, posited the origins of the planets by the cooling of the nebular gases that had surrounded the sun. Evangelicals, adopting Laplace's theory, argued that in such a view, light would be apparent on the first day, but from the perspective of earth the sun would not appear until the gases had cleared by day four.[18]

Geological catastrophism proved more congenial to Bible-believing scientists and theologians. Genesis 1:2 (". . . the earth was without form and void . . .") might indicate a catastrophe after the original creation in Genesis 1:1. Moreover, catastrophism was compatible with the six "days" of the Genesis account in proposing dramatic progressive steps in the formation of the earth. "Geology and the Bible must kiss and embrace each other," proclaimed Southern Presbyterian theologian James Henley Thornwell in 1857, "and this youngest daughter of science will be found . . . bringing her votive offerings to the Prince of Peace."[19]

Such words of triumph, of which there were many, were spoken too soon. Within two years an even younger daughter of modern science appeared who eventually would destroy the tranquility of Christian academia. The initial evangelical reactions to Darwinism, however, were not as strident nor as generally negative as might be imagined. In fact, with the exception of Harvard's Louis Agassiz, virtually every American Protestant zoologist and botanist accepted some form of evolution by the early 1870s. Generations of reconciling the Bible to the latest geological theories had prepared them well for such accommodations. What they insisted on, contrary to Darwin's own view, was that evolutionary development was compatible with purposeful design.[20]

On the other hand, a few theologians, notably the conservative

Presbyterian Charles Hodge of Princeton, argued that such accommodations were nonsense. Darwin, said Hodge, believed in a universe governed by chance, and not by design. To deny design in nature, Hodge asserted, was to deny common sense and to deny God.[21] In the South, views such as Hodge's often prevailed as evidenced by the removal of a number of professors (notably James Woodrow from Columbia, S.C., Theological Seminary in 1886) for their pro-evolutionary views. Perhaps the mentality of the lost cause, bringing southern resistance to all that was Yankee and modern, contributed to making evolution a southern symbol for apostasy.

In the rest of the country, however, the debates over evolution were much like those that had divided evangelical responses to geological theories. Twenty years after the publication of *On the Origin of Species,* one might have thought that a form of theistic evolutionism was well on its way to acceptance among Bible-believing American Protestants—at least among its scientists. Even in theologically very conservative circles opposition to all sorts of evolutionism was not necessarily a test of the faith. No better evidence of this can be found that the fact that in *The Fundamentals,* the volumes published from 1910 to 1915 that gave the name to the fundamentalist movement, two of the three major discussions of science and religion allowed that the earth was far older than suggested by literal readings of Genesis, and that limited forms of evolution were compatible with design and theism.[22]

Such middle positions, common in the early evangelical responses to Darwinism, were by now, however, losing support and failing to produce a new generation of scientific advocates. Unlike the era of reconciliation of the Bible and geology, the situation was now favoring more extreme choices. Many factors, not all of them scientific considerations, contributed to this remarkable polarization. The United States, following a half step behind major European countries, was going through a process of rapid secularization. The scientific community, often eager to free itself from the embarrassments of the narrow evangelical concerns of the previous generation, was proclaiming with its own evangelistic zeal that natural science was uncovering the mysteries of the universe. The new social and psychological sciences joined in this chorus, proclaiming that explanations of natural causes were sufficient to ensure human progress. "Science" in such views had become strictly naturalistic, excluding by definition theological considerations. By the early 1900s this orthodoxy had so far prevailed that the doors of respectable academia were rapidly closing to scientists or social scientists who might directly interject their religious beliefs into a classroom. A major social revolution—vast immigration dispelling illusions of an Anglo-Saxon Protestant consensus—reinforced these trends. The actual pluralism of American society made it, as a matter of equity, inappropriate to promote sectarian religious views in public institutions. The alleged neutrality of the new sciences was a theoretical model that (despite a shaky philosophical base in the post-Kantian world) better fit the demands of a pluralistic society than did approaches based on religious commitments.

Further driving middle positions from the discussion was the advent of negative Biblical criticism. Coming to America from Germany roughly at the same time as Darwinism, these new naturalistic explanations of the origins of the Bible supported the general trend to shift the focus of intellectual inquiry from questions of eternal truths to questions of origins, process, and development. Moreover, the new views undermined confidence in the Bible among many educated people. Reconciling the Bible to modern science was to them no longer an issue. They concluded simply that it could not be done. Even many theologians took this tact, developing the new liberal or modernist theologies that proclaimed that the essence of Christianity was in personal experience or in ethical teachings to which scientific arguments about the accuracy of the Bible were irrelevant.

Contrary to many expectations, however, traditional evangelical Christianity in America did not, like the gas light or the town pump, disappear in the face of scientific advance. Its numerous adherents were faced, nonetheless, with a most difficult situation intellectually. Much of the academic community was telling them that if they were to continue to be modern scientific people they must give up traditional beliefs about the Bible. Some of their own adherents said much the same thing. If one accepted biological evolution—that keystone of modern scientific orthodoxy—faith in the Bible would have to go. Popular evangelists who specialized in exploiting native anti-intellectualism, of course, made the most of such themes. "Whom do you trust? God or the Ph.D.'s?" But even for the serious-minded evangelicals these extremes looked more and more to be the only choices. Determined not to abandon trust in the full authority of the Bible, many of them chose for the traditional faith and accepted being intellectual outcasts.

A new cultural factor accelerated these trends. World War I precipitated in America a widespread sense of cultural crisis. The mobilization, the massive national war effort, and the ensuing "jazz age" brought acute awareness of how far secularization had progressed. The Victorian era was at an end. It was not, however, going down with a fight, as the passage of the national prohibition amendment best testifies. The war itself, despite its actual effects, was in a sense a crusade to save Victorian civilization. Americans hoped to wrest the values of democratic civilization from the onslaughts of the Huns, who had been corrupted by secular philosophies. "If you turn hell upside down," proclaimed evangelist Billy Sunday in 1918, "you will find 'Made in Germany' stamped on the bottom."[23]

In this supercharged atmosphere the evolution question took on a new dimension that pushed people further toward extremes. Biblicist opponents of Darwinism were by now well aware that the issue of evolution in modern thought and culture went far beyond matters of biology. Evolution could also mean a general philosophy of development by solely natural causes. During the war the American propaganda machine, shameless in its zeal to represent the war as the cause of civilization versus barbarism, suggested that German moral abnormality stemmed from Nietzche's "might is right" philosophy. Antievolutionists seized on this theme. Once "German *Kultur* is identified with evolution," declared evangelist Howard W. Kellogg in an address at the Bible Institute of Los Angeles in 1918, "the truth begins to be told" that Germany is "a monster plotting world domination, the wreck of civilization and the destruction of Christianity itself."[24]

The deep sense of cultural crisis that followed the war in America fostered the continuance of such themes. The same thing could happen here. German Biblical criticism and German theology had led the nation away from the evangelical faith. In place of the

Bible they had substituted a secular humanist philosophy that gave "the strong and fit the scientific right to destroy the weak and the unfit."[25]

Despite these broadsides against science without God, Bryan and the other fundamentalist critics always stressed that their views of creation were fully scientific. This indeed was one of their central themes. "It is not scientific truth to which Christians object," he declared in a speech prepared for the Scopes trial, "for true science is classified knowledge and nothing can be scientific unless it is true."[26] "Evolution," on the other hand, "is not truth; it is merely hypothesis—it is millions of guesses strung together."[27]

When fundamentalists made such remarks, as they often did, they were falling back on the strict Baconianism that had been such a prominent part of their nineteenth-century evangelical heritage. Not all nineteenth-century evangelicals, of course, had been consistent Baconians, as some of the accommodations of the Bible to geological speculations indicate. But the strands of nineteenth-century evangelicalism that most influenced the intellectual stance of fundamentalism were determinedly Baconian in their outlooks.

One of these strands was the Princeton theology, a conservative Presbyterian tradition that contributed substantially to anti-evolutionary thinking in the South. At Princeton the whole approach to theology was frankly modeled on natural science. "If the object of . . . [natural science,]" said Princeton's Charles Hodge, "be to arrange and systematize the facts of the external world, and to ascertain the laws by which they are determined; the object of . . . [theology] is to systematize the facts of the Bible, and ascertain the principles or general truths which those facts involve."[28] Particularly important in this view, then, was that the Bible be viewed as containing sure statements of hard facts.[29] Princeton theologians, consistent with this emphasis, became some of America's most ardent defenders of the doctrine of Biblical "inerrancy," the view that all the affirmations of Scripture were errorless on matters of fact. This doctrine became particularly prominent among conservative evangelicals after 1880 and soon became a leading test of faith in the fundamentalist movement.

Especially important for appreciating this concern for science and hard facts in fundamentalism is to see that such concerns antedated

Darwinism and pervaded the whole way of thinking of some conservative evangelicals. Their bias for "scientific" facts versus speculative hypothesis appeared not only in their reactions to Darwinism, but also in their principles for interpreting the Bible itself.

The ways in which these intellectual tendencies influenced fundamentalist Biblical interpretations are especially apparent in another leading strain of much of fundamentalist thought—an elaborate scheme of Biblical interpretation called dispensationalism.[30] Many fundamentalists today are dispensationalists, and some understanding of dispensationalist thinking is an important clue to seeing why they are so insistent on literal interpretations of Genesis. Dispensationalism is a method of Biblical interpretation, developed largely in the nineteenth century, that is frankly based on finding the most literal interpretations of Scripture possible. "Literal where possible" is a slogan of the movement. Particularly important are interpretations of Biblical prophecies, which dispensationalists take to refer to literal historical events. For instance, they long have predicted that the Jews would return to Israel, as the Old Testament prophesies. They also claim that the Bible predicts a series of spectacular historical events and battles that will bring the present era to a close and inaugurate a literal thousand-year reign of Jesus on earth at Jerusalem. This millennial reign will be the last of seven dispensations or distinct eras in world history. The present era, "the church age," is a time of decline and apostasy (a view that helps account for fundamentalists' perennial alarm over the state of culture).

For the dispensationalist fundamentalist, then, the Bible is among other things a prophetic puzzle. Moreover, they consider the exactitude of Biblical statements to be crucial to properly piecing together the scheme of history revealed. It is important, for instance, that the events bringing the end of our era last exactly seven years and that the millennial reign of Christ on earth be exactly 1,000 years. The inerrancy of Scripture in all its statements is accordingly an absolutely essential dogma relating to the entire dispensationalist fundamentalist world view. Moreover, this version of inerrancy carries with it a principle of interpretation. Not only does the Bible not err in any of its assertions, but its assertions are to be interpreted as literal and precise statements of historical fact whenever that is possible.

The student of the Bible then ought to be a scientific thinker who in good Baconian fashion classifies these sure and precise facts revealed in Scripture. So, for instance, C. I. Scofield, author of the immensely influential *Scofield Reference Bible* (1909), emphasized that identifying the seven dispensations of history was the result simply of properly classifying Biblical data. It was, he said in a phrase from the King James Bible, "rightly dividing the word of truth."[31] Arthur T. Pierson, another leading promoter of the dispensationalist-fundamentalist movement, was philosophically more explicit:

> I like Biblical theology that does not start with the superficial Aristotelian method of reason, that does not begin with an hypothesis, and then wrap the facts and the philosophy to fit the crook of our dogma, but a Baconian system, which first gathers the teachings of the word of God, and then seeks to deduce some general law upon which the facts can be arranged.[32]

Not surprisingly, the same views that characterized fundamentalists' interpretations of the Bible survived in their views of natural science itself. So said Harry Rimmer, a dispensationalist and a precursor of the modern creation-science movement, in a 1935 statement:

> A science is a correlated body of absolute knowledge. When knowledge of a certain subject has been gained by observation, proved by demonstration, and refined by experience, we gather the known and proved facts of that subject and correlate them. . . .[33]

This tradition, though not necessarily tied directly with dispensationalism, is carried on by contemporary defenders of creation-science. Duane Gish, a leading spokesman for that movement, says, "For a theory to qualify as a scientific theory, it must be supported by events, processes, or properties which can be observed, and the theory must be useful in predicting the outcome of future natural phenomena or laboratory experiments."[34] By this standard, as they understand it, the hypotheses of evolution simply do not qualify as good science.

One might suppose that this approach would simply lead to the conclusion that one's scientific investigations into the question of origins can not be done without a commitment of faith shaping the investigations.[35] Henry Morris, for instance, sometimes says this. "If man wishes to know anything at all about Creation . . . ," he argues, "his sole source of true information is that of divine revelation."[36] Nonetheless, Morris also insists that we can discuss the question of origins "scientifically and objectively" and "with no references to the Bible or to religious doctrine."[37] The 1981 Arkansas creation-science law reflected the same combination of views when it said, for example, "Treatment of either evolution-science or creation-science shall be limited to scientific evidences for each model and inferences from these scientific evidences. . . ." If, however, the creation-science model originates in divine revelation, it is difficult to see how one can be talking about this model without thereby referring to a religiously derived belief which shapes very much how one views the scientific evidence. Just the reference to "creation-science" itself entails at least an implicit teaching about a creator. Creation-scientists may be correct that exclusively naturalistic evolutionism represents a philosophical commitment that amounts to a virtual religious belief. If so, the same applies to their own scientific investigations. Recently, however, they seem to be trying to have it both ways—speaking of faith commitments or "models" and simultaneously speaking as though science involved simply objective observations of the evidence that could be taught, as the Arkansas law put it, without "any religious instruction or references to religious writings."[38]

Whatever the creation-scientists' own understandings of the relationships between these two strands in their thought, outsiders may be better able to understand fundamentalist views of science if they press the idea of alternative models to a more consistent conclusion. To those not already committed to the truth of the Bible, creation-scientists' scientific views appear to be sheer nonsense. The temptation, however, has been simply to reply in kind, with what amounts to an evolutionists' version of Baconianism, or at least objectivism, as though an objective look at "the facts" will prove the truth of evolution to any candid observer who can think clearly. A well-articulated version of this approach argues, for instance, that so-called scientific-creationism "is not science." Evolution on the other hand is "a fact."[39] Such approaches, while compelling to secularists, are singularly uncompelling to fundamentalists, who are likely to respond by repeating their own well-worn contrast between "the theory of evolution" and "the facts of science," or by saying once again that "evolution does not even qualify as a scientific theory. . . ."[40] The gap and the lack of communication between the two groups could hardly be greater.

One does not have to accept Thomas Kuhn's entire epistemology, nor even his whole account of the structure of scientific revolutions, to recognize that his scenario of "paradigm conflicts" applies here. Two groups, each regarding themselves as preemiently scientific, view their subject through the lenses of conflicting orthodox paradigms. Neither has a monopoly on intelligence or clear thinking. Rather, because of their basic divergence in starting presuppositions, they differ on what counts as "the facts." What is called "the facts" by one side, appears to be wildly speculative theory to the other. So the two sides, because of their differing presuppositions, models, or paradigms, have differing languages. Communication is nearly impossible and each party thinks that members of the other are virtually crazy or irremediably perverse. Neither thinks the other is doing "science" at all.[41]

To better understand fundamentalist views of science, we should review the historical territory we have covered and reflect further on what happened to create these diametrically opposed presuppositions, models, or paradigms. In the nineteenth-century American evangelical world view (which still dominated American higher education little more than a century ago), the Bible stood at the center. It was an authority as high and as sure as the best scientific or deductive reasoning. In fact, the Bible furnished the assumptions on which implicitly scientific inquiry was based. For instance, one of the reasons that the marvels of nature seemed to be irrefutable evidence of purposeful design was that scientific inquiry took place in a framework of assumptions that had originally grown up within a Christian world view where purposeful design was taken for granted. The Bible and science accordingly appeared to support each other fully as sure sources of facts. Facts, as we have seen, were regarded as eternal truths objectively discoverable by either natural or special revelation.

The intellectual revolution of the nineteenth century destroyed for many people several of the most fundamental assumptions on which this world view rested. As we have seen, this revolution was far more than a revolution in biological theory. It involved a profound change in the prevailing basic philosophical assumptions. The fundamentalist outlook preserves essentially Enlightenment and pre-Kantian philosophical categories. Truth is fixed and eternal and something to be discovered either by scientific inquiry or by looking at some other reliable source such as the Bible. Much of the rest of modern thought, however, had gradually come to view the human mind as imposing its categories on reality. Perception itself in this view is an interpretive process. Truth, moreover, is relative to the observer and to the community or culture of the inquirer. Speculative theorizing is essential, since human thought in any case involves such imposing of one's constructs on reality.

Not only did this philosophical revolution undermine the concept of eternal truths generally, it also supplied a new model for explaining reality. Everything, including religion itself, could be explained by reducing it to natural causes in process of development. The proper locus of intellectual inquiry became, often by definition, to explain development in terms of observable natural forces. Appeals to the supernatural were hence a priori written out of fields of "scientific" knowledge. By this standard, of course, the Bible had to go. So far from being at the center of intellectual life as it had often been a generation or two before, it now was eliminated as an authority altogether. So the philosophical revolution, which by itself was enough to create huge paradigm conflicts, was accompanied by a profound practical revolution in assumptions about proper sources of authority.

No wonder then that with all these immense differences in foundational assumptions, fundamentalists and naturalistic nonfundamentalists can not even communicate. Despite some commonality in their traditions in conceiving of what "science" is, the two conceptions operate in frameworks of such diametrically opposed other assumptions that debates over the scientific "evidence" are futile.

An additional point should be added in our attempts to understand the fundamentalist paradigm. Not only do they continue to operate in a framework of assumptions that allows for eternal truths, the supernatural, and the authority of the Bible, but they also have very distinctive assumptions about how the Bible is to be interpreted. As we have seen, these views follow roughly the principle "literal where possible." Such hermeneutical principles are not derived from the Bible itself, but from philosophical assumptions that appear to be closely related to the Enlightenment Baconianism of their tradition—which lends itself toward a strong preference for definite and precise statements of fact.

It is perhaps at this point that the structure of fundamentalist assumptions allows an opening for discussion and criticism in fundamentalists' own terms. Since their hermeneutical principle is not derived from the Bible itself but apparently comes from some alien philosophical sources, this principle may be questioned without threatening any of their religiously based commitments and assumptions (which are almost certainly unassailable by argument). As in every human noetic structure, some of one's basic assumptions support more of the weight of the entire structure than do others. So, for example, one cannot attack a fundamentalist's belief that "the Bible is true" without threatening to collapse his or her entire intellectual world. On the other hand, the hermeneutical principle "literal where possible" may be removable from the structure without threatening everything else. One can still believe that "the Bible is true," God has created the world, He has revealed eternal truths, the Bible reports many real historical events, and the like, even if one approaches the interpretation of the Bible with some other hermeneutic.

As a matter of fact, ever since the advent of Darwinism, many Bible-believing evangelical Christians have been willing to make just such an adjustment. Much as the early evangelical respondents to geological findings often said that the "days" in Genesis 1 need not be interpreted quite as literally as to mean 24-hour days, so evangelicals since Darwin have offered a wide variety of principles of interpretation that suggest that the first chapters of Genesis are not making the scientific claims they might seem to if taken as literally as possible. Just as the "days" can be seen as poetical ways of speaking of aeons (as even William Jennings Bryan thought),[42] so the order of the days might be poetical, and so forth and so on. The point of Genesis, such views reiterate, is not to tell us the details of *how* God created, but to assure us *that* God created the universe and the human race. Such views, which have many variations, are of course quite compatible with much of modern biological theory, even if incompatible with a sheer naturalism that would allow no room for the guiding hand of a purposeful creator.[43]

Once such points of commonality are established between Bible-believing Christians and the rest of the scientific community some fruitful discussion of science/religion questions might ensue. Both sides should be able to recognize, for instance, that the biological and geological data will not settle the questions of the relation of a creator to the origins of the universe and humanity. Such questions are almost always settled by a person's prior commitments and not by an objective look at "the evidence." Furthermore, it is difficult to imagine what sort of evidence would conclusively settle the matter anyway. The relationship of the supernatural to the natural processes in questions of origins involves essentially the same questions as the issue of the relationship of the supernatural to the natural in almost every other sphere of experience.[44] For example, religious experiences can often be accounted for largely in terms of natural causes. Yet such accounts do not at all settle the question of whether a supernatural agent might be acting through such natural means. So with the question of origins. The fact that we may be able to describe fairly completely the natural processes at work does not settle the question of whether a divine being might be using those processes as means of creating a certain type of universe. Once this is recognized—that the scientific evidence is not going to settle the religious issue—useful discussions between those who believe in divine creation and those who do not may ensue. Neither side should suppose that investigation of the technical issues will provide the key for settling the more ultimate issues of life and meaning.

NOTES

1. Maynard Shipley, *The War on Modern Science: A Short History of Fundamentalist Attacks on Evolution and Modernism* (New York, 1927), pp. 3–4.
2. H. L. Mencken, *Prejudices: Fifth Series* (New York, 1926), from selection in Henry May, ed., *The Discontent of the Intellectuals: A Problem of the Twenties* (Chicago, 1963), p. 29.
3. Quotations from Sunday and Bryan are from Richard Hofstadter,

Anti-intellectualism in American Life (New York, 1966 [1962]), pp. 122 and 125.

4. *The World's Most Famous Court Trial: State of Tennessee v. John Thomas Scopes* (Cincinnati, 1925), p. 297.
5. John Dewey, *A Common Faith* (New Haven, 1934), p. 63.
6. This observation is borrowed from the valuable work of Ernest R. Sandeen, *The Roots of Fundamentalism: British and American Millenarianism 1800–1930* (Chicago, 1970). See also his "Fundamentalism and American Identity," *The Annals of the American Academy of Political and Social Science* 387 (January, 1970), pp. 56-65, which describes the institutional basis for the fundamentalist resiliency.
7. This prevalent American understanding of Baconianism is best explained in Theodore Dwight Bozeman, *Protestants in an Age of Science: The Baconian Ideal and Antebellum American Religious Thought* (Chapel Hill, 1977).
8. Henry M. Morris, *Many Infallible Proofs: Practical and Useful Evidences of Christianity* (San Diego, 1974), p. 249.
9. This view is explicit in Tim LaHaye, *The Battle for the Mind* (Old Tappan, N.J., 1980). Richard Hofstadter, in *The Paranoid Style in American Politics, and Other Essays* (New York, 1963) and *Anti-intellectualism in American Life,* makes perceptive comments on this theme. I have discussed it further in *Fundamentalism and American Culture.*
10. This dominance is documented in Bozeman, *Protestants in an Age of Science;* George H. Daniels, *American Science in the Age of Jackson* (New York, 1968); and Herbert Hovenkamp, *Science and Religion in America, 1800–1860* (Philadelphia, 1978).
11. James Marsh, Introduction to Samuel Taylor Coleridge, *Aids to Reflection* (London, 1840 [1829]), p. 40.
12. Samuel Tyler, *Discourse of the Baconian Philosophy,* 2nd ed. (New York, 1850), p. 15, quoted in Bozeman, *Protestants,* p. 128.
13. Mark Hopkins, *Evidences of Christianity* (Boston, 1876 [1846]), p. 39. I have discussed these arguments in more detail in "The Collapse of American Evangelical Academia," in *Faith and Rationality.* Nicholas Wolterstofb ed. (Notre Dame). In Press.
14. Common Sense philosophy is discussed in S. A. Grave, *The Scottish Philosophy of Common Sense* (Oxford, 1960) and in the works by Bozeman, Daniels, and Marsden cited above.
15. Richard Rorty, *Philosophy and the Mirror of Nature* (Princeton, 1979), *passim.*
16. Hopkins, *Evidences,* pp. 97–98.

17. These responses are summarized in Hovenkamp, *Science and Religion,* pp. 119–45.
18. Ronald L. Numbers, *Creation by Natural Law: Laplace's Nebular Hypothesis in American Thought* (Seattle, 1977) is a fascinating account of reactions to this theory.
19. Quoted in Bozeman, *Protestants,* p. 121.
20. This interpretation closely follows that of Numbers, *Creation by Natural Law,* pp. 104–118, and that of the very fine comprehensive study of the variety of reactions, James R. Moore, *The Post-Darwinian Controversies: A Study of the Protestant Struggle to come to Terms with Darwin in Great Britain and America, 1870–1900* (Cambridge, 1979).
21. Charles Hodge, *What is Darwinism?* (London, 1874), pp. 169 and 173.
22. James Orr, "Science and the Christian Faith," *The Fundamentals: A Testimony to the Truth* (Chicago, 1910–15), IV, pp. 91–104; George F. Wright, "The Passing of Evolution," VII, pp. 5–20. Wright had been a protege of Harvard's Christian evolutionist Asa Gray. See William James Morison, "George Frederick Wright: in Defense of Darwinism and Fundamentalism, 1838–1921," Ph.D. dissertation, Vanderbilt, 1971. Orr and Wright vigorously attack the extravagant claims of evolutionists to explain the origins of life entirely without reference to design.
23. Quoted in Ray H. Abrams, *Preachers Present Arms* (New York, 1933), p. 79.
24. Howard W. Kellogg, " 'Kultur'—Applied Evolution," *The King's Business* (February 1919), p. 155.
25. A. C. Dixon, "The Roots of Modern Evils" (1922) quoted from selection in Willard B. Gatewood, Jr., ed., *Controversy in the Twenties: Fundamentalism, Modernism, and Evolution* (Nashville, 1969), p. 121.
26. William Jennings Bryan, "The Fundamentals," *The Forum,* LXX (July 1923), pp. 1675–80 from selection in Gatewood, pp. 137–38. This selection is among other things an attack on "theistic evolution," illustrating the tendency to drive out the middle positions.
27. *The World's Most Famous Court Trial,* p. 323. Cf. William Jennings Bryan, *In His Image* (New York, 1922), pp. 86–135, especially p. 94.
28. Charles Hodge, *Systematic Theology,* I (New York, 1874), p. 18.
29. James Moore agrees that "the conviction that ultimate certainty is the desirable and attainable product of inductive inference" was

one of the major philosophical premises in opposition to Darwinism. *Post-Darwinian Controversies,* p. 205.

30. Ernest Sandeen, *The Roots of Fundamentalism: British and American Millenarianism, 1800–1930* (Chicago, 1970) argues that dispensationalism (a species of millenarianism), sometimes allied with Princeton theology, was the leading strain in fundamentalism. Today most fundamentalists are dispensationalists. Some fundamentalists, such as William Jennings Bryan, were neither dispensationalists nor in the Princeton tradition. One can find, however, the same analogies of theology to natural science in other strands of nineteenth-century evangelical background—e.g., in the thought of Charles Finney, whose outlook resembled Bryan's in many respects. Conservatives in some confessional traditions, such as the Missouri Synod Lutheran, also sometimes hold views of facts, truth, and the inerrancy of Scripture such as those here described.
31. C. I. Scofield, *"Rightly Dividing the World of Truth"* (Revell paper edition, n.d. [1896]), p. 3.
32. Arthur T. Pierson, "The Coming of the Lord," *Addresses on the Second Coming of the Lord: Delivered at the Prophetic Conference, Allegheny, Pa. December 3–6, 1895* (Pittsburgh, 1895), p. 82.
33. Harry Rimmer, *The Theory of Evolution and the Facts of Science* (Grand Rapids, 1935).
34. Duane T. Gish, *Evolution: The Fossils Say No!* (San Diego, 1972), p. 2. Gish adds, "An additional limitation usually imposed is that the theory must be capable of falsification," pp. 2–3. Cf. the statement by Henry Morris, *Many Infallible Proofs,* p. 249, quoted early in the present paper.
35. Ronald L. Numbers, "The Creationists," from a forthcoming collection of essays on the historical relations of Christianity and science, edited by David C. Lindberg and Ronald L. Numbers, provides an excellent history of the development of creation-science thought in the twentieth century. He suggests that they have abandoned their Baconianism for categories borrowed from Thomas Kuhn (their speaking of "models") and have adopted Karl Popper's views concerning falsifiability (see previous note). However, I am not convinced that they adopt these views with any consistency. Rather, as argued below, they retain a good bit of Baconianism as well. Unlike Kuhn, they do not see the presence of a model as subverting an objective look at the facts.

    I am indebted to Professor Numbers for his work and for his helpful comments on the present essay.

36. Henry M. Morris, *Studies in the Bible and Science* (Philadelphia, 1966), p. 114. Gish, *Evolution,* p. 8, says of creation, "Neither can it qualify, according to the above criteria, as a scientific theory, since creation would have been unobservable, and would as a theory be non-falsifiable." He holds, however, that in the scientific realm either creation or evolution may serve as a "model to explain and correlate the evidence related to origins," *ibid.*
37. Henry M. Morris, ed., *Scientific Creationism* (general edition) (San Diego, 1974), pp. 9 and 3.
38. Act 590 of 1981, State of Arkansas, 73rd General Assembly, section 2.
39. Stephen Jay Gould, "Evolution as Fact and Theory," *Discovery* (May 1981), pp. 34–35.
40. Gish, *Evolution,* p. 2.
41. Thomas S. Kuhn, *The Structure of Scientific Revolutions* (Chicago, 1962), *passim.*
42. *The World's Most Famous Court Trial,* p. 302. Bryan said that it did not make any difference to him whether creation took six 24-hour days or "600,000,000 years."
43. James R. Moore, *The Post-Darwinian Controversies,* provides a fine history of the development of these views. Today they are represented in the work of many evangelical scientists in the American Scientific Affiliation, which since 1949 has published the *American Scientific Affiliation Journal.* Such evangelical scientists differ among themselves on the extent of the compatibility of evolutionary development and the Bible.
44. Moore, *Post-Darwinian Controversies,* shows that this point was made by a number of conservative Christian thinkers, notably Calvinists such as Asa Gray of Harvard and his protégé, George Frederick Wright. Augustinians had for fifteen hundred years dealt with the question of the relation of God's all-controlling providence to secondary natural means. In principle, natural selection raised no new issues. See Moore, pp. 252–351.

STEPHEN JAY GOULD

EVOLUTION AS FACT AND THEORY

Kirtley Mather, one of my dearest friends and a pillar of both science and Christian religion in America, died last year at age 89. The difference of half a century in our ages evaporated before our common interests. The most curious thing we shared was a battle we each fought at the same age. For Kirtley had gone to Dayton, Tennessee with Clarence Darrow to testify for evolution at the Scopes trial of 1925. When I think that we are enmeshed again in the same struggle for one of the best documented, most compelling and exciting concepts in all of science, I don't know whether to laugh or cry.

According to idealized, but rarely operating, principles of scientific discourse, the arousal of dormant issues should reflect fresh data giving renewed life to abandoned notions. Outsiders to the current debate may therefore be excused for suspecting that creationists have come up with something new, or that evolutionists have generated some serious internal trouble. But nothing has changed; the creationists have not a single new fact or argument. Darrow and Bryan were at least more entertaining than we lesser antagonists today. The rise of creationism is politics pure and simple; it represents one issue (and by no means the major concern) of the resurgent evangelical right in America. Arguments that seemed kooky just a decade ago have reentered the mainstream.

From *Discover,* May 1981.

Creationism Is Not Science

The basic creationist attack falls apart on two general counts before we even reach the supposed factual details of their complaints against evolution. First, they play upon a vernacular misunderstanding of the word "theory" to convey the false impression that we evolutionists are covering up the rotten core of our edifice. Secondly, they misuse a popular philosophy of science (Popper's principle of falsification) to argue that they are behaving scientifically in attacking evolution. Yet the same philosophy demonstrates that their own belief is not science, and that "scientific creationism" is therefore meaningless and self-contradictory, a superb example of what Orwell called "newspeak."

In the American vernacular, "theory" often means "imperfect fact"—part of a hierarchy of confidence running downhill from fact to theory to hypothesis to guess. Thus the power of the creationist argument: evolution is "only" a theory and intense debate now rages about many aspects of the theory. If evolution is worse than a fact, and scientists can't even make up their minds about the theory, then what confidence can we have in it? Indeed, President Reagan echoed this argument before an evangelical group in Dallas when he said (in what I devoutly hope was campaign rhetoric): "Well, it is a theory. It is a scientific theory only, and it has in recent years been challenged in the world of science—that is, not believed in the scientific community to be as infallible as it once was."

Well, evolution is a theory. It is also a fact. And facts and theories are different things, not rungs in a hierarchy of increasing certainty. Facts are the world's data. Theories are structures of ideas that explain and interpret facts. Facts don't go away when scientists debate rival theories to explain them. Einstein's theory of gravitation replaced Newton's in this century, but apples didn't suspend themselves in midair, pending the outcome. And humans evolved from ape-like ancestors whether they did so by Darwin's proposed mechanism or by some other yet to be discovered.

Moreover, "fact" doesn't mean "absolute certainty"; there ain't no such animal in an exciting and complex world. The final proofs of logic and mathematics flow deductively from stated premises and achieve certainty only because they are *not* about the empirical world. Evolutionists make no claim for perpetual truth, though creationists often do (and then attack us falsely for a style of argument that they themselves favor). In science, "fact" can only mean "confirmed to such a degree that it would be perverse to withhold provisional assent." I suppose that apples might start to rise tomorrow, but the possibility does not merit equal time in physics classrooms.

Evolutionists have been clear about this distinction of fact and theory from the very beginning, if only because we have always acknowledged how far we are from completely understanding the mechanisms (theory) by which evolution (fact) occurred. Darwin continually emphasized the difference between his two great and separate accomplishments: establishing the fact of evolution, and proposing a theory—natural selection—to explain the mechanism of evolution. He wrote in the *Descent of Man:*

> I had two distinct objects in view; firstly, to show that species had not been separately created, and secondly, that natural selection had been the chief agent of change. . . . Hence if I have erred in . . . having exaggerated its [natural selection's] power . . . I have at least, as I hope, done good service in aiding to overthrow the dogma of separate creations.

Thus, Darwin acknowledged the provisional nature of natural selection, while affirming the fact of evolution. The fruitful theoretical debate that Darwin initiated has never ceased. From the 1940's through the 1960's, Darwin's own theory of natural selection did achieve a temporary hegemony that it never enjoyed in Darwin's own lifetime. But renewed debate characterizes our decade and, while no biologist questions the importance of natural selection, many now doubt its ubiquity. In particular, many evolutionists now argue that substantial amounts of genetic change may not be subject to natural selection and may spread through populations at random. Others are challenging Darwin's link of natural selection with gradual, imperceptible change through all intermediary degrees; they are arguing that most evolutionary events may occur far more rapidly than Darwin envisioned.

Scientists regard debates on fundamental issues of theory as a sign of intellectual health and a source of excitement. Science—and how else can I say it—is most fun when it plays with interesting ideas, examines their implications, and recognizes that old information might be explained in surprisingly new ways. Evolutionary theory is now enjoying this uncommon vigor. Yet, amidst all this turmoil, no biologist has been led to doubt the fact that evolution occurred; we are debating *how* it happened. We are all trying to explain the same thing: the tree of evolutionary descent linking all organisms by ties of genealogy. Creationists pervert and caricature this debate by conveniently neglecting the common conviction that underlies it, and by falsely suggesting that evolutionists now doubt the very phenomenon we are struggling to understand.

As their second false argument, creationists claim that "the dogma of separate creations," as Darwin characterized it a century ago, is a scientific theory meriting equal time with evolution in high school biology curricula. Philosopher Karl Popper has argued for decades that the primary criterion of science is the falsifiability of its theories. We can never prove absolutely, but we can falsify. A set of ideas that cannot, in principle, be falsified is not science.

The entire creationist argument involves little more than a rhetorical attempt to falsify evolution by presenting supposed contradictions among its supporters. Their brand of creationism, they claim, is "scientific" because it follows the Popperian model in trying to demolish evolution. Yet, as Isaiah said to Hezekiah: "set thine house in order." Popper's argument must apply in both direction. One doesn't become a scientist by the simple act of trying to falsify another scientific system; one has to present an alternative system that also meets Popper's criterion—it too must be falsifiable in principle.

"Scientific creationism" is a self-contradictory, nonsense phrase precisely because it cannot be falsified. I can envision observations and experiments that would disprove any evolutionary theory I know, but I cannot imagine what potential data could lead creationists to abandon their beliefs. Unbeatable systems are dogma, not science. Lest I seem harsh or rhetorical I quote creationism's leading intellectual, Duane Gish, Ph.D., from his recent (1978) book: *Evolution? The Fossils Say No!*

> By creation we mean the bringing into being by a supernatural Creator of the basic kinds of plants and animals by the process of

> sudden, or fiat, creation. We do not know how the Creator created, what processes He used, *for He used processes which are not now operating anywhere in the natural universe* [Gish's italics]. This is why we refer to creation as special creation. We cannot discover by scientific investigations anything about the creative processes used by the Creator.

Pray tell, Dr. Gish, in the light of your last sentence, what then is "scientific" creationism?

The Fact of Evolution

Our confidence that evolution occurred centers upon three general arguments. First, we have abundant, direct, observational evidence of evolution in action, both from the field and laboratory. It ranges from countless experiments on change in nearly everything about fruit flies subjected to artificial selection in the laboratory to the famous British moths that turned black when industrial soot darkened the trees upon which they rest. (The moths gain protection from sharp-sighted bird predators by blending into the background.) I will not dwell on these data here, because creationists do not deny them; how could they? Creationists have tightened their act. They now argue that God only created "basic kinds" (see Gish quote above), and allowed for limited evolutionary meandering within them. Thus, toy poodles and Great Danes come from the dog kind, and moths can change color, but nature cannot convert a dog to a cat or a monkey to a man.

The second and third arguments—the case for major changes—do not involve direct observation of evolution in action. They rest upon inference but are no less secure for that reason. Major evolutionary change requires too much time for direct observation on the scale of recorded human history. All historical sciences rest upon inference, and evolution is no different from geology, cosmology, or human history in this respect. In principle, we cannot observe processes that operated in the past. We must infer them from results that still survive: living and fossil organisms for evolution, documents and artifacts for human history, strata and topography for geology.

The second argument strikes many people as ironic, for they feel that evolution should be most elegantly displayed in the nearly perfect adaptation expressed by some organisms—the camber of a gull's wing, or butterflies that cannot be seen in ground litter because they mimic leaves so precisely. But perfection could be imposed by a wise creator or evolved by natural selection. Perfection covers the tracks of past history. And past history—the evidence of descent—is our mark of evolution.

Evolution lies exposed in the *imperfections* that record a history of descent. Why should the embryos of whalebone whales develop teeth only to reabsorb them later, unless they evolved from toothed ancestors? Why should a rat run, a bat fly, a porpoise swim, and I type this essay with structures built of the same bones unless we all inherited them from a common ancestor? An engineer, starting from scratch, could design better limbs in each case. Why should all the large native mammals of Australia be marsupials, unless they descended from a common ancestor isolated on this island continent? Marsupials are not "better" or ideally suited for Australia; many have been wiped out by placental mammals imported by man from other continents. This principle of imperfection extends to all historical sciences. When we recognize the etymology of September, October, November, and December (7th, 8th, 9th, and 10th), we know that two additional items must have been added to an original calendar of 10 months.

The third argument relies upon transitions in the fossil record. Preserved transitions are not common, and should not be according to our understanding of evolution (see next section), but they are not entirely wanting as creationists often claim. The lower jaw of reptiles contains several bones, that of mammals only one. The non-mammalian jaw bones are reduced, step by step, in mammalian ancestors until they become tiny nubbins located at the back of the jaw. The "hammer" and "anvil" bones of the mammalian ear are descendants of these nubbins. How could such a transition be accomplished? the creationists ask; surely a bone is either entirely in the jaw or in the ear. Yet paleontologists have discovered at least two transitional lineages with a double jaw joint—one composed of the old quadrate and articular bones (soon to become the hammer and anvil), the other of the squamosal and dentary bones (as in modern mammals). For that matter, what better transitional form could we desire than the oldest human, *Australopithecus afarensis,* with its ape-like palate, its human upright stance and a cranial capacity larger than any ape's of the same body size, but a full 1000 cc below ours. If God made each of the half dozen human species discovered in ancient rocks, why did he create in an unbroken temporal sequence of progressively modern features—increasing cranial capacity, a reduced face and teeth, larger body size. Did he create to mimic evolution and test our faith thereby?

An Example of Creationist Argument

Faced with these facts of evolution and the philosophical bankruptcy of their own position, creationists continually rely upon distortion and innuendo to buttress their rhetorical claim. If I sound sharp or bitter, indeed I am—for I have become a major target of these practices.

I count myself among the evolutionists who argue for a jerky or episodic, rather than a smoothly gradual, change of pace. In 1972, my colleague Niles Eldredge and I developed the theory of punctuated equilibrium. We argued that two outstanding facts of the fossil record—geologically "sudden" origin of new species and failure to change thereafter (stasis)—reflect the predictions of evolutionary theory, not the imperfections of the fossil record. In most theories, small isolated populations are the source of new species, and the process of speciation takes thousands or tens of thousands of years. This amount of time, so long when measured against our lives, is a geological microsecond. It represents much less than one percent of the average lifespan for a fossil invertebrate species—more than 10 million years. Large, widespread, and well-established species, on the other hand, are not expected to change very much. We believe that the inertia of large populations explains the stasis of most fossil species over millions of years.

We proposed the theory of punctuated equilibrium largely to provide a different explanation for pervasive trends in the fossil record. Trends, we argued, cannot be attributed to gradual transformation within lineages, but must arise from the differential success of certain kinds of species. A trend, we argued, is more like climbing a flight of stairs (punctuations and stasis) than rolling up an inclined plane.

Since we proposed punctuated equilibria to explain trends, it is infuriating to be quoted again and again by creationists—whether through design or stupidity, I do not know—as admitting that the fossil record includes no transitional forms. The punctuations occur at the level of species; directional trends (on the staircase model) are rife at the higher level of transitions within major groups. Yet a pamphlet entitled: "Harvard Scientists Agree Evolution Is a Hoax" states: "The facts of punctuated equilibrium which Gould and Eldredge . . . are forcing Darwinists to swallow fit the picture that Bryan insisted on, and which God has revealed to us in the Bible."

Continuing the distortion, several creationists have equated the theory of punctuated equilibrium with a caricature of Goldschmidt's belief that major transitions are also accomplished suddenly by means of "hopeful monsters." (I am attracted to some aspects of the non-caricatured version, but Goldschmidt's theory still has nothing to do with punctuated equilibrium.) Creationist Luther Sunderland talks of the "punctuated equilibrium hopeful monster theory" and tells his hopeful readers that "it amounts to tacit admission that anti-evolutionists are correct in asserting there is no fossil evidence supporting the theory that all life is connected to a common ancestor." Duane Gish writes: "According to Goldschmidt, and now apparently according to Gould, a reptile laid an egg from which the first bird, feathers and all, was produced." Any evolutionist who believed such nonsense would rightly be laughed off the intellectual stage; yet the only theory that could ever envision such a scenario for the evolution of birds is creationism—God acts in the egg.

Conclusion

I am both angry at and amused by the creationists; but mostly I am deeply sad. Sad for many reasons. Sad because so many people who respond to creationist appeals are troubled for the right reason, but venting their anger at the wrong target. It is true that scientists have often been dogmatic and elitist. It is true that we have often allowed the white-coated advertising image to represent us—"scientists say that brand x cures bunions ten times faster than . . ." We have not fought it as we should because we derive benefits from appearing as a new priesthood. It is also true that faceless, bureaucratic state power intrudes more and more into our lives and removes choices and options that should belong to individuals and communities. I can understand that evolution in a mandated state curriculum might be seen as one more insult on all these grounds. But the culprit is not, and cannot be, evolution or any other fact of the natural world. Identify and fight your legitimate enemies by all means, but we are not among them.

I am sad because the practical result of this brouhaha will not be expanded coverage to include creationism (that would also make me sad), but the reduction or excision of evolution from high school curricula. Evolution is one of the half dozen "great ideas" developed by science. It speaks to the profound issues of genealogy that fascinate all of us—the "Roots" phenomenon writ large. Where did we come from? When did life arise? How did it develop? How are organisms related? It forces us to think, ponder, and wonder. Shall we deprive millions of this knowledge and once again teach biology as a set of dull and unconnected facts, without the thread that weaves diverse material into a supple unity?

Most of all I am saddened by a trend I am just beginning to discern among my colleagues. I sense that some now wish to mute the healthy debate about theory that has brought new life to evolutionary biology. It provides grist for creationist mills, they say, even if only by distortion. Perhaps we should lay low and rally round the flag of strict Darwinism, at least for the moment—a kind of old-time religion on our part. But we should borrow another metaphor and recognize that we too have to tread a straight and narrow path, surrounded by roads to perdition. For if we ever begin to suppress our search to understand nature, to quench our own intellectual excitement in a misguided effort to present a united front where it does not and should not exist, then we are truly lost.

STEPHEN JAY GOULD

CREATIONISM: GENESIS VS. GEOLOGY

G. K. Chesterton once mused over Noah's dinnertime conversations during those long nights on a vast and tempestuous sea:

> And Noah he often said to his wife
> when he sat down to dine,
> "I don't care where the water goes if
> it doesn't get into the wine."

Noah's insouciance has not been matched by defenders of his famous flood. For centuries, fundamentalists have tried very hard to find a place for the subsiding torrents. They have struggled even more valiantly to devise a source for all that water. Our modern oceans, extensive as they are, will not override Mt. Everest. One seventeenth-century searcher said: "I can as soon believe that a man would be drowned in his own spittle as that the world should be deluged by the water in it."

With the advent of creationism, a solution to this old dilemma has been put forward. In *The Genesis Flood* (1961), the founding document of the creationist movement, John Whitcomb and Henry Morris seek guidance from Genesis 1:6–7, which states that God created the firmament and then slid it into place amidst the waters, thus dividing "the waters which were under the firmament from the waters which were above the firmament: and it was so." The waters under the firmament include seas and interior fluid that may rise in volcanic eruptions. But what are the waters above the firma-

From *The Atlantic Monthly,* September 1982. Copyright © 1982 by the Atlantic Monthly Company, Boston, Mass. Reprinted with permission.

ment? Whitcomb and Morris reason that Moses cannot refer here to transient rain clouds, because he also tells us (Genesis 2:5) that "the Lord God had not caused it to rain upon the earth." The authors therefore imagine that the earth, in those palmy days, was surrounded by a gigantic canopy of water vapor (which, being invisible, did not obscure the light of Genesis 1:3). "These upper waters," Whitcomb and Morris write, "were therefore placed in that position by divine creativity, not by the normal processes of the hydrological cycle of the present day." Upwelling from the depths together with the liquefaction, puncturing, and descent of the celestial canopy produced more than enough water for Noah's world-wide flood.

Fanciful solutions often generate a cascade of additional difficulties. In this case, Morris, a hydraulic engineer by training, and Whitcomb invoke a divine assist to gather the waters into their canopy, but then can't find a natural way to get them down. So they invoke a miracle: God put the water there in the first place; let him then release it.

> The simple fact of the matter is that one cannot have *any* kind of a Genesis Flood without acknowledging the presence of supernatural elements. . . . It is obvious that the opening of the "windows of heaven" in order to allow "the waters which were above the firmament" to fall upon the earth, and the breaking up of "all the fountains of the great deep" were supernatural acts of God.

Since we usually define science, at least in part, as a system of explanation that relies upon invariant natural laws, this charmingly direct invocation of miracles (suspensions of natural law) would seem to negate the central claims of the modern creationist movement—that creationism is not religion but a scientific alternative to evolution; that creationism has been disregarded by scientists because they are a fanatical and dogmatic lot who cannot appreciate new advances; and that creationists must therefore seek legislative redress in their attempts to force a "balanced treatment" for both creationism and evolution in the science classrooms of our public schools.

Legislative history has driven creationists to this strategy of claiming scientific status for their religious view. The older laws, which banned the teaching of evolution outright and led to John Scopes's conviction in 1925, were overturned by the United States Supreme Court in 1968, but not before they had exerted a chilling effect upon teaching for forty years. (Evolution is the indispensable organizing principle of the life sciences, but I did not hear the word in my 1956 high school biology class. New York City, to be sure, suffered no restrictive ordinances, but publishers, following the principle of the "least common denominator" as a sales strategy, tailored the national editions of their textbooks to the few states that considered it criminal to place an ape on the family escutcheon.) A second attempt to mandate equal time for frankly religious views of life's history passed the Tennessee state legislature in the 1970s but failed a constitutional challenge in the court. This judicial blocking left only one legislative path open—the claim that creationism is a science.

The third strategy had some initial success, and "balanced treatment" acts to equate "evolution science" and "creation science" in classrooms passed the Arkansas and Louisiana legislatures in 1981. The Arkansas law was challenged by the ACLU in 1981, on behalf of local plaintiffs (including twelve practicing theologians who felt more threatened by the bill than many scientists did). Federal Judge William R. Overton heard the Arkansas case in Little Rock last December. I spent the better part of a day on the stand, a witness for the prosecution, testifying primarily about how the fossil record refutes "flood geology" and supports evolution.

On January 5, Judge Overton delivered his eloquent opinion, declaring the Arkansas act unconstitutional because so-called "creation science" is only a version of Genesis read literally—a partisan (and narrowly sectarian) religious view, barred from public-school classrooms by the First Amendment. Legal language is often incomprehensible, but sometimes it is charming, and I enjoyed the wording of Overton's decision: ". . . judgment is hereby entered in favor of the plaintiffs and against the defendants. The relief prayed for is granted."

Support for Overton's equation of "creation science" with strident and sectarian fundamentalism comes from two sources. First, the leading creationists themselves released some frank private documents in response to plaintiffs' subpoenas. Overton's long list of citations seems to brand the claim for scientific creationism as simple hypocrisy. For example, Paul Ellwanger, the tireless advocate and drafter of the "model bill" that became Arkansas Act 590 of 1981, the law challenged by the ACLU, says in a letter to a state legislator that "I view this whole battle as one between God and anti-God forces, though I know there are a large number of evolutionists who believe in God. . . . it behooves Satan to do all he can to thwart our efforts . . ." In another letter, he refers to "the idea of killing evolution instead of playing these debating games that we've been playing for nigh over a decade already"—a reasonably clear statement of the creationists' ultimate aims, and an identification of their appeals for "equal time," "the American way of fairness," and "presenting them both and letting the kids decide" as just so much rhetoric.

The second source of evidence of the bill's unconstitutionality lies in the logic and character of creationist arguments themselves. The flood story is central to all creationist systems. It also has elicited the only specific and testable theory the creationists have offered; for the rest, they have only railed against evolutionary claims. The flood story was explicitly cited as one of the six defining characteristics of "creation science" in Arkansas Act 590: "explanation of the earth's geology by catastrophism, including the occurrence of a worldwide flood."

Creationism reveals its nonscientific character in two ways: its central tenets cannot be tested and its peripheral claims, which can be tested, have been proven false. At its core, the creationist account rests on "singularities"—that is to say, on miracles. The creationist God is not the noble clockwinder of Newton and Boyle, who set the laws of nature properly at the beginning of time and then released direct control in full confidence that his initial decisions would require no revision. He is, instead, a constant presence, who suspends his own laws when necessary to make the new or destroy the old. Since science can treat only natural phenomena occurring in a context of invariant natural law, the constant invocation of miracles places creationism in another realm.

We have already seen how Whitcomb and Morris remove a

divine finger from the dike of heaven to flood the earth from their vapor canopy. But the miracles surrounding Noah's flood do not stop there; two other supernatural assists are required. First, God acted "to gather the animals into the Ark." (The Bible tells us [Genesis 6:20] that they found their own way.) Second, God intervened to keep the animals "under control during the year of the Flood." Whitcomb and Morris provide a long disquisition on hibernation and suspect that some divinely ordained state of suspended animation relieved Noah's small and aged crew of most responsibility for feeding and cleaning (poor Noah himself was 600 years old at the time).

In candid moments, leading creationists will admit that the miraculous character of origin and destruction precludes a scientific understanding. Morris writes (and Judge Overton quotes): "God was there when it happened. We were not there. . . . Therefore, we are completely limited to what God has seen fit to tell us, and this information is in His written Word." Duane Gish, the leading creationist author, says: "We do not know how the Creator created, what processes He used, for He used processes which are not now operating anywhere in the natural universe. . . . We cannot discover by scientific investigation anything about the creative processes used by God." When pressed about these quotes, creationists tend to admit that they are purveying religion after all, but then claim that evolution is equally religious. Gish also says: "Creationists have repeatedly stated that neither creation nor evolution is a scientific theory (and each is equally religious)." But as Judge Overton reasoned, if creationists are merely complaining that evolution is religion, then they should be trying to eliminate it from the schools, not struggling to get their own brand of religion into science classrooms as well. And if, instead, they are asserting the validity of their own version of natural history, they must be able to prove, according to the demands of science, that creationism is scientific.

Scientific claims must be testable; we must, in principle, be able to envision a set of observations that would render them false. Miracles cannot be judged by this criterion, as Whitcomb and Morris have admitted. But is all creationist writing merely about untestable singularities? Are arguments never made in proper scientific form?

Creationists do offer some testable statements, and these are amenable to scientific analysis. Why, then, do I continue to claim that creationism isn't science? Simply because these relatively few statements have been tested and conclusively refuted. Dogmatic assent to disproved claims is not scientific behavior. Scientists are as stubborn as the rest of us, but they must be able to change their minds.

In "flood geology," we find our richest source of testable creationist claims. Creationists have been forced into this uncharacteristically vulnerable stance by a troubling fact too well known to be denied: namely, that the geological record of fossils follows a single, invariant order throughout the world. The oldest rocks contain only single-celled creatures; invertebrates dominate later strata, followed by the first fishes, then dinosaurs, and finally large mammals. One might be tempted to take a "liberal," or allegorical, view of Scripture and identify this sequence with the order of creation in Genesis 1, allowing millions or billions of years for the "days" of Moses. But creationists will admit no such reconciliation. Their fundamentalism is absolute and uncompromising. If Moses said "days," he meant periods of twenty-four hours, to the second. (Creationist literature is often less charitable to liberal theology than to evolution. As a subject for wrath, nothing matches the enemy within.)

Since God created with such alacrity, all creatures once must have lived simultaneously on the earth. How, then, did their fossil remains get sorted into an invariable order in the earth's strata? To resolve this particularly knotty problem, creationists invoke Noah's flood: all creatures were churned together in the great flood and their fossilized succession reflects the order of their settling as the waters receded. But what natural processes would produce such a predictable order from a singular chaos? The testable proposals of "flood geology" have been advanced to explain the causes of this sorting.

Whitcomb and Morris offer three suggestions. The first—hydrological—holds that denser and more streamlined objects would have descended more rapidly and should populate the bottom strata (in conventional geology, the oldest strata). The second—ecological—envisions a sorting responsive to environment. Denizens of the ocean bottom were overcome by the flood waters first, and should lie in the lower strata; inhabitants of mountaintops postponed their inevitable demise, and now adorn our upper strata. The third—anatomical or functional—argues that certain animals, by their high intelligence or superior mobility, might have struggled successfully for a time, and ended up at the top.

All three proposals have been proven false. The lower strata abound in delicate, floating creatures, as well as spherical globs. Many oceanic creatures—whales and teleost fishes in particular—appear only in upper strata, well above hordes of terrestrial forms. Clumsy sloths (not to mention hundreds of species of marine invertebrates) are restricted to strata lying well above others that serve as exclusive homes for scores of lithe and nimble small dinosaurs and pterosaurs.

The very invariance of the universal fossil sequence is the strongest argument against its production in a single gulp. Could exceptionless order possibly arise from a contemporaneous mixture by such dubious processes of sorting? Surely, somewhere, at least one courageous trilobite would have paddled on valiantly (as its colleagues succumbed) and won a place in the upper strata. Surely, on some primordial beach, a man would have suffered a heart attack and been washed into the lower strata before intelligence had a chance to plot temporary escape. But if the strata represent vast stretches of sequential time, then invariant order is an expectation, not a problem. No trilobite lies in the upper strata because they all perished 225 million years ago. No man keeps lithified company with a dinosaur, because we were still 60 million years in the future when the last dinosaur perished.

True science and religion are not in conflict. The history of approaches to Noah's flood by scientists who were also professional theologians provides an excellent example of this important truth—and also illustrates just how long ago "flood geology" was conclusively laid to rest by religious scientists. I have argued that direct invocation of miracles and unwillingness to abandon a false doctrine deprive modern creationists of their self-proclaimed status as scientists. When we examine how the great scientist-theologians of past centuries treated the flood, we note that their work is distinguished by both a conscious refusal to admit miraculous events into their explanatory schemes and a willingness to abandon preferred hypotheses in the face of geological evidence. They were scientists *and* religious leaders—and they show us why modern creationists are not scientists.

On the subject of miracles, the Reverend Thomas Burnet published his century's most famous geological treatise in the 1680s, *Telluris theoria sacra* (*The Sacred Theory of the Earth*). Burnet accepted the Bible's truth, and set out to construct a geological history that would be in accord with the events of Genesis. But he believed something else even more strongly: that, as a scientist, he must follow natural law and scrupulously avoid miracles. His story is fanciful by modern standards: the earth originally was devoid of topography, but was drying and cracking; the cracks served as escape vents for internal fluids, but rain sealed the cracks, and the earth, transformed into a gigantic pressure cooker, ruptured its surface skin; surging internal waters inundated the earth, producing Noah's flood. Bizarre, to be sure, but bizarre precisely because Burnet would not abandon natural law. It is not easy to force a preconceived story into the strictures of physical causality. Over and over again, Burnet acknowledges that his task would be much simpler if only he could invoke a miracle. Why weave such a complex tale to find water for the flood in a physically acceptable manner, when God might simply have made new water for his cataclysmic purification? Many of Burnet's colleagues urged such a course, but he rejected it as inconsistent with the methods of "natural philosophy" (the word "science" had not yet entered English usage):

> They say in short that God Almighty created waters on purpose to make the Deluge . . . And this, in a few words, is the whole account of the business. This is to cut the knot when we cannot loose it.

Burnet's God, like the deity of Newton and Boyle, was a clock-winder, not a bungler who continually perturbed his own system with later corrections.

> We think him a better Artist that makes a Clock that strikes regularly at every hour from the Springs and Wheels which he puts in the work, than he that hath so made his Clock that he must put his finger to it every hour to make it strike: And if one should contrive

> a piece of Clockwork so that it should beat all the hours, and make all its motions regularly for such a time, and that time being come, upon a signal given, or a Spring toucht, it should of its own accord fall all to pieces; would not this be look'd upon as a piece of greater Art, than if the Workman came at that time prefixt, and with a great Hammer beat it into pieces?

Flood geology was considered and tested by early-nineteenth-century geologists. They never believed that a single flood had produced all fossil-bearing strata, but they did accept and then disprove a claim that the uppermost strata contained evidence for a single, catastrophic, worldwide inundation. The science of geology arose in nations that were glaciated during the great ice ages, and glacial deposits are similar to the products of floods. During the 1820s, British geologists carried out an extensive empirical program to test whether these deposits represented the action of a single flood. The work was led by two ministers, the Reverend Adam Sedgwick (who taught Darwin his geology) and the Reverend William Buckland. Buckland initially decided that all the "superficial gravels" (as these deposits were called) represented a single event, and he published his *Reliquiae diluvianae* (*Relics of the Flood*) in 1824. However, Buckland's subsequent field work proved that the superficial gravels were not contemporaneous but represented several different events (multiple ice ages, as we now know). Geology proclaimed no worldwide flood but rather a long sequence of local events. In one of the great statements in the history of science, Sedgwick, who was Buckland's close colleague in both science and theology, publicly abandoned flood geology—and upheld empirical science—in his presidential address to the Geological Society of London in 1831.

> Having been myself a believer, and, to the best of my power, a propagator of what I now regard as a philosophic heresy, and having more than once been quoted for opinions I do not now maintain, I think it right, as one of my last acts before I quit this Chair, thus publicly to read my recantation . . .
>
> There is, I think, one great negative conclusion now incontestably established—that the vast masses of diluvial gravel, scattered almost over the surface of the earth, do not belong to one violent and transitory period . . .
>
> We ought, indeed, to have paused before we first adopted the

> diluvian theory, and referred all our old superficial gravel to the action of the Mosaic flood . . . In classing together distant unknown formations under one name; in giving them a simultaneous origin, and in determining their date, not by the organic remains we had discovered, but by those we expected hypothetically hereafter to discover, in them; we have given one more example of the passion with which the mind fastens upon general conclusions, and of the readiness with which it leaves the consideration of unconnected truths.

As I prepared to leave Little Rock last December, I went to my hotel room to gather my belongings and found a man sitting backward on my commode, pulling it apart with a plumber's wrench. He explained to me that a leak in the room below had caused part of the ceiling to collapse and he was seeking the source of the water. My commode, located just above, was the obvious candidate, but his hypothesis had failed, for my equipment was working perfectly. The plumber then proceeded to give me a fascinating disquisition on how a professional traces the pathways of water through hotel pipes and walls. The account was perfectly logical and mechanistic: it can come only from here, here, or there, flow this way or that way, and end up there, there, or here. I then asked him what he thought of the trial across the street, and he confessed his staunch creationism, including his firm belief in the miracle of Noah's flood.

As a professional, this man never doubted that water has a physical source and a mechanically constrained path of motion—and that he could use the principles of his trade to identify causes. It would be a poor (and unemployed) plumber indeed who suspected that the laws of engineering had been suspended whenever a puddle and cracked plaster bewildered him. Why should we approach the physical history of our earth any differently?

GUNTHER S. STENT

SCIENTIFIC CREATIONISM: NEMESIS OF SOCIOBIOLOGY

As many other contributors to this volume must have pointed out, the case of "Scientific Creationism" can be discussed in at least three rather different contexts: the scientific, the ethico-social, and the ethico-philosophical. Of these three, the scientific context seems to me the least productive. There is no point in engaging in any scientific discussion unless there is first an agreement on the concepts, logical operations, empirical data, and verification procedures that have to be brought to bear on the intellectual activity formulated by the ancient Greeks that we call "science." Since the Creationists will not, or rather cannot, agree on these ground rules, it is futile to engage them in scientific argument. The historical claims of the Creationists regarding the actual sequence and time scale of the events by which the living world came to be as it is today can be shown to be false, provided that empirical data such as the fossil record or protein amino acid sequences are given the evidentiary weight accorded to them under the ground rules of science. And as for the theoretical claims of the Creationists regarding the causal connections that determined the history of the living world, they are incompatible with the metaphysical concept of Natural Law on which scientific theories are based. Admittedly, under the Judeo-Christian scientific tradition God is identified as the author of Natural Law, and having created us in His image, He has given us the opportunity to fathom His Law. But He remains aloof from Nature and refrains from capriciously intervening with its operation. It is possible to be a devout Jew or Christian and believe in miracles, i.e., in divine interferences with or contraventions of Natural Law, or even to believe literally in the biblical account of creation, but it is not possible to hold such beliefs and, at the same time, lay claim to being a scientist. Thus in view of the present state of knowledge reached by the life sciences, the very term "Scientific Creationism" is an oxymoron.

The ethico-social context, however, is one in which discussion regarding Creationism is not only possible but has to be faced by scientists. For unlike the obviously nonsensical scientific claims of the Creationists, their demands for equal time in the public school curriculum cannot be dismissed out of hand. These demands confront us with the complex character of the Western moral tradition which, according to Isaiah Berlin, Machiavelli discovered to be inherently contradictory a century before Galileo opened the door to modern science. The contradiction discovered by Machiavelli is not, as has often been alleged by commentators on *The Prince* and the *Discourses,* between morality and politics but between two incompatible systems of ethics. One of these systems, which Berlin terms "Christian," envisages morality as being based on "ultimate values sought for their own sakes—values recognition of which alone enables us to speak of crimes or morally to justify and condemn anything." The other system, which Berlin terms "pagan," derives its authority from the fact that man is a social animal who lives in communities. Under the "pagan" system there are no ultimate values, only communal purpose. Or, more simply stated, the two mutually incompatible aims projected into the Western City of God are freedom and justice for the individual on the one hand, and law and order for the body politic on the other. From this insight of Machiavelli it follows, according to Berlin, "that the belief that the correct, objectively valid solution to the question of how men should live can in principle be discovered is itself, in principle, not true."

Thus, in the light of the "Christian" value of freedom and justice, the claim of the Creationists that their religious freedom is violated by the teaching in the public schools of Darwinian evolution as a universal truth is by no means groundless. After all, do not devout Christian parents have as much right to have their children in tax-supported schools protected from unilateral exposure to atheistic doctrines as atheistic parents have to have their children protected from prayer? Hence classroom teaching of Darwinism as the only account of biocosmogony would be an infringement of the religious freedom of fundamentalist Christian parents to raise their children in the faith of their choice. This argument seems valid, whether or not it is true that, as held by liberal theologians, one can be a good Christian without taking Genesis literally. After all, the fundamentalist faith *is* to take the Bible literally. But the inference that follows from admitting the justice of the fundamentalist demands is not that public school biology texts should give Genesis equal time with Darwinism. Rather it has to be concluded that no public school system can operate effectively in a heterogeneous social setting without having its curriculum prejudice the minds of the pupils against the cherished beliefs of some of its citizens. In other words, here the ultimate "Christian" ethical value of freedom and individual rights for fundamentalist parents has to give way to the "pagan" social value of mounting a pedagogically effective, modern society. Scientists should take note, however, that the reintroduction of prayer into the public schools could be justified by the same argument. For if it were judged that daily prayer is an effective pedagogical device for fostering the spiritual life and moral development of children, the "Christian" claims to freedom and justice by atheist parents would have to yield to the "pagan" social aims of the Christian majority.

In an ethico-philosophical context the resurgence of Creationism in our days, when the Scopes Monkey Trial seems as much part of an archaic and primitive American past as the Salem Witch Trials, should give biologists cause for reflection on what evolutionism has wrought. It would be ridiculous, of course, for any contemporary biologist to deny the tremendous advances brought to our understanding of the living world by the Darwinian revolution in the latter part of the 19th century. Nor can there be many biologists who would not acknowledge the immense value of the brilliant wedding of genetics and evolutionary theory represented by the development of Neo-Darwinism, in particular by the "New Synthesis," in the first half of the 20th century. This general appreciation of the contributions of Neo-Darwinism is shared even by its latter-day critics within the community of professional evolutionists, who may argue against the gradualist conception of the origin of species, propose a revision of the role of natural selection in speciation, or criticize the lack of attention paid to the role of embryological processes in accounting for evolutionary change. Having got wind of this in-house controversy, the Scientific Creationists cite it as proof that evolution is just a "theory" and not a "fact," confusing, as is their wont, the historical fact of evolution, which they deny, with the theory put forward to account for that fact, which is indeed a theory and thus subject to emendation and revision. The inability of current evolutionary theory to lay claim to being final and absolute truth does not of course imply that Scientific Creationism is a genuine alternative theory deserving equal consideration in the biology curriculum.

This being said, there are nevertheless some intellectually noxious consequences of evolutionism, or rather *hyper*evolutionism, which have become manifest in our days. And the resurgence of Scientific Creationism will have done professional biologists a good turn if it caused them to reflect on these consequences. The responsibility for current excessive claims on behalf of evolutionism does not so much lie with the leading architects of modern evolutionary thought who, from Darwin onward, were generally aware of the epistemological status of their work, but with the epigones who failed to comprehend the intrinsic limitations of the theory. First, it would appear that the idea of natural selection as the grandest of all biological principles—a unifying "law" to which all explanations must ultimately refer—has been carried too far. Apparently it has been widely forgotten, at least by biology teachers, that natural selection is foremost a diachronic or historical principle whose main explanatory value concerns biological processes that occur over periods of time that are long with respect to the life span of the individual organism, i.e., evolutionary phenomena. By contrast, natural selection has little or no standing as a synchronic principle that can be drawn on for explanations of biological processes that occur over periods of time that are short with respect to life spans, i.e., physiological phenomena. As I have noticed since I began teaching the elementary biology course at the University of California a few years ago, most students now seem to be taught in high school that natural selection, and its correlate of "adaptive value," will explain everything. Such smuggling of evolutionary principles into the physiological realm has resulted in an intellectual lethargy—actually not only among my sophomore students but even among many of my biologist colleagues—which is manifest in a willingness to accept adaptive value as an instant explanation for phenomena for which functional or mechanistic explanations are wanted. For instance, to an examination question such as "why do some eggs contain very large amounts of yolk?" many students will consider "because it gives embryos developing in such eggs a selective advantage" an adequate response.

Second, hyperevolutionism has spawned sociobiology, the troublesome ethological subdiscipline specifically concerned with the evolution of social behavior. Now whereas there is no denying that this is a reasonable scientific concern, particularly insofar as its objective is to account for the origins of the "social" behavior of insects and other lower orders, the extension of sociobiological theory to human behavior is fraught with tremendous conceptual difficulties and ominous political consequences. Here too the architects of modern evolutionary thought cannot be held directly responsible for this development. Admittedly, from Darwin onwards they have propounded the importance of behavior for evolutionary processes, particularly insofar as reproductive behavior obviously plays a central role in the process of speciation. But most of the leading contemporary students of evolution regard the sociobiological methodology with disdain and, in the words of Richard Lewontin, consider that discipline as little more than a "caricature of Darwinism." Nevertheless, the scientistic pretensions of sociobiologists seem to me to be a natural product of the modern secular, or "humanistic" culture to whose rise the successful refutation by Darwinism of the very first chapter of the Bible has made a significant contribution. For by showing that, contrary to Genesis 1:11–13, 20–31, the species presently inhabiting the biosphere were not created in three days, Darwinism has helped to erode confidence in the whole of that central document of Western culture. This erosion has made people forget that the Bible remains, as always and "humanistic" pretensions to the contrary notwithwstanding, the source of the ultimate, or according to Berlin, "Christian" values that govern jointly and in conflict with the "pagan" values our paradoxical moral life. Creationists for their part have given considerable assistance to this erosion process, since their stubborn insistence on literal rather than hermeneutic interpretation of the ancient text has only served to dramatize and make more impressive the scientific refutation of Genesis. But it is precisely from the Bible that sociobiologists could have learned some of the elements of moral philosophy, from the viewpoint of which the relevance of their theories for human behavior is most open to doubt. Even cursory study of the Bible would have made it clear to them that in the Western culture to which they belong and whose values they share, the action of persons are judged on the basis of *intent* and not, as is generally assumed in sociobiological thought regarding morally relevant behavior, of the *consequences.* Had they remembered their Bible lessons, sociobiologists would have avoided the total confusion they sowed in the ethical domain by their corrupted use of the antonymous concepts of altruism and selfishness in contexts where intent plays no role. Thus I would like to regard Scientific Creationism as the Nemesis of Sociobiology and its "Selfish Gene." It would be marvelous if these two movements that threaten to pollute the intellectual environment of modern civilized society were to consume each other and let the rest of us get on with the already difficult enough task of dealing with our two incompatible ethical systems.

KENNETH E. BOULDING

TOWARD AN EVOLUTIONARY THEOLOGY

Theology consists of those images in the human mind, and the language describing them, which are relevant to and ultimately derived from the record of human experiences and states of mind and body which can be classified as "religious." Just where we draw the line between religious and nonreligious experience is not easy to say, for there is a continuum of experience from the religious to the secular, from the ecstasies of the mystic to buying clothes. Then there is a large middle ground in the experience of art and music, of human love and sacrifice, which is not easy, and perhaps not very important, to classify. Nevertheless, there is a large area of human experience reflected not only in personal testimony, but in a very ancient and complex record, the existence of which can hardly be denied. The traces of flowers on Neanderthal graves, cave paintings, figurines and little statues of household gods, funerary objects, temples, mosques and cathedrals, holy books, a vast descriptive literature of personal religious experience in journals and other writings, the enormous literature of mythology, and even the sober descriptive writings of psychologists like William James, all suggest that there is a religious element of human experience which cannot be denied as a part of the ongoing history of the human race. Furthermore, it is fairly safe to assert that to every example of religious experience there will be attached something like a theology, an image of some transcendent reality for which the religious experience is taken as evidence, and which is regarded as justification for religious practice.

The distinction between religious experience and practice is important. Experience is something that happens inside the individual that produces some kind of change of state. This can happen either spontaneously, as a product of some interior activity of the human mind, or as a result of external stimulus or input of information. Religious practice varies all the way from the almost purely internal in meditation and prayer to the almost purely external in formalistic ritual, which is performed more for the sake of performing it than for any change of inward state that it is going to produce. Some religious practice, indeed, can be interpreted in terms of an escape from the qualms and anxieties of inwardness. An important aspect of religious practice, however, is its public character—public ritual worship and public prayer. Religion here has a strong community aspect; it brings people together in rituals of legitimation and comfort and binds them into a community. The theology which emerges out of community religion, in terms, for instance, of local gods, often derives its sanction from the urge to reify the community and to identify it with the supernatural order.

There is little doubt that religion is a peculiarly human phenomenon, although we do occasionally have the uneasy experience that we are being worshiped by an adoring dog as he looks up at us. The human mind, however, has the extraordinary capacity for language and for developing images of internally created worlds and images of things that have not been experienced directly. This capacity rapidly seems to develop a distinction between the "natural" and the "supernatural." The "natural" is the external environment revealed and interpreted through the senses, other human beings, dwellings, food, the forest, the lakes and seas—whatever it is with which we have direct personal experience. Because of language, however, we can have images, however imperfect, of the interior landscapes of other human beings. We recognize the manifestations of anger, love, or hunger in others as we manifest them in ourselves, and we find that language modifies these manifestations. Language, of course, includes nonverbal as well as verbal communication.

When we find ourselves, however, in the presence of the uncontrollable—day and night, rain and shine, wind and calm, the growth of plants and the elusiveness of animals—and when we observe the stars and the succession of the seasons, it is not surprising that we seek to go beyond our own experience into a cosmology. It seems that the human mind has an ineradicable itch to perceive patterns and is uneasy with chaos. Whether we begin with some experience of the grandeur of a sunset, or the terror of a storm, or Wordsworth's experience about Tintern Abbey, and then go on to some kind of cosmology and theology, or whether we start with a theology and from this derive experience, is hard to say. What we have is a constant interaction between theology and experience.

It is not surprising that the first theologies tended to be animistic. We observe that we can change the behavior of our neighbors by talking to them. Why cannot we similarly change the behavior of the skies, or the plants, or the animals? Furthermore, it is quite easy to accumulate evidence for animism: if we say something to our neighbor and he does not respond, we conclude that we said the wrong thing and should say something else. If we say something to the mysterious forces of nature and they do not respond, then we decide we said the wrong thing, and will do better next time. It is hard to avoid selective memory of response; we can remember the successes and forget the failures. Even today the gambler kisses the dice he is about to throw. We even sacrifice a bottle of champagne when launching a ship, no doubt to cheer up its spirits. And we make huge economic and human sacrifices in the name of national defense to appease the spirit of the nation. Even the hard-headed Japanese put up miles of *torii* to appease the spirits of business success.

Animism, however, seems to lead to the gods and to the great pantheons of Greece and Rome, India and Japan. These must surely begin as fantasy and poetry, arising out of the extraordinary human capacity for creating a vast inner landscape in which we can personify and satisfy desires, exorcise our sense of impotence, and magnify our capacities both for good and for evil. We cannot resist imagining that which is larger than life, and the fantasies, the legends, and the myths then take on a reality of their own. Do we not almost believe that Mr. Pickwick and the Cheshire cat must exist in some Platonic space, whereas the dull heroes of the novels we have forgotten clearly do not? A vivid image, indeed, can create a sense experience. Jaynes's disconcerting study of the great change that seems to have taken place in the human race about 1400 B.C., after which we no longer heard the voices or saw the apparitions of the gods, certainly has some evidence to support it. Furthermore, this capacity may not be as extinct as some think. Joan of Arc and George Fox (founder of the Society of Friends) heard voices. The vast apparatus of Lourdes is not that far from the Odyssey; Athena still appears, but in the shape of the Virgin Mary.

In the record of religious experience and theology over the last 3,000 years or so, the impact of the Jews essentially dominates a large part of the world. Three of the world's major religions (perhaps one should say, faiths) emerged out of Judaism: Christianity, Islam, and Marxism. Of the non-Judaic religions, only Hinduism and Buddhism have any status as world religions, and even these are largely limited to South and Southeast Asia. All the world religions come out of sacred books. Without writing, a religion can only be local; spoken language tends to be local. Just what creates "sacredness" in books is a real puzzle. There must have been a large number of documents which claimed to have been sacred but somehow were not selected for that quality. The selective process, indeed, may depend not only on the nature and the quality of the books themselves (although this clearly has something to do with it) but also on the social setting in which they are produced and the institutional needs and support which may lead to the survival of some and not of others. It may take an institutional church to distinguish between the canon and the apocrypha, or even the expurgated works.

The Bible, both the Old and the New Testaments, seems to be unique among the sacred books of the world because a large part of it originated in the writing down of oral tradition, often passed on among common people rather than professionals or intellectuals. The Koran and the Book of Mormon are works of single individuals. The Bhagavad Gita, and much of the scripture of Hinduism and Buddhism, is the work of an intellectual class. Parts of the Bible, of course, are the work of single individuals, such as the Epistles of the New Testament and perhaps some of the prophetic books of the Old Testament. Some of the books of the Bible are literary works, such as the Book of Job or Ecclesiastes in the Old

Testament, or Revelation in the New Testament. But a good deal of it is still "gossip written down." This explains in part the continuing power of the Bible over many centuries to create new religious experiences and institutions, though the Koran, with its very different origins, seems to have something of the same capacity. And to judge by the young Hare Krishnas at the airports, some of the Hindu scriptures still activate religious experience.

A very interesting question is whether the world scientific community should be regarded as a world religion or faith. This is, perhaps, only a matter of semantics, for the world scientific community has both many similarities to a world religion, but also very important differences. It has, in the first place, a very distinctive ethic of its own, which I am almost tempted to call "the four-fold way," as it has four essential components. The first is a high value on curiosity, which not all cultures possess. The second is a high value on veracity—that is, on not telling lies—which many other cultures also do not possess. The one thing that can get a scientist excommunicated from the scientific community is to be caught deliberately falsifying his results—that is, in telling lies. Error is often pardonable, but lying is the sin that cannot be forgiven. The third ethical principle is the high value on the testing of images of the world against the external world that they are supposed to map. Mere internal consistency is not enough, for there may be views of the world which are internally consistent, but which are, nevertheless, not true, in the sense that the real world does not conform to them. There are many methods of testing. Experiment is an important method where it is appropriate, though only perhaps a third to a half of the testing activities of science consist of experiment. Careful observation and recording, coupled with systematic analysis, is another important method, such as we have in celestial mechanics and in national economic statistics. Comparative studies of systems which are alike in many respects but differ in others is another important method; for instance, in medical research and the social sciences. Underlying all these, however, is a profound belief that the real world will speak for itself if it is asked the right questions.

The fourth principle of scientific ethics is abstention from threat, embodied in the principle that people should be persuaded only by evidence and never by threat. This, of course, is in striking opposition to the ethics of many religious organizations and of all political organizations. The politicization of science, however, such as we saw it, for instance, in the Lysenko episode in the Soviet Union, and on a much smaller scale in the antievolutionary movements in the United States, is always very damaging.

One could nominate a fifth principle—that the business of the scientific community is to diminish error rather than to discover truth. Perhaps not all scientists would agree with this. All that testing can do, however, is to detect error. We can never be sure, therefore, that error will not be detected in any presently accepted scientific proposition at some time in the future. It is this tentative and essentially agnostic view of science towards truth, even towards its own propositions, that has laid the ground for the enormous increase in knowledge that science has given us, simply because of the ability of the scientific community to change its images and paradigms of the world in response to its own experience. The same thing has happened, however, in the religious communities, as theological positions and religious practices have failed certain tests. Because of the sacredness associated with religious views of the world, however, change in them is perceived as much more threatening and is much more difficult.

Science is certainly a "phylum" of human experience, just as the world religions are. It may be a little hard to pinpoint its origins, for we certainly get what might be called "protoscience" in ancient India, Babylonia, and Greece. But science as an ongoing subculture is usually dated from Copernicus, about 450 years ago. As an evolutionary phylum should, it continually branches into new disciplines and subdisciplines, each of which pursues its own dynamic. There is a certain tendency, as there is also in religious institutions, for particular disciplines to exhaust their initial potential and enthusiasm and become rather sterile and formalistic, and then they are frequently revived by new ideas and new paradigms. Thomas Kuhn's pattern of "normal science," followed by scientific revolutions and paradigm change as the old paradigms accumulate evidence against them, has striking parallels also in the religious communities, where religions and sects become conservative and stagnate, and then are sometimes revived by new outbursts of enthusiasm and experience, often as a result of rediscovery by some charismatic prophet or enthusiast of things that had been forgotten.

Science, like religion, has suffered at times from the carry-over of inappropriate methodologies and paradigms from one field of study and experience to another. Science has expanded from astronomy, physics, and mechanics, especially celestial mechanics, symbolized especially by Newton in the seventeenth century, to economics (Adam Smith); descriptive biology (Linnaeus) in the eighteenth century; chemistry (Dalton) in the early nineteenth century; geology (Lyell) in the 1830s; evolutionary biology (Darwin) in the 1860s; experimental psychology (Wundt) in the 1870s; sociology (Durkheim) in the 1880s; electrical sciences (Clerk Maxwell) in the 1870s; radioactivity, nuclear science (Bohr and Rutherford); Einsteinian physics, quantum theory, "big bang" cosmology, molecular biology, plate tectonics in geology, information theory, genetics, the neurosciences, and so on, all in the twentieth century. Newtonian science is now seen as a very special case. In the evolutionary sciences matter and energy are now seen mainly as coders and carriers of information. In this absurdly brief sketch I have left out a good many things, but included enough to suggest the extraordinary changes in cosmology, methodology, and even in views of what constitutes ultimate reality which have taken place, especially in the twentieth century.

Although science is certainly a subculture and a phylum, characteristics which it shares with the world religions, it differs sufficiently from the religious subcultures that it is an abuse of words to call it a "religion." It is not very much like a church, for it has no congregations. The members of the subculture consist almost entirely of "priests" and "priests in training." Its "churches" are classrooms and laboratories; its altars, the laboratory benches. It even has its mandalas, such as the periodic table of elements, which are now found on classroom walls in every country of the world. It has no sacred books, apart from laboratory manuals. It has no pope, though it has professors who behave a little like bishops. The saints perhaps are the Nobel Prize winners. Charisma and discipleship exist, though they are somewhat frowned upon. The renunciation of threat also involves the renunciation of authority.

Science does not have a formal creed, but it does have discipline, or at least a collection of disciplines, theoretically imposed by the nature of the real world, and even sociologically tending to come from below rather than from above. We may define a discipline indeed as a subculture within which a young person can get promoted for pointing out that an older person has fallen into error. Science is indeed a culture of critique and selection, essentially evolutionary in its model rather than authoritative. Science is an ecosystem; it is not an organism. I have sometimes teased the American Association for the Advancement of Science by comparing it with the National Council of Churches. It is certainly ecumenical with its 280 affiliated societies, none of which talk to each other very much. It also has an ethical mission of protecting the integrity and liberty of the scientific community and of helping individual scientists who get into trouble by practicing the scientific ethic. But these are cultural resemblances—shared by other social subcultures like music and the arts—which should not blind us to the fundamental differences.

Historically, science has to be regarded as a cultural mutation out of "Western" Christian European culture. One of the real puzzles of human history is what I have sometimes called the "Needham problem" because of Joseph Needham's extraordinary work on the history of science and technology in China. This is the question as to why science did not originate in China, which in 1500 had a well-organized society with an ideological base in Confucianism, which one would not have expected to be unfriendly to the study of the material world. It is perhaps easier to understand why it did not originate in India, a land with a disorganized, conquered society and an ideological base in Hinduism, which almost prohibited interest in the material world. It is puzzling also why it did not originate in the world of Islam, which, again, certainly did not dismiss the material world as illusion, and which had preserved the Greek science and philosophy that Europe had partly lost. Maybe it was too much traumatized by the Mongol invasions and destructions. Greek and Russian Orthodox Christianity, again, perhaps was too deeply traumatized by the conquests of the Tartars and the Turks.

Whatever the reason, there is no doubt that Copernicus was a Polish Catholic and Galileo an Italian one. The Lutheran Reformation came not long after Copernicus, and science then tended to shift into Protestant Europe, which was more sympathetic to it. Charles II of England might almost be called the Constantine of science. It was under his reign that the Royal Society was founded, which represented certainly the establishment of legitimacy. Right up to the beginning of the nineteenth century most scientists were Christians of a sort. Priestley was a Unitarian minister, Newton a sort of Unitarian, Dalton was a Quaker. I have not really been able to identify who was the first atheist scientist. It may well have been Laplace, who is supposed to have replied to Napoleon's question as to whether he found God in his equations of the solar system, "I have no need of that hypothesis." On the whole, however, the atheists and agnostics of the Enlightenment, like Voltaire, Rousseau, and Tom Paine, tended to be philosophers and literary types rather than scientists.

I have speculated as to why Christian culture was congenial to science even though the Catholic Church, especially, was very hostile toward it in its early days, as poor Bruno and Galileo witnessed. Scientists of the seventeenth and eighteenth centuries were either people of independent means like Boyle, lightly worked government officials like Newton or Adam Smith, associated with universities for other reasons (like philosophy), or were royal or aristocratic protégés, like Tycho Brahe. It was not until the nineteenth century that science as such really established itself in universities, and even there in a setting that was considerably dominated by clerics.

It is not wholly surprising that Christian theology and culture provided some of the preconditions of the rise of science. Christianity was essentially a working-class religion—born in a stable, founded by a carpenter, propagated by a tentmaker and a fisherman—and this gave it a profound interest in the real world. It is officially a religion of love, and the practice of threat, as in the Spanish Inquisition, at least sat uncomfortably in its basic doctrines. The love of humanity which at least its sacred documents preached can easily spill over into that love of the natural world out of which comes curiosity and the passion to find out about it. The Christian Church also arose in its early days not from philosophers and theologians, but from the record of some extraordinary things that very ordinary people believed had happened to them. It may seem a little fanciful to compare the record of the Resurrection of Jesus to the Michelson-Morley experiment, but a belief in either of them changes the paradigm of the world. Each of them suggests indeed that there is another world beyond the cozy experience of ordinary daily life. The Michelson-Morley experiment, of course, is more repeatable—and I presume has been repeated—but while science may be legitimately suspicious of a unique and nonrepeatable experience, it cannot be ruled out a priori. There is evidence, for instance, for a spontaneous nuclear explosion in Gabon some two million years ago, which seems to have been unique in the history of the planet as far as we know.

Even if we accept the view of science as a mutation out of Catholic Christian culture, continuing on into Protestant Christian culture, we should not be surprised to find that from the very beginning severe tensions existed between the scientific culture and the culture out of which it sprang, and that these continue. Christian cosmology is still in large part Judaic. That means that it is both creationist and homocentric. It stands in awe before the wonder, complexity, and splendor of the universe and says, "How could this come to be?" It looks at the human race and sees it as a creator, again of an awesome variety of artifacts—cities, bridges, houses and clothing, pots and pans, jewelry and works of art, poems and stories, legends and fantasies—and says, "How did these things come to be?" The answer is very clear: We made them. And how did the universe come to be? God made it. "When I consider thy heavens, the work of thy fingers, the moon and the stars, which thou hast ordained; / What is man, that thou art mindful of him? and the son of man, that thou visitest him? / For thou hast made him a little lower than the angels, and hast crowned him with glory and honor, / thou madest him to have dominion over the works of thy hands; thou has put all things under his feet." (Psalms 8: 3–6).

The key, of course, is "thou visitest him." The Bible, by and large, is a record of people who felt that they were visited in the inner recesses of the human experience, sometimes in times of stress, sometimes in the intense quiet of which the mind is also capable. The experience of the "visitor," a still small voice that may not even say very much, is attested by innumerable records, writings, and indeed it is hard to explain a good deal of human history without it. What it corresponds to, "the real world," we cannot be sure, especially as we understand consciousness so little. It may be some quirk of the vast electrical and chemical activity of the human brain that produces an illusion of visitation, but then the only reason why we believe that all our experience is not just a quirk of the brain, a dream and illusion, is that it has pattern, coherence, and testability. Religious experience exhibits this too, though in a more uncertain mode. The experience of "visitation" is by no means confined to the religious experience. It is found also in art and in science. Mathematicians and scientists have often recorded the experience of a sudden flash of insight about a problem they have been wrestling with, perhaps for a long time without success. Darwin records how the idea of natural selection came to him as he was "reading Malthus for pleasure one evening." Poets have the experience of being "visited" by a poem which they do not write, but which writes itself, having been living in their head, as it were, all the time until the moment comes to be born.

The constantly changing and developing image of the universe which we find in the scientific community has been a continuous challenge to the simpler and more metaphorical folk images which emerge out of the religious experience. This conflict started even with Copernicus, who destroyed the Ptolemaic image on which, for instance, the geography of Dante's *Inferno* and *Paradiso* were based. Suddenly the universe becomes much colder and less homocentric. In the vast universe of modern astronomy, earth seems to be nowhere in particular, and there seems a high probability that in such a vast universe, human intelligences perhaps superior to our own have already developed, although we have no evidence of this. What astronomy and Copernicus did for space, geology and Darwin did for time. In this enormous universe, ten or twenty billion years old in time, twenty billion light years across in space, we may well say, "What is man that anyone is mindful of him?" The Lord God of Israel hardly seems big enough to take on this enormous enterprise. The "big bang" theory, of course, suggests a kind of deism, but it only postulates that the universe was created at a given moment. The days of Genesis, indeed, have now shrunk into nanoseconds, but the creator of the "big bang" is a very long way from the divine visitor, the still small voice.

If the "big bang" theory seems to lead to a super deity who set the whole thing up in the first nanosecond and has not visited it since, evolution seems to point to an even bleaker and more meaningless universe in which an impersonal natural selection sorts out random mutations and genetic structure, and where, even more than Laplace, we have no need of the hypothesis of God. There are passages from Tennyson's "In Memoriam" which are a classic example of what might be called the "evolutionary angst," and the amazing thing is that this was written a whole generation before Darwin's *Origin of Species.* In the modern world it is perhaps the poets who are the real prophets.[1]

In dealing with something as large and complex as the universe as it spreads out through space and time, scientists need to remember that they should be agnostic and that their language should exhibit a greater humility than it frequently does. I thought the creationists indeed rather improved the biology texts in California by making them a little more humble. It would be nice if the creationists could be a little more humble too! The record of the past, even of the fairly immediate past, is extremely imperfect. The record is not only a very small sample of the reality, but is a highly biased sample, biased by durability, for only durable records survive. Hence, we are trying to piece together an enormous pattern, most of which is missing. Any pattern that we think we perceive has a quite strong presumption that it might be wrong.

As science moves into the study of increasingly complex systems (and certainly the most complex system with which we are acquainted is the human brain), it loses its earlier simplicity and innocence and itself becomes more complex. The Newtonian universe of matter and energy will no longer do. The universe also contains information. Evolution is very largely a process in information, and even beyond information into what I have called "know-how" and "know-what," or even "know-whether." In the evolutionary process matter and energy are significant mainly as limiting factors and as coders of information. Furthermore, it is clear that something is at work that involves the recreation of potential, that the great world of human artifacts is not the only example of creation. All biological organisms are created by their genetic structure coded in DNA. Every time an egg is fertilized, biological potential is created. Every organism is a planned economy from the moment of the fertilization of the egg until death. Like all planned economies, the plan does not always come off, but it is still there in the form of potential.

Furthermore, we detect in the evolutionary record at least three or four of "time's arrows"—that is, directionalities of change toward complexity. There is little doubt that we are more complex than the amoeba, just measured in terms of bits. Then there is a movement towards control, towards cybernetics, in the development of the senses, of locomotion, of warm-bloodedness, of increasingly complex systems of homeostasis as time goes on. And there is a movement toward something that might be called "cleverness," knowledge in some sense—that is, structures inside the organism which have some kind of one-to-one mapping on structures outside it. Finally, there is consciousness, of which we may be the only really good example, which is a very important source of human knowledge, of images of the past and future and of ourselves, but which we still understand very little and cannot begin to simulate.

It is clear that the evolutionary process has prejudices. The same could be said of that continuation of evolution in the field of human artifacts, which is human history. Here, too, there are movements toward complexity, cybernetics, cleverness, and consciousness. Error is a little less stable than truth. Perhaps in some sense evil is a little less stable than goodness. As social consciousness increases, we become aware not only of ourselves and of our societies, but of the whole ongoing stream of human history, and may have choices toward betterment or toward worsening.

Theology tends to be in the language of metaphor. In the very act of worship and prayer we confess our own ignorance and impotence. Religion all too often produces an unbecoming arrogance, especially in charismatic and popular preachers. Nevertheless the metaphors of religion provide, as it were, an excuse for practices of the spiritual life which large numbers of human beings have found of great value. These valuations are not universal. Many human beings do very well without a spiritual life, just as there are those who do without music, poetry, and art. But that the spiritual life is part of the potential of human beings, as is the enjoyment of music, poetry, and art, can hardly be denied.

It seems trite to say that an evolutionary theology will have to evolve, but this means that it cannot really be ordered in advance. The human mind is an ecosystem inhabited by innumerable mental species, with constantly changing niches, even though the pattern also has a lot of stability. We can expect new species coming out of new mutations, new insights, new revelations. What these will be, nobody can predict. It is highly probable, however, that an evolutionary theology will emerge, to provide language and metaphor to accommodate and utilize the enormously enriched images of the universe which science has given us, at the same time recognizing that science is just another set of species in the great ecosystem and that it too constantly changes and accommodates.

It would certainly be premature, and indeed presumptuous, on the part of an amateur in the field to try to state the tenets of an evolutionary theology. It is likely indeed that there will be many evolutionary theologies, each coming out of a particular tradition, discipline, or "phylum." There is a precedent for this in the scientific community, where the impact of evolutionary ideas on economics, as expressed, for instance, in my own work on evolutionary economics,[2] is by no means the same as their impact, shall we say, on chemistry in the work of Prigogine,[3] or on paleontology at different levels, or on genetics or sociobiology. Each field and discipline is itself a phylum which is capable of receiving grafts of new ideas which will interact in different ways with the existing structure.

Nevertheless, it may be interesting to speculate tentatively about some possible characteristics of evolutionary theology, conceived as an attempt to interpret varieties of religious experience in language which at least resonates to the world view of evolution as it develops in the scientific community. Within the Judeo-Christian tradition, an evolutionary theology would not be comfortable with a rigid and literalistic interpretation of scripture, but would regard the Bible and other writings of the past as the paleontologist regards the rocks and fossils—as an imperfect record of real events which constitute evidence from which tentative images of the real events may be constructed by careful thought and analysis. An evolutionary theology also, while recognizing the large and ancient record of religious experience in prayer, worship, meditation, even ecstasy and personal transformation, and recognizing that these experiences gain their validity by being directed towards an object beyond the individual, will be chary of idolatry, which might almost be defined as premature identification of the object of religious practice and experience, whether this is a physical or a mental idol. "Seeing in a glass darkly," as St. Paul says, does not cut us off from wonder or adoration, praise or mystery, and may prevent us falling into bigotry and arrogance. It is not surprising to find this same problem in the secular world, such as in nationalism and Marxism. Both the "cult of personality," as with Stalin and Mao, and the "cult of literalism" in the interpretation of doctrine, have disastrous effects on the quality of life of the society.

Creation and revelation are concepts which cannot be cast aside in any theological interpretation of experience and yet which present real difficulties. Science in its present state must be agnostic about a creator, or it would transcend the limitations of its epistemological field. A creator almost by definition is outside the system of the perceived universe. The record of the past does suggest, however, that there are moments of creation in the development of evolutionary potential, moments before which something did not exist and after which it did. The current cosmology of the "big bang" certainly suggests a moment in time before which the universe in some sense did not exist and after which it did, though mainly in the form of potential. Similarly, there was a time before which there was no life on earth, after which there was; another time before which there were no vertebrates, after which there were; another time before which there were no mammals, after which there were; another time before which there were no humans, after which there were. These moments of creation, if we can call them such, of major evolutionary potential, are not, of course, reproducible or subject to laboratory investigation. They are too rare.

We are also aware of creation in the biosphere, as when an egg is fertilized and a new member of a species starts its life, and no matter what the National Academy of Sciences may say, I am pretty well convinced that before the moment at which my egg was fertilized, something called "me" did not exist, and afterward it did, again largely in the form of potential. These, I think, can legitimately be called "antientropic events," even though they may do no more than segregate thermodynamic entropy, and evolution is inexplicable without them. We are conscious in our own lives also of moments of creation, whether this is a poem; or a work of art; or a marriage before which a family did not exist, after which it did; or in the religious life, moments of conversion and conviction, of creation of a new identity, and so on.

We can perhaps distinguish between creation, the moment when a new individual comes into being (and death, a kind of negative creation), and relevation, which is the moment when a new evolutionary potential comes into being, and it is surprisingly hard to find a more secular word for this concept. It is usually very difficult to identify these moments, either of creation or of revelation, at the time when they occur. I am sure it never occurred to Pontius Pilate that his name would appear in millions of church services for 2,000 years! It is often hard to identify the moment of revelation, but certainly in the life of Julius Caesar there was no Christian Church; at the time of Nero there certainly was. In 500 A.D. there was no Islamic religion; in 700 A.D. there certainly was, and so on. A post-Newtonian science should have no difficulty with these rare events, discontinuities, and irreversible patterns in time. Newtonian mechanics must now be seen as a very special case.

On the other hand, an evolutionary theology would also recognize that there is an ecology of revelation, that what creates evolutionary potential in one time and place will not at others. We build our identities around different traditions, we belong to different phyla; there is nothing wrong with this, provided that in the white blaze of truth we bow our heads, humble our hearts, and keep silent. An evolutionary theology, however, must see revelation as continuing and not closed, for the present moment is just a point on the line of the little-known past and the unknown future.

Whether an evolutionary theology could resolve the conflict between the scientific and the religious phyla, I do not know, and I am by no means sure that it should. This tension properly managed can be creative to all parties. It is perhaps significant that in the recent creationist trial in Arkansas, most of the witnesses for the creationists were Ph.D.s in the sciences, and for the opposition were churchmen. I certainly rejoice in the verdict, for the attempt to impose the content of education by the threat of legal sanctions is a danger to education, contrary to the basic ethic of science that people should be persuaded by evidence and not by threat, and a danger to the integrity of religious experience, which must also follow this same ethic if it is not to be corrupted. Nevertheless, if the creationists make the scientists a little more humble, particularly in their teaching, and induce them to give more emphasis to the tentative character of any propositions about the past and to emphasize that science deals with evidence, not truth, something good will have been accomplished. If the scientists can make the creationists a little more humble, it might be a more difficult but even greater achievement. It is a curious paradox that it is the business of both science and religion to seek and not to be afraid of finding, and yet not to be too cocksure about what either thinks it has found.

NOTES

1. Alfred Tennyson, *In Memoriam,* LV, verse 4, 1850.
2. K. E. Boulding, *Evolutionary Economics* (Beverly Hills, Calif.: Sage Publications, 1981).
3. Grégoire Nicolis and Ilya Prigogine, *Self-Organization in Nonequilibrium Systems: From Dissipative Structures to Order through Fluctuations* (New York: Wiley-Interscience, 1977).

GARRETT HA

"SCIENTIFIC CREATIONISM" —MARKETING DECEPTION AS TRUTH

The seven-part TV series *The Voyage of Charles Darwin* ended in a reenactment of the 1860 Huxley-Wilberforce debate, in which Dr. Samuel Wilberforce, bishop of Oxford, attacked Thomas Henry Huxley for upholding Darwin's views but was thoroughly trounced. A television viewer might well have concluded that Darwinism had triumphed. How wrong he would have been!

Among scientists, it is true, the Darwinian theory did pass from triumph to triumph in the years after the debate to become the only view seriously entertained by professional biologists. The idea of natural selection now suffuses every branch of biology. There, Darwin has won.

But in the public arena, things are quite otherwise. Sixty-five years after Huxley-Wilberforce, the trial of John T. Scopes, a high school teacher, revealed an enormous resistance to Darwin's ideas among Fundamentalist Protestants. To the dismay of both parties in the dispute, this celebrated 1925 "monkey trial," in which Scopes was accused of teaching the theory of evolution in Dayton, Tennessee, was ultimately decided on purely technical grounds. Scopes was first convicted and fined $100, but on appeal he was acquitted on the technicality that the fine had been excessive. Within a few years, other trials around the country determined that state laws could not mandate the teaching of the biblical story of creation nor forbid the teaching of evolution in the public schools.

From *The Dial,* vol. 1, no. 1, September 1980.

Both violated the First Amendment of the Constitution, which established the separation of Church and State.

In the 1860 debate, evolutionists won the battle; in the following century, they nearly lost the war. By the time of the centenary of the *Origin of Species,* in 1959, the vast majority of high school biology texts had resolved the dispute simply by suppressing both special creation and evolution. The word "evolution" was usually omitted, with the flabby word "development" standing in its place. Natural selection was scarcely touched upon. A high school student in 1960 would generally have had no inkling of the importance of Darwin in the intellectual history of humanity.

The public resurrection of Darwinism came, curiously, from space. In October 1957, the Soviet Union launched *Sputnik I,* the first artificial earth satellite. By beating us out in the race to space, the Soviets shattered American complacency about our technological superiority. There arose an immediate outcry for greater emphasis on the teaching of science in the high schools. As biologists took up their portion of the educational burden, they became aware of how disastrously school administrators and textbook publishers had sabotaged biology. A feisty geneticist, Nobel Prize winner H. J. Muller, protested in an article entitled "One Hundred Years Without Darwin Is Enough." In response, the Biological Sciences Curriculum Study, the official arm of the biology teaching profession, put out three different high school textbooks, each of them assigning a major role to evolution and natural selection. When the state board of education in Texas asked for a special edition that would mitigate these frightening ideas, BSCS refused to compromise.

In human affairs as in Newtonian physics, action provokes reaction. Within a few years, Fundamentalists had developed a new attack, which ran around the end of the First Amendment. Knowing that they could not insert an explicitly religious view into the school curricula, they called their view scientific, christening it "scientific creationism." Their plea that it be included in the curricula had a surface plausibility. No human being *was* present at the origin of life on earth, nor did anyone actually observe and record the evolution of one species into another millions of years ago. Therefore (said the creationists), it is just as scientific to believe that all existing species were created in an instant in exactly the same forms that they now appear as it is to suppose that they evolved. Scientific creationists do not ask that their theory displace Darwin's in the schools. They ask only for equal time.

Are scientific creationists concerned primarily with science or with religion? In a presentation to the California Board of Education, one of their spokespersons said, "Creation in scientific terms is *not* a religious or philosophical belief." At the same time, an appeal for funds made by the Creation Science Research Center, in San Diego, bragged that it intended "to take advantage of the tremendous opportunity that God has given us . . . to reach the 63 million children in the United States with the scientific teaching of Biblical creationism."

Even at the religious level the creationists' view is a biased one. The only creation story they mention is the one in Genesis (in which there are actually two stories—the version in the first chapter being so different from that in the second chapter that biblical scholars believe they were written hundreds of years apart). Why do they not mention the belief of Hindus that the world began with the creation of the cosmic egg? What about the Babylonians' belief that there was not a single creationist god but two cosmic parents?

Many outsiders see the creationists' call for fair play as little more than a legal ploy. A close reading of Fundamentalist literature by social scientist Dorothy Nelkin, of Cornell University, led her to believe that these earnest people are most deeply disturbed by what they regard as the moral disintegration of our society—rising crime rates, profligate sexuality, breakdown of the family, undermining of authority, and so on. Darwin may be only the scapegoat.

Because many of the views of Fundamentalists are widely shared, creationists have considerable support among those who couldn't care less about the creation-versus-evolution argument. During the past generation, Americans have become ever more concerned about fair play toward minorities. Protecting minorities increases diversity, which is regarded as a positive good. Scientists have long insisted that truth cannot be determined by majority vote: Galileo, after all, was in his day a minority—or "a majority of one," to use Thoreau's inspired phrase. We worship fair play; we are intolerant of dogmatism.

So in town meetings and in public debates, scientific creationists have proved formidable opponents. Scientists have not found it easy to explain to creationist supporters why a view held by a sizable minority should be forcibly excluded from the public schools.

To see what is involved, let us adopt a tactic discovered long ago by the mathematicians: When one question stumps you, ask another. That is, ask a related question whose answer throws light on the first.

Let our other question be this: Why don't we teach astrology in the schools? Astrology holds that the course of each human life is determined to a considerable degree by the position of the stars in the sky at the exact moment of the individual's birth. Belief in it, in one variant or another, has probably been held by most of the people on earth. Even today, some universities in India offer degrees in the subject. Yet American believers do not pressure boards of education to add their subject to the curriculum. If believers in astrology became as well organized as the creationists, it is hard to see how their demands could be withstood. Our emotions concerning this issue have not been aroused; we can objectively examine the issues. On what grounds might scientists object to the inclusion of astrology in the public schools?

The reason for not calling astrology a science is simple: Its assertions cannot be proved false.

There is a widespread belief among the public that the statements of science are *provable.* Scientists and philosophers now agree this is wrong. No scientific statement is ever fully proved. Science is made up of statements that *may* be proved false but that have not, in fact, been proved false by the most rigorous tests. Those that are, in principle, not falsifiable are waterproof hypotheses, and they are beyond the pale.

Let's see why astrology is not science. Over 1,500 years ago, Saint Augustine cited what he regarded as a definitive disproof of astrology. He knew of two babies who were born at the same time, one to a wealthy couple and the other to a slave woman. When these babies grew up—surprise!—the child born to wealth became wealthy, and the slave's child became a slave. Since they had been born at the same instant, it was obvious, said Saint Augustine, that the astrological hypothesis was nonsense.

Did Saint Augustine prevail? He did not. Astrologers had a very simple response to his "disproof," which they continue to repeat to the present day. It is this: No two babies are ever born at *exactly* the same instant. Therefore, their astrological signs are different, and their futures must differ as well. Insistence on the word "exactly" converts the astrological position into a waterproof hypothesis.

Should astrology be taught in public schools? Not as science. On this scientists must be adamant. The total exclusion of doctrines based on waterproof statements is one of the few dogmas of science. If the public wants to have astrology taught as part of some other course—history? sociology?—that is a matter about which a scientist, *as a scientist,* has nothing to say.

Having shown that astrology is not scientific, we can return to our principal question: Is scientific creationism scientific? Curiously, a complete answer to this question was worked out more than a century ago in a brief dispute that has, by a quirk of history, been almost completely forgotten. The idea of evolution is much older than Darwinism. What Darwin contributed was a believable mechanism to account for evolution. Fifteen years before the *Origin of Species,* an anonymous volume, *Vestiges of the Natural History of Creation,* espoused the evolutionary view. Scientifically, *Vestiges* was, in the opinion of scientists both then and now, a poor thing, but it was very popular; it went through ten editions before the *Origin of Species* was published.

Many religious people saw evolution as a threat to morality and religion. One of the most disturbed of these was Philip Gosse, a minister in the Fundamentalist group called the Plymouth Brethren. Gosse was not only a minister but also a naturalist (a common combination in Victorian England). During the 1850s, Darwin consulted him on many matters, though without ever revealing the heretical trend of his thought.

Gosse, upset by *Vestiges,* set out to demolish completely all theories of evolution. He began with geology. Geologists explain the strata of the rocks by physical principles, deducing that it must have taken millions of years to deposit layer upon layer of sedimentary rocks. There is no way to reconcile this deduction with the religious belief that the world began in 4004 B.C., as proclaimed in the seventeenth century by James Ussher, archbishop of Armagh. But Gosse thought he had found a way. His book, published two years before the *Origin,* was entitled *Omphalos.* The name is significant: It is Greek for "belly button."

Consider Adam and Eve, said Gosse. Did they have navels? Since the navel is a vestige of the link between the fetus and the placenta, one could argue that they had no navels, since Adam was created from dust and Eve was created from Adam's rib. But one could also argue that the first human had to have a navel; it is inconceivable that God (a perfect being) would create imperfect creatures. Adam's and Eve's navels were not evidence of a pre-existing being (namely a mother) but were merely what one would expect in God-created creatures.

Gosse explained the stratification of the rocks by the same logic. Strata are not evidence of processes occurring over millions of years; they are merely what one would expect to find in a perfect world. The strata and their fossils were all created on day three (see Genesis) as a materialization of God's thought. The fossils are merely artifacts that God was pleased to place among the strata when he created the world. The deductions of the geologist and the biologist fall to ground, and the Bible stands supreme as the revelation of truth. So said Gosse.

Gosse expected *Omphalos* to be attacked by scientists. It was. He was not prepared for the bitter denunciation by the religious community. Asked to write a review of *Omphalos,* his friend Charles Kingsley, a minister and the author of *Westward Ho!,* refused. He wrote a letter to Gosse explaining why.

"You have given," Kingsley said, "the 'vestiges of creation theory' the best shove forward which it has ever had. I have a special dislike for that book; but, honestly, I felt my heart melting towards it as I read *Omphalos.*

"Shall I tell you the truth? It is best. Your book is the first that ever made me doubt [the doctrine of absolute creation], and I fear it will make hundreds do so. Your book tends to prove this—that if we accept the fact of absolute creation, God becomes God-the-Sometime-Deceiver. I do not mean merely in the case of fossils which *pretend* to be the bones of dead animals; but in . . . your newly created Adam's navel, you make God tell a lie. It is not my reason, but my *conscience* which revolts here . . . I cannot . . . believe that God has written on the rocks one enormous and superfluous lie for all mankind.

"To this painful dilemma you have brought me, and will, I fear, bring hundreds. It will not make me throw away my Bible. I trust and hope. I know in whom I have believed, and can trust Him to bring my faith safe through this puzzle, as He has through others; but for the young I do fear. I would not for a thousand pounds put your book into my children's hands."

Gosse, abandoned by churchmen, gave up theorizing and returned to merely observing nature. As a popularizer of nature, his position in science education is an honorable one. His *Evenings at the Microscope* persuaded many an English gentleman to take up the microscope as a hobby.

Returning to the present, we note that there has been no improvement in the arguments for creation since *Omphalos.* Of course we now have the ingenious "radioactive clock" method of dating strata and fossils, but this can be explained away as easily as Adam's belly button. If an Archeozoic crystal has more lead and less uranium than one formed during the Cenozoic Era, it is merely because God set the two clocks at different times when he started both of them ticking in 4004 B.C. So say the creationists.

Neither scientist nor scientific creationist can suggest any deduction from the creation hypothesis that can be proved false, now or in the future. But the hypothesis of evolution *is* falsifiable by a thousand conceivable observations, for example, finding *Australopithecus* bones in strata from the Mesozoic Era. Evolution, therefore, might be a false hypothesis. But creationism can never be proved false.

The Reverend Charles Kingsley was closer to the truth than perhaps he knew when he said it was not his reason but his conscience that made him reject the waterproof belly button argument. In some abstract sense, sicence may (as some claim) be value free, but the practitioners of science often become very emotional when they are confronted with waterproof hypotheses. They exhibit what can only be called moral indignation—or the sort of contemptuousness a professional gambler would express if he were asked to play poker with twos, threes, fours, fives, and one-eyed jacks wild. Grown men don't play such games.

There is a paradox in the present Mexican standoff between scientists and scientific creationists. Bible supporters want Genesis taught because (they say) it is scientific; evolutionists want waterproof hypotheses excluded because (they feel) they are intellectually immoral. Small wonder for confusion.

Actually, all of the arguments given here *could* be included in public schools and with considerable educational benefit. That such material is not included has many explanations. The principal one is no doubt this: It is always easier to teach facts than arguments. It is particularly difficult to *examine* for an understanding of arguments. Teachers—some of them—are lazy. So are some students. Classes—most of them—are large; this militates against teaching subtle arguments. A pluralistic society like ours makes it easier to run away from a controversy than to deal with it fairly and openly.

One wonders: When the second centenary of the *Origin of Species* rolls around, in the year 2059, will the theory of evolution through natural selection be universally accepted? Evidences of natural selection are everywhere: in the unwanted appearance of DDT-resistant insects and antibiotic-resistant disease germs as well as in the wanted development of domestic plant and animal varieties in response to breeding programs in which man defines the selective criteria. But these evidences are nothing to a person who does not reject waterproof hypotheses.

Our social world is a chaotic one. It is understandable that many sincere people should seek emotional refuge in a waterproof hypothesis like that of instantaneous creation. Broadening the support for Darwin's view depends not so much on accumulating more scientific evidence as it does on getting more people to understand the nature of science itself.

LAURIE R. GODFREY

SCIENTIFIC CREATIONISM: THE ART OF DISTORTION

Where is the science in "scientific creationism"?

In 1963 American historian Richard Hofstadter wrote that "today the evolutionary controversy seems as remote as the Homeric era." The Biological Sciences Curriculum Project, supported in part by federal funds, was preparing secondary school texts that openly presented evolution as the foundation of biology. And George McCready Price, an outspoken leader of the protest against evolution in the days of the Scopes "monkey trial" and author of numerous antievolutionary tomes, including *The Phantom of Organic Evolution* (1924), *A History of Some Scientific Blunders* (1930), *The Modern Flood Theory of Geology* (1935), and *Genesis Vindicated* (1941), died at the age of 92. But 1963 was also the year that the Creation Research Society—and with it, organized "scientific creationism"—was born.

The Creation Research Society was founded by a group of ten men led by Walter E. Lammerts and William J. Tinkle. Many of these men were disaffected members of the American Scientific Affiliation, a theistic organization founded in 1941 and devoted to the reconciliation of science and evangelical Christianity. The increasing domination of the organization by evolutionists disturbed those who wanted it to oppose evolutionism. The "team of ten" vowed to work, through what they regarded as scientific endeavors, for a revival of belief in special creation as described in the King

This article is a revision of "The Flood of Antievolutionism," which appeared in *Natural History,* vol. 90, no. 6, pp. 4–10.

James version of the Bible. While they held populist William Jennings Bryan, the Scopes prosecutor, in high esteem, the new activists were creationists of a different ilk.

Bryan had mocked his scientific opponents: "You believe in the age of rocks; I believe in the Rock of Ages." He had preached to the masses, "I would rather begin with God and reason down than begin with a piece of dirt and reason up." But the new creationists profess no disdain for science. College-educated fundamentalist Christians with a strong commitment to both science (particularly in the form of technology and engineering) and to a literal interpretation of the Bible, they have set out to convince the public that "true science" supports the creation model of world and life origins. Denying that they are trying to bring religion into the public schools, they are seeking to have their model taught as science.

By the end of 1980, seventeen years after Hofstadter had pronounced the evolution controversy dead, "two-model" scientific education bills—which would require public schools to present creation as a scientific model alongside evolution—had been introduced and debated in the state legislatures of Florida, Georgia, Illinois, Iowa, Kentucky, Louisiana, Minnesota, New York, South Carolina, Tennessee, and Washington and were being introduced elsewhere. In addition, various local school boards around the country had passed resolutions that made concessions to creationist pressure. The membership of the Creation Research Society, based in Ann Arbor, had grown to 2,500. Sister organizations such as the Bible Science Association (Minneapolis), The Creation Social Science and Humanities Society (Wichita), the Institute for Creation Research and the Creation Science Research Center (San Diego) had been formed to defend scientific creationism and promote the teaching of creation on a par with evolution. In 1981 two states, Arkansas and Louisiana, passed versions of the creationists' "two-model" bill, one of which was overturned by Federal District Court Judge William Overton very early the following year.*

Led by virtually the same nucleus of antievolutionists, these organizations have become efficient factories of purportedly scientific

* The Louisiana bill has also been overturned.

antievolutionary propaganda. Conventions, as well as debates, textbooks, and films, are the means to the political end of building a constituency. The Institute for Creation Research (ICR) now offers college- and graduate-level programs as well as summer institutes (offering optional college credit) on creationism; distributes antievolutionary books, pamphlets, and cassettes; and sponsors creation/evolution debates and nationally distributed weekly radio broadcasts. And the ICR also funds research: to find evidence of Noah's ark and a global flood; evidence of coexisting humans, trilobites, and dinosaurs; and proof of a recent creation of the universe and the planet Earth (the earth is assumed to be roughly 10,000 years old). The Creation Research Society developed the first "creation science" biology textbook meant for use in public secondary schools, and since 1964 the society has published a quarterly journal dealing with evidence that supports a literal interpretation of the Bible.

The scientific creationists make no attempt to hide the proselytizing role of the various research organizations. Emphasis Creation 1980 was a midwestern convention of scientific creationists sponsored jointly by the ICR and the Bible Science Association. The Director of ICR, Henry Morris, gave explicit instructions, which appeared in the newsletter of the ICR's Midwest Center:

> The purpose of such a convention should not be to provide a forum where various creationists get together to present papers arguing for their own particular interpretations on details of science or Scripture. Rather, it should seek to reach as large and general an audience as possible with carefully chosen papers (and other activities) by qualified speakers who will make the greatest impact for the creationist cause in general.

The newsletter went on to list acceptable and unacceptable topics. The former included refutations of evolutionism; legal, political, and educational aspects of teaching creation in schools; scientific evidence for a recent creation of the earth and universe; and "flood geology," which attributes a wide range of fossil-bearing geologic deposits to a single catastrophic global event, the Noachian deluge. Unacceptable topics included plate tectonics and continental drift (listed among others as areas of questionable or peripheral significance to creationism) and all "highly technical and specialized treatments of individual problems."

Field or laboratory research represents a very minor charge of scientific creationists. Most efforts are directed toward rewriting the discoveries and interpretations of evolutionists. In this endeavor, numerous evolutionists are portrayed as scientists who have all the evidence to disprove evolution (and support creation) at their fingertips, but who are either too stubborn or too deeply indoctrinated in evolutionary dogma to appreciate it. Arguments of anthropologists, biologists, chemists, geologists, astronomers, physicists, and engineers are reinterpreted or taken out of context. In this way, as I will show below, creationists manage, among other things, to convert arguments about the pattern and process of evolutionary change into arguments about the existence of change.

The primary tactic of the scientific creationists is to find controversy, disagreement, and weakness in evolutionary theory—by no means a difficult task. Having demonstrated problems with various aspects of evolutionary theory (some fabricated, some real), the creationists then conclude that we must accept the Judeo-Christian biblical account of creation as the only possible, logical alternative. Thus scientific creationism proceeds by constructing an artificial dichotomy between two models—evolution and creation—both incorrectly represented as monolithic.

Actually, various evolutionary explanations are possible, and numerous models, both Darwinian and non-Darwinian, have been posed. They have in common the notion that the earth's life forms are related by common ancestry, whether or not they have since achieved reproductive isolation. Evolutionists agree that the evidence supports this premise of genetic continuity, although as scientists they do not rule out the logical possibility that life could have arisen independently on more than one occasion on the earth or in the universe.

Creationism, on the other hand, is based on the idea that reproductive isolation often signals the absence of common ancestry. Given genetic discontinuity, numerous creation-based explanations are nevertheless possible: witness the global diversity of creation myths. Ignoring this diversity, however, scientific creationists begin with one specific and detailed explanation of the universe and require its acceptance on faith as a prerequisite of membership in their various research organizations. The Statement of Belief of the Creation Research Society begins: "The Bible is the written Word of God, and because we believe it to be inspired throughout, all of its assertions are historically and scientifically true in all of the original autographs." The scientific creationists do not pose and test alternative creation models. Doing science is not the business of scientific creationists; destroying the public credibility of evolution is their real goal. "New evidence," the press is told, reveals "major weaknesses" in evolution. Oddly, the creationist tactic of discovering controversies within evolutionary biology amounts to discovering that evolutionary biologists are guilty of doing science—posing, testing, and debating alternative explanations.

Biologists seek answers to many diverse and intriguing questions about evolutionary processes. They debate the rate and pattern of evolutionary change, its directionality or lack of directionality, and most importantly, its causal mechanisms. Some evolutionary biologists defend Darwin more ardently than others; some have accounted for aspects of the biological record of change in ways Darwin never considered, and thus favor non-Darwinian causal mechanisms for at least some aspects of evolutionary change. But whatever the causes, it is abundantly clear that organisms have evolved. Thus we must separate the concept of evolution from those theories which attempt to explain evolutionary change, including Darwinian natural selection. The latter was first widely accepted as the primary mechanism of evolutionary change not in Darwin's day but well afterwards, in the 1930s and 1940s when the "neo-Darwinian" Modern Synthesis was born (Fisher, 1930; Dobzhansky, 1937; Simpson, 1944 and 1948). In recent decades neo-Darwinism has come under renewed scrutiny. While the 1970s and 1980s might be characterized as a period of excitement, vigorous growth and polarization in the field of evolutionary biology (see, for example, Eldredge, 1971; Eldredge and Gould, 1972; Ayala, 1975; Hallam, 1977; Schwartz and Rollins, 1979; Stanley, 1979, 1981; Kimura, 1979; Gould, 1980; Cronin et al., 1981; Brace, 1981; Stebbins and Ayala, 1981), the debates of the 70s were never about the occurrence of evolution but about its processes.

Creationists are decidedly uninterested in the merits of any of these arguments and, indeed, in the issues themselves. But they are intrested in convincing the public that the concept of evolution is utterly bankrupt. By citing and falsely *reinterpreting* excerpts from a vast and complex scientific literature, they have built for their constituency a false (though superficially plausible) picture of what the issues really are. Furthermore, since most laypersons equate Darwinism with evolution, critics of neo-Darwinism are especially vulnerable to misrepresentation. But the non-Darwinian arguments so often cited by creationists can hardly be taken as support for creationism by anyone who bothers to discover what they are really about.

One scientific debate in particular, that between the neocatastrophists (or punctuationalists) and the phyletic gradualists, has fueled the fires of scientific creationism. In 1972 Niles Eldredge of the American Museum of Natural History and Harvard paleontologist Stephen Jay Gould launched their new theory of evolution by "punctuated equilibria." Evolution, they claimed, proceeds by rapid fits and starts, punctuating long periods of relative stasis. Drawing from the work of other paleontologists and neontologists, Eldredge, Gould, and later, Steven Stanley of Johns Hopkins developed the implications of a punctuational model of evolutionary change (Eldredge and Gould. 1972; Gould and Eldredge, 1977; Stanley, 1979, 1981). In so doing, they challenged the assumption that most evolutionary change occurs as a by-product of slow, ceaseless natural selection acting on variation in well-established populations.

While they have not abandoned the concept of natural selection as an important evolutionary process, the punctuationalists have reinterpreted its role. Central to their argument is the view that most evolutionary change occurs in association with speciation, that is, the formation of independent species by the splitting of lineages into reproductively isolated populations. Drawing from the work of Ernst Mayr (1954, 1963) on founder populations, they argue that speciation may be achieved rapidly in small, geographically isolated populations and that, in such populations, chance (as well as natural selection) can exert much greater influence on genetic change in form than is otherwise possible. They further argue that rapid or dramatic evolutionary changes cannot occur in the *absence* of speciation. The mechanisms and importance of speciation lie at the heart of the debate between the punctuationalists and their opponents. Unlike the phyletic gradualists, the punctuationalists conclude that in macroevolution (evolution viewed in the long range and on a large scale) an episodic pattern of change is the expectation rather than an exception to the rule.

A second important issue in evolution that has attracted the attention of creationists is the question of the relative importance of chance as a factor in evolutionary change. Using computer simulations, David Raup and his colleagues at Chicago's Field Museum of Natural History have argued that chance is very important in macroevolution as well as microevolution (Raup, 1976, 1977, 1981, in press; Raup and Gould, 1974). Raup believes that many genetic changes that do not greatly affect "fitness" may survive or fail to survive owing to chance. Such evolution by chance is called neutral, or non-Darwinian, evolution. The debate in evolutionary biology is over its relative importance, not its existence.

It is hard to imagine creationists drawn to the arguments of Eldredge, Gould, and Raup, since they are antithetical to creationist tenets. First, the question of the genetics of speciation, which is central to the theory of the punctuationalist school, is foreign to creationism. "Speciation" is rarely part of the creationist vocabulary; "special creation" is used instead. Creationists claim that each life form was created as a separate "kind" and that nature permits variation only within such created kinds. They conveniently downplay speciation when they ridicule evolutionists for sometimes reconstructing ancestors which resemble organisms that are still extant. For example, Morris (1974) considers that the discovery of a living crossopterygian fish effectively refutes the suggestion that amphibians may have evolved from crossopterygian fishes. Of course, it does nothing of the kind. This evolutionary scenario requires only that speciation has occurred.

Creationists rely on the fact that most people have some commonsense notion of biological discontinuity and can use this to intuit "kind." They insist that kinds are not species; yet they seem sometimes to use the notion "created kind" as a substitute for the biologists' species concept. When pressed, creationists maintain that a "kind" may be recognized by the "criterion of interfertility"—

based on the notion that God's creatures brought forth progeny "after their kind." (Note also Byron Nelson's definition of "kind" as "natural species," on pp. 156–173 of *After its Kind,* revised ed., 1967.) If so, reproductive isolation should designate separate kinds. But creationists must acknowledge the existence of speciation (that is, multiplication of species) if the term "created kind" is to remain unassailable, if vacuous. Only by *defining* any variation which can be shown to have evolved *in a short time span* as variation "within kinds," whether or not reproductive isolation has been achieved, can creationists maintain that variation only *exists* within "kinds."

In fact, the definition of a "created kind" is one of the most confused issues in the creationist literature (see discussion in Cracraft, 1983). Frank Lewis Marsh (1978) who deems himself discoverer of a "principle of limitation of variation within kinds" cites Theodosius Dobzhansky's *Genetics of the Evolutionary Process* to establish the existence of discontinuities which, he implies, separate "created kinds." But Dobzhansky (1970) is describing a *hierarchical* pattern of discontinuities with speciation and evolutionary divergence as its fundamental cause. Such a hierarchical pattern is a direct prediction of the evolutionary hypothesis (a fact which few creationists acknowledge).

Punctuationalists theorize that rapid evolutionary change associated with speciation is the dominant cause of macroevolutionary change; Steven Stanley (1979) calls this "quantum speciation." Quantum speciation is a known biological phenomenon (see Stanley, 1979)—what is *not* known is its relative importance for macroevolution. Creationists acknowledge speciation only as a by-product of "variation within kinds," and they insist that it results in no significant morphological change. They trivialize the voluminous literature on the mechanisms of speciation (slow and rapid) and thus effectively deny the existence of the very process which punctuationalists take as the prime cause of macroevolution.

Nothing about punctuationalism supports the creationist viewpoint. Punctuationalists simply maintain that while much evolutionary change is very slow or static, very rapid "jumps" can occur naturally and these are the important stuff of macroevolutionary change. Genetically, such jumps are as comprehensible as slow phyletic changes. Indeed, whether they are perceived as jumps at all depends upon one's expectations concerning the scale and pace of evolutionary change. As Gould has written (*Natural History*, August 1979):

> New species usually arise, not by the slow and steady transformation of entire ancestral populations, but by the splitting off of small populations from an unaltered ancestral stock. The frequency and speed of such speciation is among the hottest topics in evolutionary theory today, but I think that most of my colleagues would advocate ranges of hundreds or thousands of years for the origin of most species by splitting. This may seem like a long time in the framework of our lives, but it is a geologic instant, usually represented in the fossil record by a single bedding plane, not a long stratigraphic sequence.

Second, "chance" is also foreign to creationism. One Florida-based organization puts out a flier that reflects the widespread creationist notion that nothing (or nearly nothing) ever happens by chance: "Evolution demands what has not, and cannot happen, even with careful planning—much less by total accident!" It is, of course, a misstatement of evolution to claim that this body of theory argues that change comes about "by total accident," for selection is not a random process. Yet the non-Darwinian school ascribes to chance a much more central role than is admitted by other evolutionary biologists. Ironically, in their effort to show disagreement among evolutionists, the creationists are citing the work of paleontologists whose arguments are, in many ways, the most antithetical to creationism.

One reason creationists are able to exploit the current debates among evolutionists is that certain key phrases have entirely different meanings for paleontologists and for creationists (or their constituency). When such phrases are lifted from the work of evolutionists and inserted into creationist literature, they acquire new meaning simply because of differences in assumed knowledge. For example, the "neocatastrophism" of paleontology (widely quoted in support of creationist catastrophism) has nothing to do with either creation or a great flood. But creationists automatically associate the term "catastrophism" with the concept of the Noachian deluge.

Creationist Gary Parker wrote an essay on neocatastrophism

that was circulated in the October 1980 issue of *Acts & Facts,* the free monthly newsletter of the ICR. Reading his article, one cannot avoid the conclusion that Raup and Gould consider the creation model tenable, if not actually preferable to evolutionism. Here is a passage from Parker's essay:

> "Well, we are now about 120 years after Darwin," writes David Raup of Chicago's famous Field Museum, "and the knowledge of the fossil record has been greatly expanded." [Parker cites a 1979 article by Raup.] Did this wealth of new data produce the "missing links" the Darwinists hoped to find? ". . . ironically," says Raup, "we have even fewer examples of evolutionary transition than we had in Darwin's time. By this I mean that some of the classic cases of darwinian change in the fossil record, such as the evolution of the horse in North America, have had to be discarded or modified as a result of more detailed information." Rather than forging links in the hypothetical evolutionary chain, the wealth of fossil data has served to sharpen the boundaries between the created kinds. As Gould says, our ability to classify both living and fossil species distinctly and using the same criteria "fit splendidly with creationist tenets." "But how," he asks, "could a division of the organic world into discrete entities be justified by an evolutionary theory that proclaimed ceaseless change as the fundamental fact of nature?" [Parker cites a 1979 *Natural History* article by Gould.] ". . . we still have a record which *does* show change," says Raup, "but one that can hardly be looked upon as the most reasonable consequence of natural selection." The change we see is simply variation within the created kinds, plus extinction.

The arguments Parker presents outside as well as inside quotation marks seem to be those of Raup and Gould. Given these selected tidbits, there is no way to interpret the statements of Raup and Gould except within the framework of the creation model. The reader is not told what Raup and Gould are arguing but is left instead to surmise, incorrectly, that evolution itself is under attack. Furthermore, Parker has chosen to cite titles that seem to support such an interpretation. Raup's article is called "Conflicts between Darwin and Paleontology." Gould's is entitled "A Quahog Is a Quahog."

Those familiar with Raup's research will not be surprised to find that his article is actually a treatise concerning problems with Darwinian gradualism. Raup first deals with the complex, uneven record of evolutionary change. His point, quoted more fully, is that "some of the classic cases of darwinian change in the fossil record, such as the evolution of the horse in North America, have had to be discarded or modified as a result of more detailed information—what appeared to be a nice simple progression when relatively few data were available now appears to be much more complex and much less gradualistic." Raup goes on to discuss the potential of chance processes to bring about apparently patterned evolutionary change—in particular, the extinction of lineages.

Gould's article is also about problems with Darwinian gradualism. It takes to task those biologists and anthropologists who argue that species boundaries are artifacts of the human capacity to classify and construct artificial divisions. Gould argues, as Ernst Mayr did years before, that species are real biological entities, but he does not suggest that they are genealogically unrelated to one another or that they cannot give rise to new species.

Gould and his colleagues are widely cited by creationists in their effort to establish that the fossil record documents "no transitions." To creationists this is taken to mean that there are no evolutionary links between "created kinds." But Gould, Eldredge and Stanley are talking about the failure of the fossil record to document *fine-scale* transitions between pairs of species, and its dramatic documentation of rapid evolutionary bursts involving multiple speciation events—so-called adaptive radiations. They are not talking about any failure of the fossil record to document the existence of intermediate forms (to the contrary, there are so many intermediates for many well-preserved taxa that it is notoriously difficult to identify true ancestors even when the fossil record is very complete). Nor are Gould, Eldredge, and Stanley talking about any failure of the fossil record to document large-scale trends, which *do* exist, however jerky they may be. Furthermore, fine-scale transitions are *not* absent from the fossil record but are merely *underrepresented.* Eldredge, Gould, and Stanley reason that this is the unsurprising consequence of known mechanisms of speciation. Additionally, certain ecological conditions may favor speciation and rapid evolution, so new taxa may appear abruptly in the fossil record in association with adaptive radiation. Since creationists acknowledge that fine-scale transitions (including those resulting in reproductive isolation) exist, and since the fossil record clearly documents large-scale "transitions," it would seem that the creationists have no case. Indeed, they do not. Their case is an artifact of misrepresentation to the lay public of exactly *what* the fossil record fails to document.

Despite the attempts of scientific creationists to play up the signs of controversy among evolutionists, there is actually widespread agreement in scientific circles that the evidence overwhelmingly supports evolutionism. Confirmation has sometimes taken unexpected forms, as in the high correlation between the degree of biochemical difference between pairs of species and the amount of paleontological time since their apparent separation.

There is agreement that the *pattern* of origin of taxa in the paleontological record strongly supports genetic continuity and, therefore, evolution. The punctuationalists' concept of evolutionary stasis has been misused by creationists to argue against such a pattern, but evolutionary stasis contradicts only strict gradualism, not evolution. The fact is, the genus *Homo* does not occur in the Mesozoic alongside the brontosaurus, as the creationists claim; if it did, we would indeed have to question our evolutionary assumptions.

Scientists do ask questions about the pattern of evolutionary change. In particular, does the fossil record bear witness to the slow, continuous, gradual change envisioned by Darwin and supported by neo-Darwinists? Although still a matter of considerable debate, some form of punctuationalism is gaining increasing support among evolutionists (but see Stebbins and Ayala, 1981; Cronin et al., 1981; and Brace, 1981 for counterarguments). Scientists also ask questions about the process or mechanism of evolutionary change: for example, given a pattern of punctuational change, is Darwinian natural selection the best explanation for macroevolutionary trends?

The current debate is complicated because the concept of natural selection embraced by Darwinians has changed with the introduction of population genetics. Steven Stanley's concept of species selection (the differential survival of species) is part of natural selection as formulated by Darwin and some modern biologists, but not as formulated by population geneticists focusing on selection operating within populations. Therefore, when Eldredge, Gould, and Stanley proclaim natural selection to be an inadequate explanation of macroevolutionary change, it is important to realize that they are talking about natural selection as mathematized, reformulated, and restricted to populational variation by population geneticists in the 1930s.

When a creationist such as Parker describes the putative failure of natural selection, however, it is to an audience that simplistically equates natural selection with evolution—an audience that does not know the difference between natural selection and species selection. Most students of scientific creationism know little about the debate between the phyletic gradualists and punctuationalists or that between proponents of Darwinian (nonrandom) and non-Darwinian (random) processes of change. And they will not learn what the debates are about from Parker and his colleagues.

"It's so utterly infuriating to find oneself quoted, consciously incorrectly, by creationists," Gould has said. "None of this controversy within evolutionary theory should give any comfort, not the slightest iota, to any creationist." Yet the scientific creationists, by misrepresenting the ongoing work of evolutionists, have helped the antievolutionary cause to gain more momentum than ever before in the twentieth century. Scientific creationists are widely viewed as learned scholars with impressive credentials, and more and more people are being persuaded that staggering evidence is on their side. Many scientists are baffled that such poor science can be so easily swallowed, and that creation is being taught as science in some schools around the country. Scientific creationism may be poor science, but it is powerful politics. And politically, it may succeed.

BIBLIOGRAPHY

Ayala, Francisco J. (ed.). 1975. *Molecular Evolution.* Sunderland, Mass.: Sinauer.

Brace, C. Loring. 1981. Tales of the phylogenetic woods: the evolution and significance of evolutionary trees. *American Journal of Physical Anthropology* 56(4):411–29.

Cracraft, Joel. 1983. Systematics, comparative biology, and the case against creationism. In L. Godfrey (ed.), *Scientists Confront Creationism.* New York: W. W. Norton, pp. 163–69.

Cronin, J. E., N. T. Boaz, C. B. Stringer, and Y. Rak. 1981. Tempo and mode in hominid evolution. *Nature* 292:113–22.

Dobzhansky, Theodosius. 1937. *Genetics and the Origin of Species.* 1st ed. New York: Columbia Univ. Press.

———. 1970. *Genetics of the Evolutionary Process.* New York: Columbia Univ. Press.

Eldredge, Niles. 1971. The allopatric model and phylogeny in Paleozoic invertebrates. *Evolution* 25:156–67.

Eldredge, Niles, and Stephen J. Gould. 1972. Punctuated equilibria: an alternative to phyletic gradualism. In T. J. M. Schopf (ed.), *Models in Paleobiology.* San Francisco: Freeman, Cooper, pp. 82–115.

Fisher, Ronald A. 1930. *The Genetical Theory of Natural Selection.* Oxford: Clarendon Press.

Gould, Stephen J. 1979. A quahog is a quahog. *Natural History* 888(8):18–26.

———. 1980. Is a new and general theory of evolution emerging? *Paleobiology* 6:119–30.

Gould, Stephen J., and Niles Eldredge. 1977. Punctuated equilibria: the tempo and mode of evolution reconsidered. *Paleobiology* 3(2): 115–51.

Hallam, A. (ed.). 1977. *Patterns of Evolution.* Amsterdam: Elsevier.

Hofstadter, Richard. 1963. *Anti-intellectualism in American Life.* New York: A. Knopf.

Kimura, Motoo. 1979. The neutral theory of molecular evolution. *Scientific American* 241(5):94–104.

Marsh, Frank Lewis. 1978. Variation and fixity among living things. A new biological principle. *Creation Research Society Quarterly* 15: 115–18.

Mayr, Ernst. 1954. Change of genetic environment and evolution. In J. Huxley, A. C. Hardy and E. B. Ford (eds.), *Evolution as a Process.* London: Allen and Unwin, pp. 157–80.

———. 1963. *Animal Species and Evolution.* Cambridge, Mass.: Harvard Univ. Press.

Morris, Henry M. (ed.). 1974. *Scientific Creationism.* San Diego: Creation-Life Publishers.

Nelson, Byron. 1967. *After Its Kind.* rev. ed. Minneapolis, Minn.: Bethany Fellowship, Inc.

Price, George McCready. 1924. *The Phantom of Organic Evolution.* New York: Fleming H. Revell Co.

———. 1930. *A History of Some Scientific Blunders.* New York: Fleming H. Revell Co.

———. 1935. *The Modern Flood Theory of Geology.* New York: Fleming H. Revell Co.

———. 1941. *Genesis Vindicated.* Takoma Park, Md. and Washington, D.C.: Review and Herald Publ. Assoc.

Raup, David M. 1976. Species diversity in the Phanerozoic: an interpretation. *Paleobiology* 2(4):289–97.

———. 1977. Stochastic models in evolutionary paleontology. In A. Hallam (ed.), *Patterns of Evolution.* Amsterdam: Elsevier, pp. 59–78.

———. 1979. Conflicts between Darwin and paleontology. *Bulletin of the Field Museum of Natural History* 50:22–29. Chicago.

———. 1981. Extinction: bad genes or bad luck? *Acta Geologica Hispanica* 16:25–33.

———. In press. The role of chance in evolution. In L. Godfrey (ed.), *A Century After Darwin: Issues in Evolution.* Boston: Allyn and Bacon.

Raup, David M., and S. J. Gould. 1974. Stochastic simulation and evolution of morphology—towards a nomothetic paleontology. *Systematic Zoology* 23:305–22.

Schwartz, Jeffrey H., and Harold B. Rollins (eds.). 1979. Models and methodologies in evolutionary theory. *Bulletin of Carnegie Museum of Natural History* 13. Pittsburgh, Pa.

Simpson, George Gaylord. 1944. *Tempo and Mode in Evolution.* 1st ed. New York: Columbia Univ. Press.

———. 1949. *The Meaning of Evolution.* New Haven: Yale Univ. Press.

Stanley, Steven M. 1979. *Macroevolution: Pattern and Process.* San Francisco: W. H. Freeman.

———. 1981. *The New Evolutionary Timetable: Fossils, Genes and the Origin of Species.* New York: Basic Books.

Stebbins, G. Ledyard, and Francisco J. Ayala. 1981. Is a new evolutionary synthesis necessary? *Science* 213:967–71.

ISAAC ASIMOV

THE "THREAT" OF CREATIONISM

Scientists thought it was settled.

The universe, they had decided, is about 20 billion years old, and Earth itself is 4.5 billion years old. Simple forms of life came into being more than three billion years ago, having formed spontaneously from nonliving matter. They grew more complex through slow evolutionary processes and the first hominid ancestors of humanity appeared more than four million years ago. *Homo sapiens* itself—the present human species, people like you and me—has walked the earth for at least 50,000 years.

But apparently it isn't settled. There are Americans who believe that the earth is only about 6,000 years old; that human beings and all other species were brought into existence by a divine Creator as eternally separate varieties of beings; and that there has been no evolutionary process.

They are creationists—they call themselves "scientific" creationists—and they are a growing power in the land, demanding that schools be forced to teach their views. State legislatures, mindful of votes, are beginning to succumb to the pressure. In perhaps 15 states, bills have been introduced, putting forth the creationist point of view, and in others, strong movements are gaining momentum. In Arkansas, a law requiring that the teaching of creationism receive equal time was passed this spring and is scheduled to go into effect in September 1982, though the American Civil Liberties

From *The New York Times Magazine,* 14 June 1981. Copyright © 1981 by *The New York Times* Company. Reprinted by permission.

Union has filed suit on behalf of a group of clergymen, teachers, and parents to overturn it. And a California father named Kelly Segraves, the director of the Creation-Science Research Center, sued to have public-school science classes taught that there are other theories of creation besides evolution, and that one of them was the Biblical version. The suit came to trial in March, and the judge ruled that educators must distribute a policy statement to schools and textbook publishers explaining that the theory of evolution should not be seen as "the ultimate cause of origins." Even in New York, the Board of Education has delayed since January in making a final decision, expected this month, on whether schools will be required to include the teaching of creationism in their curriculums.

The Rev. Jerry Falwell, the head of the Moral Majority, who supports the creationist view from his television pulpit, claims that he has 17 million to 25 million viewers (though Arbitron places the figure at a much more modest 1.6 million). But there are 66 electronic ministries which have a total audience of about 20 million. And in parts of the country where the Fundamentalists predominate—the so-called Bible Belt—creationists are in the majority.

They make up a fervid and dedicated group, convinced beyond argument of both their rightness and righteousness. Faced with an apathetic and falsely secure majority, smaller groups have used intense pressure and forceful campaigning—as the creationists do—and have succeeded in disrupting and taking over whole societies.

Yet, though creationists seem to accept the literal truth of the Biblical story of creation, this does not mean that all religious people are creationists. There are millions of Catholics, Protestants, and Jews who think of the Bible as a source of spiritual truth and accept much of it as symbolically rather than literally true. They do not consider the Bible to be a textbook of science, even in intent, and have no problem teaching evolution in their secular institutions.

To those who are trained in science, creationism seems like a bad dream, a sudden reliving of a nightmare, a renewed march of an army of the night risen to challenge free thought and enlightenment.

The scientific evidence for the age of the earth and for the evolutionary development of life seems overwhelming to scientists. How can anyone question it? What are the arguments the creationists use? What is the "science" that makes their views "scientific"? Here are some of them:

- The argument from analogy.

A watch implies a watchmaker, say the creationists. If you were to find a beautifully intricate watch in the desert, far from habitation, you would be sure that it had been fashioned by human hands and somehow left there. It would pass the bounds of credibility that it had simply formed, spontaneously, from the sands of the desert.

By analogy, then, if you consider humanity, life, Earth, and the universe, all infinitely more intricate than a watch, you can believe far less easily that it "just happened." It, too, like the watch, must have been fashioned, but by more-than-human hands—in short by a divine Creator.

This argument seems unanswerable, and it has been used (even though not often explicitly expressed) ever since the dawn of consciousness. To have explained to prescientific human beings that the wind and the rain and the sun follow the laws of nature and do so blindly and without a guiding hand would have been utterly unconvincing to them. In fact, it might well have gotten you stoned to death as a blasphemer.

There are many aspects of the universe that still cannot be explained satisfactorily by science; but ignorance implies only ignorance that may someday be conquered. To surrender to ignorance and call it God has always been premature, and it remains premature today.

In short, the complexity of the universe—and one's inability to explain it in full—is not in itself an argument for a Creator.

- The argument from general consent.

Some creationists point out that belief in a Creator is general among all peoples and all cultures. Surely this unanimous craving hints at a great truth. There would be no unanimous belief in a lie.

General belief, however, is not really surprising. Nearly every people on earth that considers the existence of the world assumes it to have been created by a god or gods. And each group invents full details for the story. No two creation tales are alike. The Greeks, the Norsemen, the Japanese, the Hindus, the American Indians, and so on and so on all have their own creation myths, and all of these are recognized by Americans of Judeo-Christian heritage as "just myths."

The ancient Hebrews also had a creation tale—two of them, in fact. There is a primitive Adam-and-Eve-in-Paradise story, with man created first, then animals, then woman. There is also a poetic tale of God fashioning the universe in six days, with animals preceding man, and man and woman created together.

These Hebrew myths are not inherently more credible than any of the others, but they are our myths. General consent, of course, proves nothing: There can be a unanimous belief in something that isn't so. The universal opinion over thousands of years that the earth was flat never flattened its spherical shape by one inch.

- The argument by belittlement.

Creationists frequently stress the fact that evolution is "only a theory," giving the impression that a theory is an idle guess. A scientist, one gathers, arising one morning with nothing particular to do, decides that perhaps the moon is made of Roquefort cheese and instantly advances the Roquefort-cheese theory.

A theory (as the word is used by scientists) is a detailed description of some facet of the universe's workings that is based on long observation and, where possible, experiment. It is the result of careful reasoning from those observations and experiments and has survived the critical study of scientists generally.

For example, we have the description of the cellular nature of living organisms (the "cell theory"); of objects attracting each other according to a fixed rule (the "theory of gravitation"); of energy behaving in discrete bits (the "quantum theory"); of light traveling through a vacuum at a fixed measurable velocity (the "theory of relativity"), and so on.

All are theories; all are firmly founded; all are accepted as valid descriptions of this or that aspect of the universe. They are neither guesses nor speculations. And no theory is better founded, more closely examined, more critically argued and more thoroughly accepted, than the theory of evolution. If it is "only" a theory, that is all it has to be.

Creationism, on the other hand, is not a theory. There is no evidence, in the scientific sense, that supports it. Creationism, or at least the particular variety accepted by many Americans, is an expression of early Middle Eastern legend. It is fairly described as "only a myth."

- The argument from imperfection.

Creationists, in recent years, have stressed the "scientific" background of their beliefs. They point out that there are scientists who base their creationist beliefs on a careful study of geology, paleontology, and biology and produce "textbooks" that embody those beliefs.

Virtually the whole scientific corpus of creationism, however, consists of the pointing out of imperfections in the evolutionary view. The creationists insist, for example, that evolutionists cannot show true transition states between species in the fossil evidence; that age determinations through radioactive breakdown are uncertain; that alternate interpretations of this or that piece of evidence are possible and so on.

Because the evolutionary view is not perfect and is not agreed upon in every detail by all scientists, creationists argue that evolution is false and that scientists, in supporting evolution, are basing their views on blind faith and dogmatism.

To an extent, the creationists are right here: The details of evolution are not perfectly known. Scientists have been adjusting and modifying Charles Darwin's suggestions since he advanced his theory of the origin of species through natural selection back in 1859. After all, much has been learned about the fossil record and about physiology, microbiology, biochemistry, ethology, and various other branches of life science in the last 125 years, and it is to be expected that we can improve on Darwin. In fact, we have improved on him.

Nor is the process finished. It can never be, as long as human beings continue to question and to strive for better answers.

The details of evolutionary theory are in dispute precisely because scientists are not devotees of blind faith and dogmatism. They do not accept even as great a thinker as Darwin without question, nor do they accept any idea, new or old, without thorough argument. Even after accepting an idea, they stand ready to overrule it, if appropriate new evidence arrives. If, however, we grant that a theory is imperfect and that details remain in dispute, does that disprove the theory as a whole?

Consider. I drive a car, and you drive a car. I do not know exactly how an engine works. Perhaps you do not either. And it may be that our hazy and approximate ideas of the workings of an automobile are in conflict. Must we then conclude from this disagreement that an automobile does not run, or that it does not exist? Or, if our senses force us to conclude that an automobile does exist and run, does that mean it is pulled by an invisible horse, since our engine theory is imperfect?

However much scientists argue their differing beliefs in details of evolutionary theory, or in the interpretation of the necessarily imperfect fossil record, they firmly accept the evolutionary process itself.

- The argument from distorted science.

Creationists have learned enough scientific terminology to use it in their attempts to disprove evolution. They do this in numerous ways, but the most common example, at least in the mail I receive, is the repeated assertion that the second law of thermodynamics demonstrates the evolutionary process to be impossible.

In kindergarten terms, the second law of thermodynamics says that all spontaneous change is in the direction of increasing disorder—that is, in a "downhill" direction. There can be no spontaneous buildup of the complex from the simple, therefore, because that would be moving "uphill." According to the creationist argument, since, by the evolutionary process, complex forms of life evolve from simple forms, that process defies the second law, so creationism must be true.

Such an argument implies that this clearly visible fallacy is somehow invisible to scientists, who must therefore be flying in the face of the second law through sheer perversity.

Scientists, however, do know about the second law and they are not blind. It's just that an argument based on kindergarten terms is suitable only for kindergartens.

To lift the argument a notch above the kindergarten level, the second law of thermodynamics applies to a "closed system"—that is, to a system that does not gain energy from without, or lose energy to the outside. The only truly closed system we know of is the universe as a whole.

Within a closed system, there are subsystems that can gain complexity spontaneously, provided there is a greater loss of complexity in another interlocking subsystem. The overall change then is a complexity loss in line with the dictates of the second law.

Evolution can proceed and build up the complex from the simple, thus moving uphill, without violating the second law, as long as another interlocking part of the system—the sun, which delivers energy to the earth continually—moves downhill (as it does) at a much faster rate than evolution moves uphill.

If the sun were to cease shining, evolution would stop and so, eventually, would life.

Unfortunately, the second law is a subtle concept which most people are not accustomed to dealing with, and it is not easy to see the fallacy in the creationist distortion.

There are many other "scientific" arguments used by creationists, some taking quite clever advantage of present areas of dispute in evolutionary theory, but every one of them is as disingenuous as the second-law argument.

The "scientific" arguments are organized into special creationist textbooks, which have all the surface appearance of the real thing, and which school systems are being heavily pressured to accept. They are written by people who have not made any mark as scientists, and, while they discuss geology, paleontology and biology with correct scientific terminology, they are devoted almost entirely to raising doubts over the legitimacy of the evidence and reasoning underlying evolutionary thinking on the assumption that this leaves creationism as the only possible alternative.

Evidence actually in favor of creationism is not presented, of course, because none exists other than the word of the Bible, which it is current creationist strategy not to use.

- The argument from irrelevance.

Some creationists put all matters of scientific evidence to one side and consider all such things irrelevant. The Creator, they say, brought life and the earth and the entire universe into being 6,000 years ago or so, complete with all the evidence for an eons-long evolutionary development. The fossil record, the decaying radioactivity, the receding galaxies were all created as they are, and the evidence they present is an illusion.

Of course, this argument is itself irrelevant, for it can neither be proved nor disproved. It is not an argument, actually, but a statement. I can say that the entire universe was created two minutes ago, complete with all its history books describing a nonexistent past in detail, and with every living person equipped with a full memory: you, for instance, in the process of reading this article in midstream with a memory of what you had read in the beginning—which you had not really read.

What kind of a Creator would product a universe containing so intricate an illusion? It would mean that the Creator formed a universe that contained human beings whom He had endowed with the faculty of curiosity and the ability to reason. He supplied those human beings with an enormous amount of subtle and cleverly consistent evidence designed to mislead them and cause them to be convinced that the universe was created 20 billion years ago and developed by evolutionary processes that included the creation and development of life on Earth.

Why?

Does the Creator take pleasure in fooling us? Does it amuse Him to watch us go wrong? Is it part of a test to see if human beings will deny their senses and their reason in order to cling to myth? Can it be that the Creator is a cruel and malicious prankster, with a vicious and adolescent sense of humor?

- The argument from authority.

The Bible says that God created the world in six days, and the Bible is the inspired word of God. To the average creationist this is all that counts. All other arguments are merely a tedious way of countering the propaganda of all those wicked humanists, agnostics, and atheists who are not satisfied with the clear word of the Lord.

The creationist leaders do not actually use that argument because that would make their argument a religious one, and they would not be able to use it in fighting a secular school system. They have to borrow the clothing of science, no matter how badly it fits, and call themselves "scientific" creationists. They also speak only of the "Creator," and never mention that this Creator is the God of the Bible.

We cannot, however, take this sheep's clothing seriously. However much the creationist leaders might hammer away at their "scientific" and "philosophical" points, they would be helpless and a laughing stock if that were all they had.

It is religion that recruits their squadrons. Tens of millions of Americans, who neither know nor understand the actual arguments for—or even against—evolution, march in the army of the night with their Bibles held high. And they are a strong and frightening force, impervious to, and immunized against, the feeble lance of mere reason.

Even if I am right and the evolutionists' case is very strong, have not creationists, whatever the emptiness of their case, a right to be heard?

If their case is empty, isn't it perfectly safe to discuss it since the emptiness would then be apparent?

Why, then, are evolutionists so reluctant to have creationism taught in the public schools on an equal basis with evolutionary theory? Can it be that the evolutionists are not as confident of their case as they pretend. Are they afraid to allow youngsters a clear choice?

First, the creationists are somewhat less than honest in their demand for equal time. It is not their views that are repressed: Schools are by no means the only place in which the dispute between creationism and evolutionary theory is played out.

There are the churches, for instance, which are a much more serious influence on most Americans than the schools are. To be sure, many churches are quite liberal, have made their peace with science and find it easy to live with scientific advance—even with evolution. But many of the less modish and citified churches are bastions of creationism.

The influence of the church is naturally felt in the home, in the newspapers, and in all of surrounding society. It makes itself felt in the nation as a whole, even in religiously liberal areas, in thousands of subtle ways: in the nature of holiday observance, in expressions of patriotic fervor, even in total irrelevancies. In 1968, for example, a team of astronauts circling the moon were instructed to read the first few verses of Genesis as though NASA felt it had to placate the public lest they rage against the violation of the firmament. At the present time, even the current President of the United States has expressed his creationist sympathies.

It is only in school that American youngsters in general are ever likely to hear any reasoned exposition of the evolutionary viewpoint. They might find such a viewpoint in books, magazines, newspapers, or even, on occasion, on television. But church and family can easily censor printed matter or television. Only the school is beyond their control.

But only just barely beyond. Even though schools are now allowed to teach evolution, teachers are beginning to be apologetic about it, knowing full well their jobs are at the mercy of school boards upon which creationists are a stronger and stronger influence.

Then, too, in schools, students are not required to believe what they learn about evolution—merely to parrot it back on tests. If they fail to do so, their punishment is nothing more than the loss of a few points on a test or two.

In the creationist churches, however, the congregation is required to believe. Impressionable youngsters, taught that they will go to hell if they listen to the evolutionary doctrine, are not likely to listen in comfort or to believe if they do.

Therefore, creationists, who control the church and the society they live in and who face the public school as the only place where evolution is even briefly mentioned in a possibly favorable way, find they cannot stand even so minuscule a competition and demand "equal time."

Do you suppose their devotion to "fairness" is such that they will give equal time to evolution in their churches?

Second, the real danger is the manner in which creationists want their "equal time."

In the scientific world, there is free and open competition of ideas, and even a scientist whose suggestions are not accepted is nevertheless free to continue to argue his case.

In this free and open competition of ideas, creationism has clearly lost. It has been losing, in fact, since the time of Copernicus four and a half centuries ago. But creationists, placing myth above reason, refuse to accept the decision and are now calling on the Government to force their views on the schools in lieu of the free expression of ideas. Teachers must be forced to present creationism as though it has equal intellectual respectability with evolutionary doctrine.

What a precedent this sets.

If the Government can mobilize its policemen and its prisons to make certain that teachers give creationism equal time, they can next use force to make sure that teachers declare creationism the victor so that evolution will be evicted from the classroom altogether.

We will have established the full groundwork, in other words, for legally enforced ignorance and for totalitarian thought control.

And what if the creationists win? They might, you know, for there are millions who, faced with the choice between science and their interpretation of the Bible, will choose the Bible and reject science, regardless of the evidence.

This is not entirely because of a traditional and unthinking reverence for the literal words of the Bible; there is also a pervasive uneasiness—even an actual fear—of science that will drive even those who care little for Fundamentalism into the arms of the creationists. For one thing, science is uncertain. Theories are subject to revision; observations are open to a variety of interpretations, and scientists quarrel among themselves. This is disillusioning for those untrained in the scientific method, who thus turn to the rigid certainty of the Bible instead. There is something comfortable about a view that allows for no deviation and that spares you the painful necessity of having to think.

Second, science is complex and chilling. The mathematical language of science is understood by very few. The vistas it presents are scary—an enormous universe ruled by chance and impersonal rules, empty and uncaring, ungraspable and vertiginous. How comfortable to turn instead to a small world, only a few thousand years old, and under God's personal and immediate care; a world in which you are His peculiar concern and where He will not consign you to hell if you are careful to follow every word of the Bible as interpreted for you by your television preacher.

Third, science is dangerous. There is no question but that poison gas, genetic engineering, and nuclear weapons and power stations are terrifying. It may be that civilization is falling apart and the world we know is coming to an end. In that case, why not turn to religion and look forward to the Day of Judgment, in which you and your fellow believers will be lifted into eternal bliss and have the added joy of watching the scoffers and disbelievers writhe forever in torment.

So why might they not win?

There are numerous cases of societies in which the armies of the night have ridden triumphantly over minorities in order to establish a powerful orthodoxy which dictates official thought. Invariably, the triumphant ride is toward long-range disaster.

Spain dominated Europe and the world in the 16th century, but in Spain orthodoxy came first, and all divergence of opinion was ruthlessly suppressed. The result was that Spain settled back into blankness and did not share in the scientific, technological and commercial ferment that bubbled up in other nations of Western Europe. Spain remained an intellectual backwater for centuries.

In the late 17th century, France in the name of orthodoxy revoked the Edict of Nantes and drove out many thousands of Huguenots, who added their intellectual vigor to lands of refuge such as Great Britain, the Netherlands, and Prussia, while France was permanently weakened.

In more recent times, Germany hounded out the Jewish scientists of Europe. They arrived in the United States and contributed immeasurably to scientific advancement here, while Germany lost so heavily that there is no telling how long it will take it to regain its former scientific eminence. The Soviet Union, in its fascination with Lysenko, destroyed its geneticists, and set back its biological sciences for decades. China, during the Cultural Revolution, turned against Western science and is still laboring to overcome the devastation that resulted.

Are we now, with all these examples before us, to ride backward into the past under the same tattered banner of orthodoxy? With creationism in the saddle, American science will wither. We will raise a generation of ignoramuses ill-equipped to run the industry of tomorrow, much less to generate the new advances of the days after tomorrow.

We will inevitably recede into the backwater of civilization, and those nations that retain open scientific thought will take over the leadership of the world and the cutting edge of human advancement.

I don't suppose that the creationists really plan the decline of the United States, but their loudly expressed patriotism is as simple-minded as their "science." If they succeed, they will, in their folly, achieve the opposite of what they say they wish.

SIDNEY W. FOX

CREATIONISM AND EVOLUTIONARY PROTOBIOGENESIS

I. Introduction

A. HISTORY OF CREATIONISM

Between 1921 and 1929, twenty states of the Union considered forty-five legislative actions designed to prohibit the teaching of evolution. These efforts were instigated by fundamentalists acting through pressure groups (Wilhelm, 1978). Of the forty-five actions, three became law. The continued teaching of evolution did not suffer from massive legal interference, but the treatment of evolution in textbooks took a big step backward (Grabiner & Miller, 1974).

In the thirty-plus years of 1930–1963, no new attempts to impose legal restrictions were made. Instead, attempts to repeal previous legislation were instituted. None of these were successful. Both in the stormy period of the twenties and in the relative calmness of the three following decades, restrictive attempts concerned the Darwinian theory of evolution but did not include attacks on theories of original life. The latter had no support at that time from experiments; the "theory" was yet in the era of pure armchair speculation (Lehninger, 1970).

In 1963 (Biological Sciences Curriculum Study), the American Institute of Biological Sciences introduced the (Blue Version) high school textbook *Biological Science: Molecules to Man.* This volume presented the ideas of Oparin, the Miller-Urey experiment, and the proteinoid experiments relating to the first life. Four photomicrographs of proteinoid microspheres dominate the illustrations in these passages.

In the later 1960s, the Institute for Creation Research came into existence. The fundamentalists began to use the new name of creationists. Their tactics as a group had changed; they now had a larger target since the phrase "origin of life" has obviously deeper scriptural significance than "evolution of life." It is not clear, however, whether the creationists were a group intent on making waves or whether they chose to ride the crest of a wave of the general antiscientific swell of the 1960s.

B. STATE OF THE EXPERIMENTAL ART

Some readers may be sufficiently familiar with the state of the art that they will prefer to read first the analysis of objections that constitute the bulk of this chapter. Others may wish to obtain the extensive orientation available in a book (Fox & Dose, 1977) or a thorough chapter (Fox, 1978). Others may choose to read first the section State of the Experimental Art later in this chapter.

C. CREATIONISTIC AND SCIENTIFIC VIEWS

The biblical story of life's beginning deals with the range of life that we can see—grass, herbs, fruit trees, fish, whales, fowl, cattle, creeping things, beasts, and man. Biology, in its early descriptive phases, has dealt also with what we can see. Darwin's theory of evolution has dealt in a less direct manner with visible forms. Thus, Darwin's theory of evolution and the Book of Genesis are alternative explanations for visible life and its variety. The scientific understanding is derived from detailed, repeatable observations of natural phenomena; the biblical narrative is rationalized as a series of supernatural acts not capable of repetition by man. The significance of unicellular organisms in origin and early evolution (perhaps for 1½ billion years of life's term) could not be assessed by authors of the Book of Genesis since microscopes had not been developed at the time.

Although the Institute for Creation Research in San Diego does not speak for all creationists, it undoubtedly does the most speaking. It has analyzed the evolutionists' position in great detail. It is the organized voice for "scientific creationism," a term that is evidently also of its own creation. The ICR claims a scientific staff and writers and consultants, many of whom hold Ph.D. or M.S. degrees. The related Creation-Life Publishers of San Diego has made and sold numerous printings of numerous books and other materials. Relevant to this paper is their book *Scientific Creationism* (Morris, 1974) in both General and Public School editions.

The internal inconsistency in the phrase "scientific creationism" is matched by comments in the official book of the same title. For example, on page iv, the authors state:

> The purpose of *Scientific Creationism* (Public School Edition), is to treat all of the more pertinent aspects of the subject of origins and to do this solely on a scientific basis, with no references to the Bible or to religious doctrine.

Whereas on page 4 of the same book we read:

> A scientific investigator, be he ever so resourceful and brilliant, can neither observe nor repeat *origins!*

And on page 5:

> It is impossible to devise a scientific experiment to describe the creation process, or even to ascertain whether such a process *can* take place. The Creator does not create at the whim of a scientist."

By the admission of the ICR authors, creation cannot be described scientifically. Why, then, is the book called *Scientific Creationism?*

The ICR creationists are, accordingly, unintegrated in their view of what science and creation are and also have their own understanding of what constitutes research. Adoption of the term "creationism" appears to have resulted from an effort to avoid the *losing* word, "fundamentalism." Modifying the word "creationism" by the adjective "scientific" was an even graver breach of general understanding.

The creationists' position is such that, if they are to be scientific, they must, for example, explain by natural processes Huxley's image of the instant creation of a rhinoceros not only out of a half-ton of component substances (Gillespie, 1979, p. 147) but out of nothing. The view of some evolutionists that permits sudden occurrences, i.e., *stepwise* assemblies, to be part of evolution (Baltscheffsky, 1981; Fox, 1980a, 1981a) is in no way comparable. The sequence of assembly steps, plus the gradual steps of selection, occupied billions of years; thus, the total sequence was far from instantaneous. Moreover, the first step began from identifiable matter (Fox, 1981a), a kind of protein that possessed the functions that evolved to the panoply of modern proteins (Vegotsky and Fox, 1962); the latter constituted a biota of internally ordered microsystems, microassemblies, and macroassemblies thereof. The last of these could be recognized by the eye as "alive."

The creationistic view would of course be paralytic for a scientist who set out to learn, or happened to find himself on the trail of learning (as this author did), how living cells came into existence from inanimate matter. He would not undertake research for which he believed the processes were unrepeatable. Failure to learn due to making no effort to learn, then, could and has been misquoted as reenforcement for the preclusive premise that one *cannot* learn about origins by experiment.

II. Creationists' Arguments Against Origin-of-Life Experiments

The "scientific creationists" claim that "debates have conclusively demonstrated at least two facts: (a) it is possible to discuss the evidences relating to evolution versus creation in a scientific context exclusively, without reference to religious literature or doctrine; (b) there is great interest in this subject, especially among young people (Morris, 1974).

We agree with the second of these interpretations, but any objective reading of the evidence must find the statement of the first "fact" to be categorically misleading, as already indicated.

Religious doctrine and religious literature, i.e., the King James version of the Bible, have been introduced to the vast majority of children in the Judeo-Christian culture well before they hear the terms "science" or "nature" in school. The religious context already exists for them by the time they enter school, and it is reenforced by annual celebrations of Christmas and by other events which do not recognize the assumptions on which they are based.

Our knowledge of how life began was acquired in one way. This was by experimental retracement of the steps in molecular evolution. Arguments about such steps are found in the book *Scientific Creationism* (Morris, 1974). They will be answered here. In addition, other arguments on later evolutionary steps are selected for response on the basis of a discernible relationship of origins to later evolutionary steps.

The author's view of abandoning unsettled scientific questions (Fox, 1981b) to creationists is here quoted from a British journal of science:

> . . . to attack an area of science because it has *not yet* reached a given stage, and then to argue a need for resorting to supernatural explanations because specific scientific answers are not reported, consolidated, or agreed upon, is what is referred to in an American idiom as a "copout."[1]

The arguments of the creationists against the accomplishments to date are in many cases reworded statements of noncreationists against the claims to date. Answering the latter thus serves a double purpose. It is perhaps well to bear in mind, however, that focusing *in extenso* on detailed arguments often causes sight to be lost of the basic American constitutional principles of the separation of church and state; this is the center of the intellectual target.

Many authors, such as Darnbrough et al. (1981a,b) and Yockey (1981), feel they can reconcile scripture and science; mostly these are not creationists in the sense of the ICR fundamentalists. The existence of varied commitments is not surprising; what is surprising is that scientifically trained authors who proselytize for religion in a magazine of science would expect that readers would regard their judgment on related scientific problems to be objective.[2]

A. *"Life has not been created in the test tube"* (Morris, 1974, p. 49ff).[3]

The ICR states, "Because of misleadingly enthusiastic newspaper accounts, many people have the impression that scientists have actually been able to 'create life in a test tube.' However, this most certainly is not the case."

The ICR statement is out of focus for at least two reasons. One is that a definition of life is not easily agreed upon, nor is a definition presented by the creationists. The other, related fact is that life arose in stages, by self-organization (Fox, 1969), which makes a single definition difficult. The self-organization theme is accepted by numerous scientists, e.g., Eigen (1971), Kuhn (1981), Matsuno (1981), although not by others such as Yockey (1981), whose numerous quotations of scripture in his critique of self-organization raises a question of the purity of his scientific premises. For ICR Ph.D.'s, who are committed to *instant creation* in advance of reviewing the data, the concept of stepwise attainment of life is also ruled out in advance.

Isaac Asimov, properly referred to in the ICR book as a "competent scientist" (Morris, 1974, p. 42), has explained the zonal origin of life. In his 1967 book *Is Anyone There?,* the author makes the point that there is no sharp boundary between life and nonlife. Asimov then states that

> the (proteinoid) microspheres, while a long way from the completely alive side of the boundary zone, are at least a small way past the nonalive side.

Of course, much new evidence and understanding have accumulated since 1967. They indicate that the usual nucleic acid-primacy of the question is outdated (Follmann, 1982) and the stepwise answer differs from that of the creationists (Fox, 1981a).

B. *"Life could not have arisen by chance three billion years ago."*

The ICR tract (Morris, 1974, p. 49) emphasizes that even though a living system may someday be made in the laboratory, that will not prove the "same thing" happened by chance three billion years ago.

True science supports rather than proves. Science will not, however, necessarily support a happening *by chance.* The origin of life can be said to have been predetermined. That determinism is seen to rest on the appropriate reactant molecules and their shapes.

We live in a deterministic universe. In contrast to the assertion of the ICR Ph.D.'s, other creationists, and the views of a number of scientists, many scientists view natural phenomena as deterministic, not chancy.

Thomas Hunt Morgan, who was perceptive enough to formulate the theory of the chromosomal gene from an experimental program that he led (Allen, 1978) was induced by his understanding of genes to recognize their deterministic nature. Morgan said (1932, p. 219),

> . . . by "the order of nature in the living world," I mean large-scale processes that take determinism for granted, because so far as the evidence goes we are dealing with phenomena of this sort.

And again (Morgan, 1932, p. 228) he said,

> . . . by laying all emphasis on the importance of the materials in which the change takes place, we get a picture of necessity rather than of chance.

In the light of this scientific view of the nature of the universe, the ICR Ph.D.'s are forcing the question into an irrelevant context by invoking chance.

They furthermore fail to discuss adequately the accomplishment that most eludes them in their 1974 *Scientific Creationism.* This was the production of proteinoid microspheres, the models for an early stage of life's emergence, which Asimov commented on in 1967. The experiments that yield these remarkable microspheres by exceedingly simple processes *have been repeated* by uncounted thousands of high school and college students.

While a fully detailed relationship of the laboratory protocells to modern cells is still being clarified, numerous expert critics had commented on the remarkable properties already catalogued for these bodies by the time of the second printing of *Scientific Creationism* (e.g., Ross, 1962; Young, 1966; Glasstone, 1968; Oparin, 1968; Calvin, 1969; Ambrose and Easty, 1970; Baker and Allen, 1971; Dowben, 1971; Korn and Korn, 1971; CRM, 1972; Brooks and Shaw, 1973; Easton, 1973; Kenyon, 1973; Fuller, 1974; Kenyon, 1974; Knight, 1974; Price, 1974; Turcotte et al., 1974; Broda, 1975; Florkin, 1975; Lehninger, 1975; Wolken, 1975; Dickerson & Geis, 1976; Mader, 1976; Winchester, 1976; Hartl, 1977; Volpe, 1977; Weinberg, 1977; Winchester, 1977; Wolfe, 1977). The simple and direct manner in which the proteinoids assemble to yield the laboratory protocells by natural processes rivals the direct simplicity of the biblical narrative for the origin of life. These aspects are not called to the attention of the readers in the creationists' writing, even though they are among the most relevant.

Characteristic of the creationists' critiques is that they tend to concentrate on what they evidently regard as a weak aspect of that which they are attacking. In assailing the absence of transitional forms in the fossil record, they largely ignore all the other kinds of evidence: comparative biochemistry, genetics, embryology, comparative anatomy, biogeography, taxonomy, neurobiology, etc. In their attack on experiments explaining the spontaneous generation of life, they similarly select what some may regard as weak points in the evidence, but they fail to present the strong aspects. These are taken up later in section V, State of the Experimental Art.

C. *"Although amino acids can be linked, the conditions used in the laboratory are not geological" (Morris, 1974, p. 50).*

The ICR Ph.D.'s state:

> *Linking of amino acids.* Sidney Fox and others have been able, by very special heating techniques and certain conditions which could never have existed on the hypothetical primeval earth to bond the amino acids together to form what he called "poteinoids." These were not in any sense the highly ordered specific proteins found in living substances, however. They were mere "blobs," with no order and no utility. Even these would quickly have been destroyed if they had ever been actually produced on the primeval earth."

Several aspects of this statement are purely gratuitous assumptions and are not honestly possible if the critics were to read the scientific papers.

First, the techniques simply are not dependent upon special heating. Warming rather than heating is sufficient. Rohlfing (1976) has shown that temperatures of 65°C or less are adequate. The polymerization occurs from a state of low water content. If the amino acids are originally present in water as in a "warm little pond" or tide pools, the water will evaporate; the residue will then polymerize (Snyder & Fox, 1975). Reaction at low temperatures are facilitated by phosphates (Harada & Fox, 1965). Osterberg & Orgel (1972) have shown that suitable phosphates would have originated at temperatures below 100°C. A likely locale for rapid events is that of a submarine or other hydrothermal vent (Fox, 1957; Corliss et al., 1981; Copeland, 1936).

A main chemical contribution is the presence of trifunctional amino acid in the mixture of amino acids. Such trifunctional amino acids, aspartic and glutamic acids, are found in extracts of lunar fines, meteorites, and bacteria-free terrestrial lava (Fox, et al., 1981).

The creationists thus fail to present criticisms of any weight against the experimental demonstration of how the right kind of inanimate matter would arise and organize itself into infrastructured cell-like beings having membranes with selective permeability, protobioelectrical activity, protometabolic competence, growth potential, and the ability to proliferate (Fox, 1981a; Ishima et al., 1981).

The laboratory conditions are all found on the surface of even the modern Earth and are not hypothetical. Although we cannot be sure of conditions on the early Earth, temperatures known to be common on the modern Earth (cf. Rohlfing, 1976) are extremely likely for the primitive Earth. Some geologists believe the surface of the early Earth was warmer. Temperature can anyhow be traded for time in accord with the laws of physical chemistry. Moreover, the production of proteinoid and microspheres are typically repeated in open glassware in the open air, whether the experiments are performed in the research laboratory by experts or at science fairs by high school students.

D. *"Proteinoids are not sufficiently like proteins to be called proteinoids."*

The term "proteinoid" was coined to signify molecules like proteins and to indicate at the same time that the polymers are not proteins in the sense of having been produced by organisms.

The term "proteinoid" was first defined in 1967 (Hayakawa et al.). The standard authority of chemical terminology, Chemical Abstracts, began to use "proteinoid" as an indexing term in 1967. Beginning in 1972, Chemical Abstracts has employed in each of its semiannual indexes a term even more suggestive of similarity between modern proteins and the thermal polymers. Chemical Abstracts indexes the polymers under *Proteins, thermal;* they were evidently the original coiners of this term. The use of thermal proteins is presumably justified on the basis that (a) the thermal polymers are indeed much like proteins and (b) any user of Chemical Abstracts would already know that modern organisms do not make proteins by heating amino acids (Fig. 1 presents the main relationships between the compounds).

The extensive data supporting the essential correctness of the terms "proteinoids" and "thermal proteins" are summarized (Fox & Dose, 1977) and can be brought up to date by any reader by references to the papers summarized in Chemical Abstracts.

E. *"The proteinoids have no order, in contrast to proteins."*

The assumption that proteinoids were not ordered in their original form is held by many noncreationist scientists as well as by Ph.D. creationists and is an assumption in accord with widespread thinking about random states and processes (Eigen, 1971; Crick et al., 1976; cf, Fox, 1981c). Some recognize that proteins in modern organisms are nonrandom for the set of molecules as a whole, this kind of nonrandomness (or order) being derivative of the action of DNA and RNA (Monod, 1971, p. 7). Monod and others, however, see protein molecules at the same time as being random within *each* molecule (Monod, 1971, p. 96), an evaluation that is derived from sequence studies (Gamow et al., 1956; Sanger and Thompson, 1953; Vegotsky and Fox, 1962). While the ICR statement does not distinguish between intramolecular and intermolecular types of order, it is in agreement with the between-molecules order found in modern proteins.

The creationists' statement that the proteinoids, i.e., protobiotic proteins, were without order is, however, belied by extensive evidence for the model protobiotic proteins produced by heating of mixtures of amino acids (Fox, 1980b). *The instructions for protobiotic proteins have been explained experimentally as arising from the amino acids themselves. This novel experimental fact is crucial to the entire understanding of the origin of life.* The fact that polyamino acids are ordered is based now on evidence from many laboratories (e.g., Fox and Harada, 1958; Rohlfing, 1967a; Oshima, 1968; Calvin, 1969; Dose and Zaki, 1971; Hennon et al., 1975;

Melius and Sheng, 1975; Nakashima et al., 1977). The ICR statement is thus a crucially incorrect statement. Noncreationist scientists, also, as indicated, have made similar statements that disagree with the facts (Fox, 1981c).

Since the order in the amino acids results from no materials other than the reactants, mixed *amino acids must order themselves.* The polyamino acids have many functions (see below); the experiments indicate that protobiotic proteins were both informational and functional. Prior nucleic acids were not required as they are in modern cellular systems. The primary dilemma in the origin of life (Crick, 1968; Monod, 1971, p. 143) is thus resolved. The resolution is orthogenetic and strengthens belief in an overview of determinism.

F. *"The proteinoids have no (biological) utility."*

This statement of the creationists cannot be made in either accuracy or honest knowledge of the scientific literature. Shortly after the ability to produce proteinlike polymers by geological heating of amino acids containing sufficient trifunctional amino acids was announced, a number of biochemists began to seek enzyme-like activities. They found a number of such activities plus also hormonal activity and photocatalytic activity (Rohlfing and Fox, 1969). The photocatalytic activity is best understood on the basis of the additional fact that pigments such as flavins and pteridines are formed during the warming of amino acids, according to Heinz and Ried (1981). The pigments must form during reactions of amino acids, as results of thermolysis.

In no case is the amplitude of activity as great as in a counterpart modern enzyme. The models are, however, models for protoenzymes that would have evolved through hierarchical steps of protein synthesis (Fox and Nakashima, 1980; Fox, 1981f) to modern enzymes. The evolutionary relationship converting weak protobiological activity to strong modern biological activity is not honored in the creationists' writing, probably because of their general failure to honor evolutionary processes. To require that a model protocell have all the properties of a modern cell in as full measure as the modern cell has proved to be an imposed requirement. Although clothed in respectable syntax, such criticisms are implicitly an argument for instant creation.

In the case of Oparin and others who have thought in his context, however, the explanation for the evaluation is another one. Oparin's one kind of experimental work was on coacervate droplets, which he regarded as model protobionts. These are made from biologically inert materials typified by gum arabic and gelatin. *Since such materials are obtained from evolved organisms, they fail to model the first organisms in a valid way.*

In order to judge the first organisms (protocells, protobionts) it was instead necessary to assemble models from polymer produced under geological conditions, not from polymers produced in evolved cells. The synthesis of prebiotic copolyamino acid has been accomplished in the body of proteinoid, and in one earlier model (sulphobes; Herrera, 1942) and in some later models. Only in the proteinoid microspheres have arrays of enzymic activity, hormonal activity, photocatalytic activity, and cytophysical activities been painstakingly catalogued (Fox and Nakashima, 1980). The inference that protocells possessed protobiological activity is now regarded as one of the crucial components of a valid theory.

Darwin's theory of evolution has been subjected to and withstood criticisms in unprecedented degree (Mayr, 1972). The critical attention paid to the proteinoid theory of origins may be comparable in amount, especially when one allows for an existence of two decades of protobiogenetic theory versus six times as long for Darwin's principle of selection.

The criticisms of the younger theory have been valuable in two ways. One has been to provide emphasis and stimulus in selection of ongoing experiments. The other is that the proponents of the theory, in a spirit of fellowship of science, have found so many criticisms to be fatuous that confidence in the theory was thereby reinforced. If unanswerable criticisms had been part of theoretical reality and had the experimenting proponents earlier missed them, the extensive efforts of scientists and creationists should have uncovered them.

G. *"Proteinoid microspheres would have been quickly destroyed on the primitive Earth"* (Morris, 1974, p. 50).

This belief is based on the assumption that the flux of ultraviolet radiation on the primitive Earth was high and very destructive due to the absence of a protective ozone layer. The ICR authors do not

Table I. Biological-Type Activities in Proteinoids

REACTION, FUNCTION OR SUBSTRATE	AUTHORS, YEAR
Hydrolysis	
p-Nitrophenyl acetate	Fox, Harada, and Rohlfing, 1962 Rohlfing and Fox, 1967 Usdin, Mitz, and Killos, 1967
p-Nitrophenyl phosphate	Oshima, 1968
Decarboxylation	
Glucuronic acid	Fox and Krampitz, 1964
Pyruvic acid	Hardebeck, Krampitz, and Wulf, 1968
Oxaloacetic acid	Rohlfing, 1967b
Amination	
α-Ketoglutaric acid	Krampitz, Baars-Diehl, Haas, and Nakashima, 1968
Deamination	
Glutamic acid	Krampitz, Haas, and Baars-Diehl, 1968
Oxidoreductions	
H_2O_2 (catalase reaction)	Dose and Zaki, 1971
H_2O_2 and hydrogen donors (peroxidase reaction)	Dose and Zaki, 1971
Synthesis with ATP (in proteinoid microparticles)	
Internucleotide bonds formed	Jungck and Fox, 1973
Peptide bonds formed	Fox, Jungck, and Nakashima, 1974 Nakashima and Fox, 1980
Photo-activated decarboxylation	
Glyoxylic acid	Wood and Hardebeck, 1972
Glucuronic acid	
Pyruvic acid	
Hormonal activity	Fox and Wang, 1968

present specifics on this point, however, so they evidently expect their readers to take their statement on faith. They do give a reference, however, in connection with their general statement; that reference explains why and how proteinoids would *not* be destroyed.

A more worthy criticism raised by noncreationists is that neither proteinoids nor microspheres would survive since the thermodynamic data indicate that peptides tend to break down thermodynamically in aqueous solution (Huffman, 1942). The energetic relationships between amino acids, peptides, and proteins or proteinoids are reviewed in Figure 1.

Determination of thermodynamic constants was a main part of this author's Ph.D. dissertation, measurements having been made under the stewardship of the late Hugh M. Huffman. One of Huffman's favorite statements was, "Thermodynamics only tells you what can't happen; it does not tell you what can happen." What evidently happens with proteinoids, which are peptides, is that they aggregate into microspheres. Both the molecular arrangements and the microspherical configurations evidently then involve stabilizing intramolecular interactions.

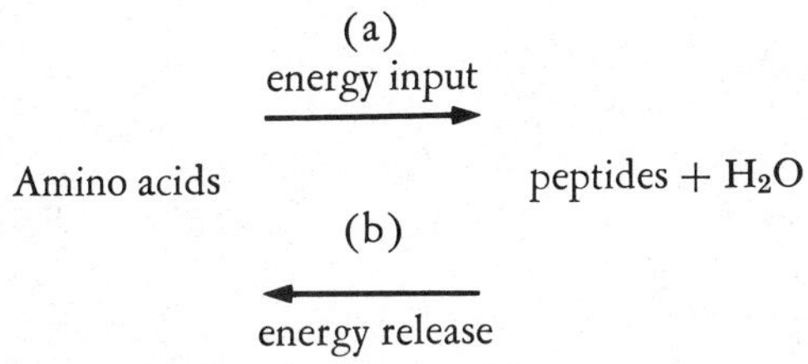

Fig. 1. The thermochemical equilibrium between amino acids and peptides.

Reaction (a) can be caused by removal of H_2O.

Reaction (b) can occur in aqueous milieu, if catalyzed.

Proteinoids (thermal proteins) are large peptides made from amino acid mixtures by heat.

Proteins are large peptides made by organisms from amino acids by energy from ATP.

Proteinoid solutions will not remain as such in the open air of a modern laboratory over a moderately large pH range due only to their serving as nutritional substrate for organisms from the air. Accordingly, Dr. L. L. Hsu (1975), while functioning as a microbiologist in our laboratory, prepared a suspension of proteinoid microspheres aseptically. These lasted six years without appreciable deterioration, at which time they were discarded. Evidently, they and the component proteinoids are stable indefinitely in the presence of water.

When illuminated in aqueous suspension for 24 hours, as they have been for experiments in the production of ATP from ADP (Fox et al., 1980), no significant destruction of the proteinoid is observed. They have, accordingly, great radiative and thermal stability.

H. *"Scientists agree that a solution is nowhere in sight."*

It may well be true that *most* scientists agree that a scientific solution to the problem of life's origin is nowhere in sight. This appraisal has been changing, however. We can see the change in illustrative comments since *Scientific Creationism* appeared as a book in 1974.

Following detailed analyses by cellularly oriented biochemists Florkin (1975) and Lehninger (1975) and his own appraisal, Dickerson (1978) stated, "The goal . . . appears to be in sight." A fuller quotation of Dickerson is presented later in this article.

More recently a review by a third biochemist (Follmann, 1982) has stated,

> It is generally accepted that protocells first assembled from abiotically produced proteinoids ("microspheres"), which then incorporated polyribonucleotides and other simple compounds (amino acids, fatty acids, sugars, heterocycles) present in the primordial organic soup.

No one, two, or three experts speak for all scientists but they come closer to doing that than do the creationists, who obviously do not allow themselves the upsetting disciplinary benefit of investigation. Experimental research could have led the creationists to internally directed, nonrandom, stepwise reactions. In any event, scientists as a whole do not agree that a solution is nowhere in sight.

III. Objections of Creationists to Other Aspects of Evolutionary Theory

The fact of evolution as change through modification by descent of the biota on this planet can no more be denied than one can deny his own senses. Each of us need only examine human offspring and their parents to attain this inference. The possibility, however, that our *understanding* of the *mechanism* of evolution is not yet adequate (cf. "random matrix") is less obvious but also cannot be denied. In other words, evolution is a fact, but how it operates is incompletely understood. The various explanations for evolution may not be as numerous as the various religions but they are of many kinds. Both our understanding of evolutionary mechanisms and of supernatural mechanism rest on systems of belief, at least in part. The challenges to these beliefs, whether from religionists, evolutionists, or protobiologists, have received consideration.

Criticisms aimed at the inferences from the proteinoid model of protobiogenesis have been answered above. Such answers also affect answers to criticisms of events inferred to have occurred later in evolutionary history. Later events must be relevant if it is true that protobiogenesis helped to set the pattern for later evolution. Discussion of these more traditional aspects of evolution is on this basis germane to a discussion dealing in the main with concepts of protobiogenesis (Salthe, 1982).

Some objections by creationists to specific aspects of postorigin evolution, and answers to them, follow:

A. *"A complex living system could not have arisen instantly by chance"* (Morris, 1974, p. 59).

The preface to this statement is an earlier assertion (Morris, 1974, p. 59) that an efficiently functioning complex system could not have arisen by random processes, but that the "evolution model nevertheless assumes all of these [living organisms] have arisen by chance and naturalism."

One aspect of the fallacy in the creationists' statement is the instant origin of complex living organisms. This view is a view that one may easily infer from the Book of Genesis, in which narrative each form of life *did* arise instantly. To a naturalist, *instant* origin is not a requirement. Indeed, a pervasive tenet of the natural theory is that life arose in steps. The inference and vision of the creationists, then, are prelimited by their basic premise.

When we consider the role of chance, which the creationists and others equate to "random processes," the experimental view is that the processes are nonrandom, i.e., *not* by chance. It is true that many evolutionary theorists reason within a context of implicit belief in chance processes and random matrix (Fox, 1981c), but that is not a component tenet in the most modern theory derived from experiments.

Accordingly, when the creationists infer, "for all practical purposes, there is no chance at all" to the formation of an integrated functioning organism, they are superposing a natural unsupported assumption on a biblically conditioned assumption. If both are wrong, as the experiments indicate them to be, the misstatements are compounded.

B. *"The probability of synthesis of a DNA molecule is nil."*

The creationists (Morris, 1974, p. 62) quote Salisbury (1971) as calculating a probability of 1 in 10^{600}; thus, "the DNA molecule could never arise by chance." With this calculated inference we agree. It is true that some scientists treat the formation of DNA as if it were an act of instant creation. However, the experiments demonstrate that DNA did not arise by chance; it arose as a product of *internally self-ordered, stepwise processes* (Fox, 1981a; Follmann, 1982).

The ICR authors selected those aspects of Salisbury's article that suited them. What they neglected to state is that Salisbury explicitly disavowed the possibility of a satisfying scientific solution to the problem of variability by invoking God, as the ICR Ph.D.'s do. Salisbury states that "Perhaps the mutations upon which natural selection acts are not really random at all." This cogent comment is not presented by ICR in their quotation of Salisbury.

A comprehensive statement supported by experimental evidence was presented three years after Salisbury's statement in the same *American Biology Teacher* magazine (Fox, 1974) on the basis of the many years of experimental demonstration of nonrandomness in the polymerization of amino acids. Again, *the nonrandomness of reaction of amino acids during simulated geothermal polymerization is crucial to all of evolution and to its understanding.*

C. *"The probability of synthesis by gradual accretion . . . seems beyond all plausibility"* (Morris, 1974, pp. 63–66).

In this argument, the creationists do confront the stepwiseness of evolution. In so doing, they consider "trial-and-error" processes, and state that (a) "each trial step would have to be immediately beneficial; there could be no failures or backward steps."

Also, they say, (b) "All would agree, surely, that a probability of 1/2 for each change would be quite optimistic," and (c) "Undoubtedly the actual probability of success is far less than that."

The next statement is a popular scenario of matrix by Dr. Larry Butler (p. 65), followed by (d) "So far, no one has accepted his challenge!"

In this series of four statements we find a mixture of unjustified assumptions, rhetoric, and failure to read literature comprehendingly.

There is no reason to accept (a). Trial steps that failed could not be expected to have left traces for billions of years. Backward steps are not relevant to evolution; they need not be mentioned.

(b) That probabilities of 1/2 are "quite optimistic" is a statement out of thin air. The *steps* that have been identified by experiment are:

1. Formation of amino acids from primordial reactants.
2. Formation of proteinoids by polymerization of sets of amino acid.
3. Formation of microstructures by contact of polymers with water.
4. Origin of the genetic apparatus within such microstructures (Fox, 1981f).

The experimental results show (c) that the probability of each step is close to unity, not 50%. For purposes of argument let us assume 90%, although the experiments indicate > 90%. The probability of reproductive, infrastuctured cell-like entities arising in steps from inorganic matter containing atoms of carbon, hydrogen, oxygen, and nitrogen is then at least $(\frac{9}{10})^4 = 65\%$. (All such "numerology," however, on either side may be considered a quantitative euphemism for ignorance, which is in turn dispelled by identification of the processes.)

In addition, thanks to the nature of geometric increases in interactions, complexity arises rapidly from simplicity (Fox, 1973). This article and the evidence it cites indicate that (d) the challenge was accepted long ago to the point described by steps 1–3 above; the challenge was met by 1967 (Asimov, 1967; Fox, 1969, 1981a; Ambrose and Easty, 1970; Florkin, 1975).

D. "*An increasing complexity of living systems is improbable*" (Morris, 1974, p. 66).

For evolution, say the creationists, "the accepted explanation, of course, is that of random mutation and natural selection." This is a correct observation of Neo-Darwinian thinking (Eden, 1967).

Many or most evolutionists agree that random mutation and natural selection constitute a valid explanation for evolution. However, not all evolutionists accept the premise of *random* mutation; some of the most profound students have not (Morgan, 1932). Wigner (1961) has analyzed some of the difficulties posed by involving a random matrix, as has Salisbury (1971). Eden (1967) has performed similar analyses and has inferred that

> The process of speciation by a mechanism of random variation in offspring is usually too imprecisely defined to be tested. When it is precisely defined it is highly implausible.

The model-building experiments that produce a simulated protocell indicate a high degree of nonrandomness at the molecular level, which is the level at which the question is meaningfully asked. Even though disagreement among natural scientists exists on this question, abandonment of the scientific argument is not called for as a result. However, and despite the earlier-mentioned protestation of no need to mention religion (Morris, 1974, p. 4), the creationists aver that all of the "complexities in the living world . . . simply reflect the omniscient, omnipotent Creator (Morris, 1974, p. 69). The scientific answer is nonrandomness in a universe that increases both in entropy and complexity (Fox, 1980b; Saunders and Ho, 1976).

IV. Objections in the Scientific Literature

Some of the objections raised by scientists to the proteinoid model are at times repeated by creationists, not necessarily the ICR creationists. There is, furthermore, little doubt that some degree of creationistic thinking enters into the personally held paradigms of many scientists (Gillespie, 1979; Fox, 1959). Some of the objections in the scientific literature will accordingly be answered here. No attempt will be made to deal with all of them. In a comparable treatment in his *Origin of Species,* Darwin (undated, pp. 153–83) said in a 30-page chapter titled Miscellaneous Objections to the Theory of Natural Selection,

> it would be useless to discuss all of them, as many have been made by writers who have not taken the trouble to understand the subject.

A. *"Temperatures of 150°–180° are required; they are rare"* (Miller & Orgel, 1974).

The temperatures cited can be used to polymerize sets of amino acids, but temperatures at that level *are not required.* The pervasiveness of the repeated misleading statement is such as to have stimulated a printed *re*affirmation that such temperatures are not required (Fox, 1980c). Rohlfing (1976) has reported making significant quantities of proteinoids in a number of weeks at 65°, and he has further reemphasized that temperature can be traded for time, a physiocochemical axiom. Moreover, since amino acid polymerization can occur rapidly in polyphosphoric acid (PPA) (Harada & Fox, 1965), it is significant that Osterberg and Orgel (1972) reported that they can make PPA from NH_4HPO_4 at 100° or less.

B. *"Proteinoids contain linkages not found in proteins"* (Temussi et al., 1976).

This criticism is usually directed at the fact that in proteinoids there are β- as well as α-aspartoyl links, γ- as well as α-glutamoyl links, and ϵ- as well as α-lysyl links. Proteinoids are in the main much like proteins, but the very name indicates that they are not proteins. What is required of proteinoids for a conceptual evolutionary overview is their progressive replacement through intermediate stages (Fox, 1976a; Pivcova et al., 1981). The essence of the evolution involving geological peptides that make cellular peptides has been demonstrated (Nakashima and Fox, 1980). With catalysis of peptide formation by lysine-rich proteinoid, it could well be that only modern proteinous structures and bonds would result. This is being investigated. The criticism is rooted in a popular anachronism demanding that a primitive form be fully like a modern form; the critic is ignoring stepwiseness in evolution.

C. *"Proteinoid microspheres on the primitive Earth would have suffered dissolution"* (Fox, 1979).

In common with some other criticisms, this one was relevant only in the early days of this research program. The proteinoid investigators were the first to recognize such a potential defect and were the first to report an experimental answer (Fox and Yuyama, 1963a). The earliest studied microspheres, of acidic proteinoid, do undergo dissolution at pH 7. However, when acidic proteinoid is combined with more basic proteinoid, the resultant microspheres persist at pHs up to 8–9 (Fox and Yuyama, 1963a) and can withstand hot water. This mixed type, moreover, is the kind that makes peptides from ATP and free amino acids (Fox and Nakashima, 1980). It is also the kind that catalyzes the synthesis of internucleotide bonds (Jungck and Fox, 1973), and it is furthermore the kind that Barghoorn et al. showed could be artificially fossilized to yield units indistinguishable from artificially fossilized algae (Francis et el., 1978).

Moreover, it is not necessary to visualize the origin of two proteinoids in two locales followed by mixing; both kinds of proteinoid arise simultaneously in one preparation if seawater salt is present (Snyder and Fox, 1975; Fox, 1979). The mixed microspheres are made from highly undersaturated solutions.

D. *"Life began only once."*

This comment is found from scientists, creationists, and people of both frames of mind (e.g., Morris, 1974; Iben, 1973; Aw,

1976). The classical requirement for scientific phenomena that they be repeatable is perhaps subject to exceptions. On this problem that exception is, however, unnecessary (Fox, 1981a). The experiments suggest, moreover, numerous repetitions of the same generative process on the primitive Earth, all alike due to molecular determinism.

E. *"The proteinoid microspheres have leaky membranes because they do not contain lipid."*

It is true that lipid membranes (lipid can be introduced into microspheres) are more discriminatory than lipid-free proteinoid membranes so far studied. Various crude proteinoids, however, contain acetone-soluble and chloroform-soluble fractions (e.g., Grote et al., 1978); these are ordinarily removed to protect storage life of proteinoids. Even with such fractions removed, the microspheres have a lipid-like barrier due to nonpolar side chains in amino acids (Lehninger, 1975, p. 1047).

Highly discriminatory membranes in early microspheres are seen as an evolutionary disadvantage (Kuhn, 1976; Fox and Nakashima, 1980). Until the evolving microspheres developed comprehensive anabolic activities, they would have required many intermediates to diffuse freely *inward* from the environment. It follows that, as early cells acquired during evolution the ability to make their own biochemical intermediates (Krebs cycle, etc.), they also developed the ability to make their own phospholipids, their own polyamino acids (now experimentally indicated), and their own nucleic acids (Jungck and Fox, 1973). Logically, evolution had to come to this anyhow in order to conform eventually to the efficient membrane activity in modern cells. But "the contemporary did not evolve from the contemporary."

Other criticisms do not so easily seem to contain some logical substance. Consistent with Darwin's observations of objections to his theory, many would not have been printed if the objectors had taken the trouble to read either that to which they were objecting, or perhaps to read their own critiques. The ones cited represent those deemed most worthy of response. All of the objections, however, serve another purpose, since all criticisms seen are systematically considered and processed in this research program. When opponents of the proteinoid theory extended the criticisms which the proponents had themselves applied, and no insurmountable difficulties were found to remain from such examination, confidence in the theory was thereby enhanced.

A more extreme evaluation is that no other theory will be generated from experiments. This thinking is based in large part on the unexpected results of experiments in molecular evolution, which stamp the pathway of evolution as narrowly self-limiting (Fox, 1981e).

The principal counterobjection to the total objections, however, is not the emptiness of the comments on the experimental fact. It is, rather, the extent to which the interpretations are anachronistically judged in a context of the modern natural world instead of in a protobiological situation.

V. State of the Experimental Art

An assessment of the state of the experimental art is demanding. This is a natural consequence of an extensive literature that has been accumulating for as long as 29 years.

The problem has been viewed as scientifically imponderable for many decades—a view that has by default strengthened the attraction of the biblical narrative of creation. When attempts are made to reason the assembly of primordial cells from data obtained analytically from modern cells, the problem is indeed imponderable. The only procedure that has yet been shown to yield a comprehensive explanation (Lehninger, 1975) for a protocell in a forward direction from a geological matrix is experimental retracement of the steps in molecular and cellular evolution (Fig. 2). This explanation has proved to be so different from the Aristotelean view that much challenge, objection, and defense has resulted. In our laboratory the investigation has consumed well over 225 man-years. In addition, other laboratories that have carried out investigations on the proteinoid track are numerous. Major ones include: Professor Klaus Dose, Department of Biochemistry, Johannes Gutenberg University, Mainz, Germany; Professor Kaoru Harada, Department of Chemistry, Tsukuba University, Tsukuba, Japan; Professor Duane L. Rohlfing, Department of Biology, University of South Carolina, Columbia, South Carolina; Professor Paul Melius, Department of Chemistry, Auburn University, Auburn, Alabama;

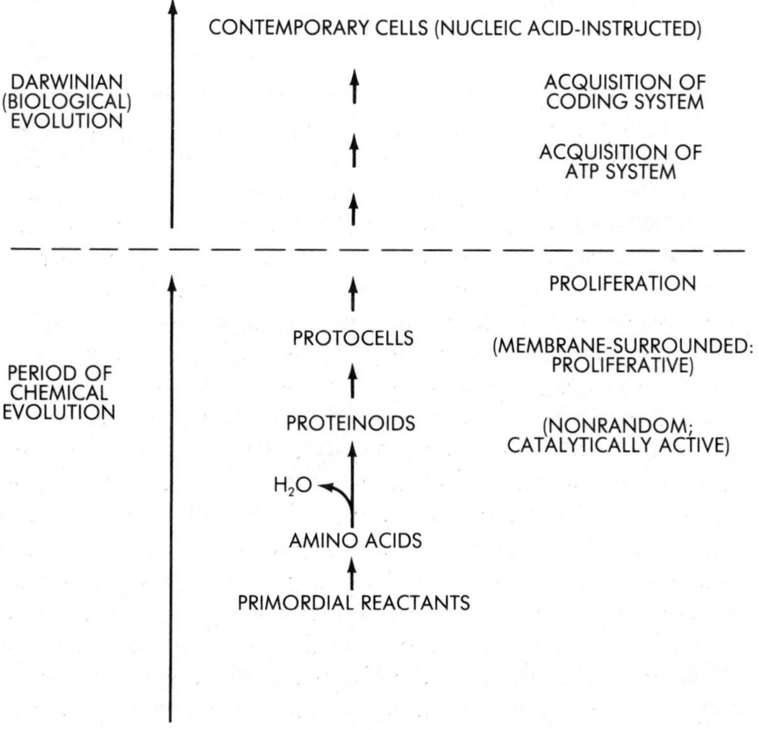

Fig. 2. Flowsheet of protobiogenesis, as slightly modified from Lehninger (1975, p. 1048). Horizontal line indicates origin of protoreproductive protocell having ordered molecular components, protometabolic activity, infrastructure including membrane, ability to grow by accretion to a set size, and electrical membrane properties.

Professor Gottfried Krampitz, Biochemistry Division, Institute for Anatomy, Friedrich Wilhelm University, Bonn, Germany; and Professor James C. Lacey, Jr., Department of Biochemistry, University of Alabama in Birmingham, Alabama.

Leading theoreticians who have drawn from the data on proteinoids are: Professor Koichiro Matsuno, Technological University of Nagaoka, Japan and Professor John R. Jungck, Department of Biology, Beloit College, Beloit, Wisconsin.

The art assessed here is very much a distillation of the experiments and interpretations of these individuals and numerous others, some of whom have studied in these laboratories. Since evaluations from so many sources require much reading, summary statements are selected.

In addition to orientation in literature, the orientation for the problem of the first cell may best be biochemical, since this is the most relevant joint discipline for the modulation from a chemical world to a biological one (geology and biophysics being also highly relevant disciplines).

A third demand is that the evaluator not be influenced by biblical indoctrination, which view is a theme of this volume. Examples of precharged evaluation have been discussed.

An additional difficulty may be that some commentators will require that full validation of experimental retracement of molecular evolution be the production of a *modern* cell. This appears not to be possible since we do not have a single definition of a modern cell. The experiments already provide definitions of primitive cells, but in their various evolutionary stages. The problem of definitions becomes increasingly complex as the retraced steps increase in number.

A. THE EVALUATION OF 1978

The state of the art was reviewed in an article in *Scientific American* in 1978 (Dickerson). In that article Dickerson said:

> The broad goal is to arrive at an intellectually satisfying account of how living forms could have emerged step by step from inanimate matter on the primitive earth. That goal appears to be in sight.

B. ADVANCES IN THREE YEARS FOLLOWING DICKERSON'S ARTICLE

1. The beginning of cellular protein synthesis (Nakashima and Fox, 1980; Fox and Nakashima, 1980; Nakashima & Fox, 1981).

This necessary step in evolution has been shown to depend upon basic amino acid residues in larger molecules. The molecules are of the kind that could become parts of protocells. These systems are capable of converting free amino acids and ATP in aqueous suspensions to small and large peptides. In a more primitive model, ATP is replaced by pyrophosphate (Nakashima and Fox, 1980).

2. Origin of the genetic mechanism and code.

Because of the experimental fact that the same kind of lysine-containing structure catalyzes formation of the internucleotide bonds of DNA and RNA (Jungck and Fox, 1973) as well as peptide bonds, a braided formation of the two kinds of informational macromolecule has been demonstrated for protocell models (Fox, 1981c). It is in such an evolving cellular context that we can now visualize the evolution of the genetic code (Fox, 1965) and can obtain some experimentally derived insight into how that occurred.

3. The first cells had protobiological activities.

The history of development of models for the first cell (protocell, protobiont, eobiont) has conditioned many to think of them as biologically inert. The historically significant protobiont models (Herrera, 1942; Oparin, 1957) have been inert. The coacervate droplets of Oparin are made from evolved polymers, a fact that disqualifies them as protobiont models (Fox, 1976b), and enzymic activity has been introduced only by incorporating modern enzymes (Oparin, 1971). Herrera's sulphobes (1942) were not reported to possess enzymic activities. One crucial significance of the proteinoid microspheres is the finding of arrays of catalytic activities present due to the content of proteinoids, polymers prepared in the same "spontaneous generation" series that yields the organized structures.

The extensive effort of numerous investigators have yielded the information summarized in Table II.

4. Establishment of bioelectrical behavior in proteinoid microspheres (Ishima et al., 1981).

This activity includes membrane potentials, spiking, and oscillatory discharge (Fox et al., 1982). These activities are induced by changing ion concentrations in bathing fluids, current injection, or illumination; but each of these has also been observed to occur endogenously ("spontaneously"), albeit not in the total absence of light.

Table II. Protobiological Characteristics of Proteinoid Microspheres

Protometabolism (Fox and Dose, 1977)
Catabolism (Rohlfing and Fox, 1969)
Anabolism
Synthesis of protein linkages (Nakashima and Fox, 1980)
Synthesis of nucleic acid linkages (Jungck and Fox, 1973)
Protogrowth (by accretion) (Fox et al., 1967)
Proliferation (protoreproduction) (Fox, 1973b; Fox et al., 1967)
Selective permeability (Fox et al., 1969)
Irritability (excitability) (Fox et al., 1982)
Conjugation (→ protocommunication) (Hsu et al., 1971)
Motility (Fox et al., 1966).

These results suggest that the first cells on Earth already had bioelectric activity. They also add to the evidence that proteinoid microspheres contain effective membranes. The earlier evidence is from electron micrographs (Fox and Dose, 1977) and from studies of selective permeability (Fox et al., 1969).

5. Extrapolation of protobiological theory to Big Bang and to behavior.

Based on inferences by others from the Big Bang studies, principles of protobiology have been back-extrapolated to primordial cosmic events (Fox, 1980b). The principles have also been extrapolated forward to psychobiological behavior (Fox, 1980d). A comprehensive theory of evolution from Big Bang to behavior has accordingly resulted. That human behavior may be governed by genetic determinism (Morgan, 1932; Wilson, 1978) is at odds with much traditional thinking, and has profound significance for societal problems such as criminal rehabilitation.

VI. Comprehensiveness within Protobiological Theory

Two leading tomes of biochemistry have reveiwed the experiments and conceptualizations of protobiochemistry. One of these, Lehninger's *Biochemistry* (1975) describes much of the proteinoid cluster of concepts and presents it as a comprehensive flowsheet. Such evaluations are a proper part of the science of biochemistry for the reason that, as Lehninger states, the question has outgrown "pure armchair speculation."

A. PROPERTIES OF PRODUCTS FROM AMINO ACIDS

While Lehninger has presented the connected experimental description and opposing armchair concepts in a balance useful for beginning students, Florkin, in his *Comprehensive Biochemistry* (1975), has been partial to proteinoids. Florkin's statements include:

> The striking properties of poteinoids have made the thermal theory of the origin of proteins very attractive.
>
> The present writer believes that the real virtue of the proteinoid-microsphere theory resides in the possibility of obtaining microspheres from mixtures of amino acids and in the unexpected properties such microspheres were proved to possess.
>
> One of these is in the production of informed proteins in the absence of the sequencing information provided in contemporary life by nucleic acids.

B. DESCRIPTION

Approximately 225–250 man-years of experimental investigation have been expended in retracement of the earliest steps in molecular evolution. No other investigative process is believed to be relevant. Indeed, the defeatism which attends this problem (Popper, 1974; Iben, 1973; Monod, 1971) is the inevitable feeling which emerges from the popular assumption that one can infer such assembly processes from Aristotelean analysis of modern organisms (Fox, 1977, 1981a).

The first comprehensive result of such work is summarized in the flowsheet of Figure 2, which is a minor modification of one appearing in Lehninger's *Biochemistry* (1975).

Figure 3 brings Figure 2 more up to date by including new details learned in the 1975–1980 period. Table I also presents a summary of functions found in proteinoid in the individual studies from a number of laboratories.

In Figure 4 is seen a scanning electron micrograph of proteinoid microspheres. This micrograph illustrates the numerousness and the almost complete uniformity of the product.

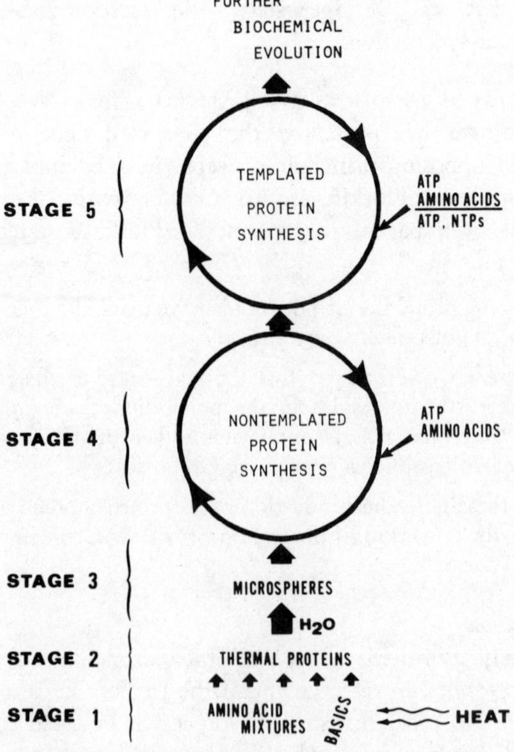

Fig. 3. Extension from Fig. 2 as inferred from experiments. Horizontal line has same meaning as in Fig. 2. This figure includes sufficient basic amino acid in mixture of stage 1. Thermal proteins (proteinoids) and microspheres, being rich in basic amino acid, are capable of converting ATP and amino acids to polypeptides (stage 4) and ATP and other nucleoside triphosphates (NTPs) to nucleic acids (stage 5). Some template activity has been demonstrated.

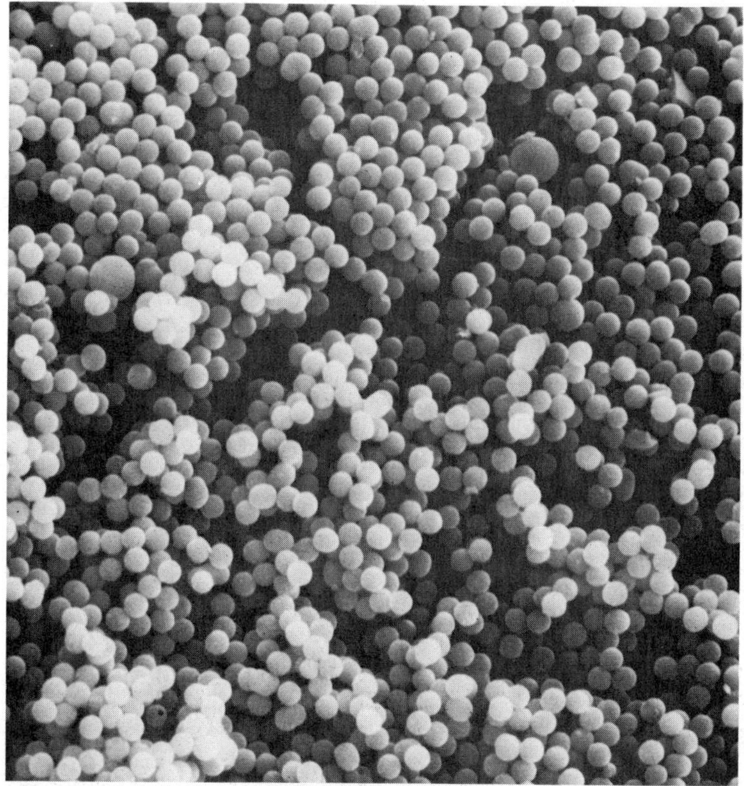

Fig. 4. Scanning electron micrograph of proteinoid microspheres, thanks to Steve Brooke. Photograph shows numerousness and highly uniform size.

Figure 5 presents examples of the protobioelectric behavior recently observed in these simulated protocells. In addition to adding to the list of Table I, these findings open a door to studying molecular neurobiology at various stages of evolution.

C. LOCALES

Understanding of the possibilities for the assembly of appropriate precursors into living systems is not confined to inferences from data obtained in the laboratory.

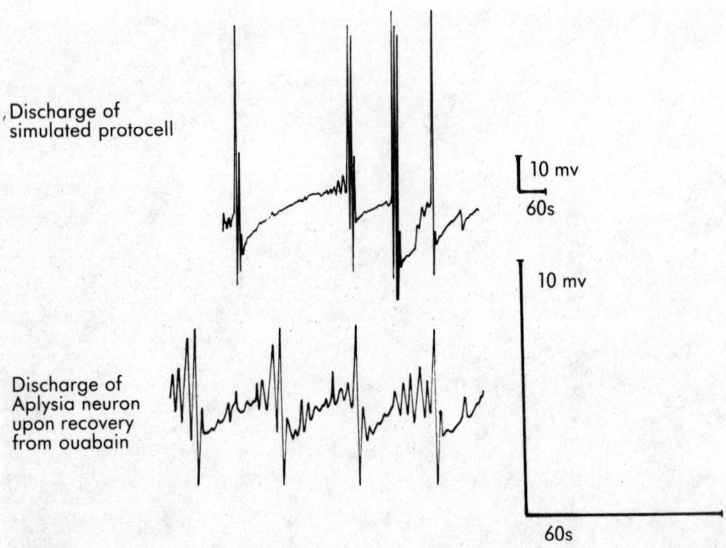

Fig. 5. Action potentials from laboratory protocell (*above*) and from *Aplysia* neuron (*below*).

Awareness of the rapidity with which appropriate monomers (amino acids) can condense to yield polymers, which in turn aggregate to cell-like structures, plus finding of evidently primitive unicellular organisms in locales such as Spirit Lake (Mount St. Helens) and the Galápagos Rift have led Corliss et al. (1981) to propose hydrothermal vents as a locale for original life. This is consistent with a proposal by Copeland (1936) that life began in thermal waters. Copeland's suggestion was prompted by his earlier examination of both hot-water algae and cold-water algae in Yellowstone Hot Springs. Copeland found that the cold-water types were, as judged by biosystematics, descendants of the hot-water types, whereas the latter had no ancestors; hence his suggestion of *de novo* generation.

This kind of result led Fox (1957) to propose submarine hydrothermal vents as likely locales. Mueller (1972) made a similar suggestion. The polymerizability of mixtures of amino acids (containing some trifunctional amino acid) on basalt, a most likely substrate, has been demonstrated (Fox, 1964). Basalt would presumably line the throats of hydrothermal vents. Further study of the metabolic capabilities of hot-springs candidates to compare with the capabilities of the laboratory protocells should help discipline these proposals.

D. MICROFOSSILS

Another kind of evidence relating the laboratory products to the field is the finding of microfossils of appearance similar to those from the laboratory. These similarities (Fox and Dose, 1977, p. 304) have prompted Keosian (1974) to state that

> Structurally, the oldest microfossils have a closer resemblance to clusters of Fox's microspheres than to primitive algae.

The possibility that some microfossils are lithified proteinoid microspheres (Fig. 6) was first suggested in 1962 (Fox and Yuyama, 1963b). Artificial fossilization led to an acknowledgment by micropaleontologists of that possibility as a probability 16 years later (Francis et al., 1978). The statement was reflected by others (Cloud and Morrison, 1979).

E. NOVELTY

The theoretical paradigm for the origin of life is being replaced by one derived from experiments. The departures in concept are simultaneously departures from some established scientific concepts, and are almost totally a departure from the creationists' paradigm. In an analytical overview, principal features of the novelty include the phenomenon of self-ordering.

1. Self-ordering of amino acids.

As already emphasized, this phenomenon is crucial to organic evolution. Without the internal self-limitation seen to result from self-ordering (self-sequencing) of amino acids, an evolutionary pathway could not have developed (Fox, 1978). The self-ordering explains also the similarity of multiple *de novo* generations.

The self-ordering of heated amino acids is manifest if the mixture of these monomers includes trifunctional types such as aspartic acid and glutamic acid (Fox and Dose, 1977). Part of the mecha-

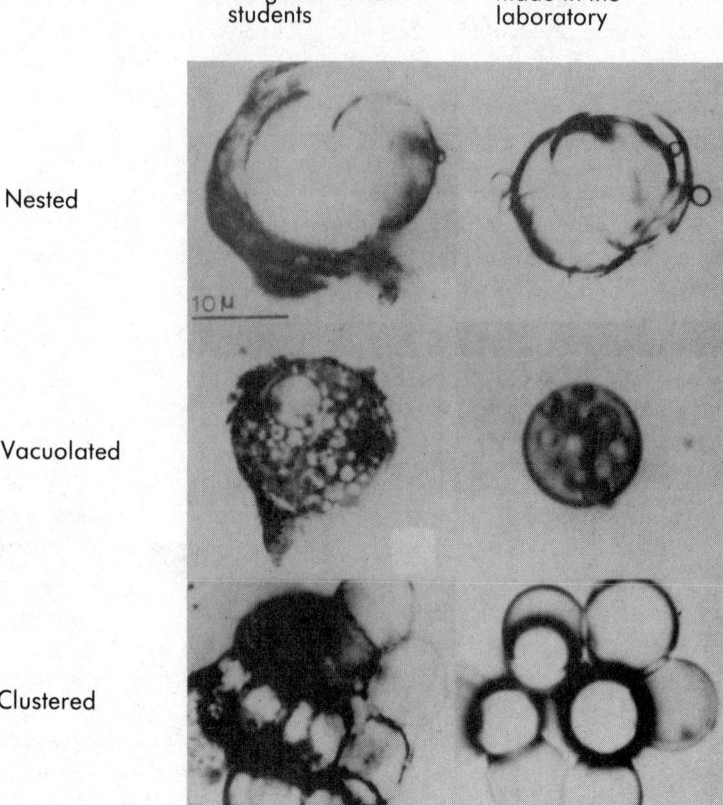

Fig. 6. Microfossils from ancient strata in left-hand column. Proteinoid microparticles in right-hand column.

nism for self-ordering has been explained as being due to the stereochemical modulation of one trifunctional amino acid, glutamic acid, on warming (Fox, 1978).

2. The self-organization of copolyamino acids to laboratory protocells.

The fundamental question of how spontaneous generation might have occurred was asked by Louis Pasteur in 1864 (Vallery-Radot, 1922) with the words (trans.):

> There is the question of so-called spontaneous generation. Can matter organize itself? In other words, are there beings that can come into the world without parents, without ancestors? That is the question to be resolved.

Copeland (1936) answered the question of parentless beings rather rigorously. The work with proteinoid suggests an affirmative answer also to the question of whether matter can organize itself; this answer designates the special kind of matter—thermal protein as the ancestor of modern protein. The fact that stable, yet dynamic, cell-like structures can assemble by interactions instead of covalent bonding is also a departure in thinking.

The proteinoids are polymers unusually rich in charged groups. While the explanation for the aggregation is not conclusive, this fact presents a plausible answer.

3. Proteins-first instead of DNA-first.

The self-ordering and the demonstrated ease of formation of protocells (Fox, 1976b) are consistent with the prior evolutionary emergence of proteins (Fox, 1981c). No experimental demonstration of DNA-first exists; it is yet being sought (Florkin, 1975; Kuhn, 1981). Experiments explaining how proteins could have yielded cells, and *then* cellular proteins and cellular nucleic acids, have been reported (Fox, 1981c).

4. The protocells had protobiological activities.

This view is novel against the background of knowledge of the inertness of a principal model for the protocell, the coacervate droplet (Oparin, 1957, 1971). Coacervate droplets are, however, made from evolved biopolymers instead of geologically generated polyamino acids. By hindsight, it is possible to state that one cannot postulate the later appearance of biological functions in a line of evolving cells for which the first ones did not possess some protobiological activity. The aggregation of protobiologically active proteinoids, however, has been shown experimentally to yield active laboratory protocells (Fox, 1980e).

5. The emergence of DNA and genetic factors in cells after (thermal) proteins and protocells.

The experiments suggest that DNA and RNA formed in protocells that had already appeared as the result of self-ordering of amino acids to give informed, aggregatable polymers. Protobiological information thus existed before protocells and before the genetic mechanism. Protein synthesis and nucleic acid synthesis, each catalyzed by lysine-rich proteinoid, would thus have been coordinated in the locale of the protocell (Fox, 1981d). The function of inheritance would have arisen after the first life cycle, which is what was inherited—when life-cycle biochemistry existed to be inherited.

6. From Big Bang to behavior.

Increasingly, study of the evolutionary sequence is leading to the kind of comprehensive picture that the Book of Genesis attempted to supply.

The proteinoid theory of origins participates in the interface with astronomical and geological evolution through fossils as explained, and in an earlier phase through ordered prebiotic matter (Fox, 1980b). The interface with biology is being broadened, partly because the proteinoid microspheres can be seen to have some primordial functions of psychobiology (Fox, 1976c, 1980d, 1982). Outstanding among the latter are communication and electrophysiological analogs. Like other natural interpretations, we cannot expect completeness in the odyssey of Big Bang to behavior, but the evolutionary sequence is more and more an overview constructed of solid and detailed explanation of the stepwise nature of all evolution.

Other modes of recognition of the state of the proteinoid art disagree with the negative evaluation of the Institute for Creation Research. They are: the fact that proteinoid as a word is part of the language, that "thermal proteins" is a standard term in Chemical Abstracts, the very large number of textbooks that present the developments, the description in text books such as that of Sherman and Sherman (1975) which state that "the laboratory-produced proteinoids demonstrate how primordial life could have arisen in the absence of parental or ancestral systems on a lifeless earth," numerous listings of the "remarkable," "unique," etc., properties of microspheres (e.g., Ambrose and Easty, 1970; Dowben, 1971; Etkin et al., 1972; Kenyon, 1973, 1974; Fuller, 1974; Lehninger, 1975), and encyclopedic listing of what is equivalent to "laboratory synthesis" of a primitive organism (Marquis, 1976).

VII. Design and Direction

The experimental findings, especially the phenomenon of self-ordering of amino acids, place in a new perspective questions of whether life began once or more and the related questions of what is either natural design and direction in evolution or the equivalent creationists' process of divine design and stewardship.

A single beginning of life is of course consistent with the concept of an act of special creation. The growing tradition in the methodology of science requires, on the other hand, that natural phenomena be repeatable. The variety of living things that we observe on this Earth can lead to the interpretation, on first viewing the question, that multiple beginnings of life (polyphylety) preceded all of the numerous lines of evolving organisms.

The experiments suggest instead, however, that protoreproductive cellular systems began innumerable times in uncounted locales on the young yet aging Earth. The self-ordering, i.e., self-limiting, nature of these processes would then have resulted in products that would have been much alike from one event of spontaneous generation to another. This would be a lateral manifestation of molecular determinism. Molecular determinism, moreover, would have been the evolutionary precursor of genetic determinism (Morgan, 1932; Wilson, 1978).

The essential feature is that, although the original processes occurred innumerable times, the products were so much alike due to the internal self-limitations of molecular stereospecificity, that they would have been of a single family type. Origins of life were thus multiple in historical occurrence but almost singular in type of product.

This kind of internal directedness has much in common with concepts of orthogenesis as well as determinism. The uniqueness of assemblies (Fox, 1980a) also has a quality that easily suggests single creations of each type, including the original, as marked by an act of aggregation (Fox, 1968). Single, or special, creations easily suggest, for an anthropomorphizing observer, a Creator who is designing and directing (Fox, 1980e). The natural equivalent of the ramified mythology of Genesis is the variety of shapes of molecules, magnified in macroevolution by the social ecology of materials able to assume myriads of individual configurations due to the directive nonrandomness of their interactions. The naturalist, accordingly, has available an understanding of how life and its variety of variegations came to be. He however must rely on the principle of *stereochemical selection* rather than on supernatural personal selection. Easy replacement of simple 2000-year-old concepts of divine genesis with an evolutionary hierarchy rooted in the shapes of molecules and their interactions may seem to be much to ask for. The material development of this situation over three billion years on this garden globe has been so ramified that we can hardly expect most members of the human species to come to accept it and its emphasis on stereochemistry without some intellectual tossing and turning.

ACKNOWLEDGMENT

Most of the experiments which have led to the views expressed have been supported by NASA Grant NGR 10-007-008, the National Foundation for Cancer Research, and Mr. David Rose.

NOTES

1. Copout = a presumption of failure.
2. In the discussion with Darnbrough et al., relayed correspondence has revealed that a technical objection (Darnbrough et al., 1981b) is based on our use of pH 11 in peptide syntheses (Fox & Nakashima, 1980). Had Darnbrough merely read one page further than that describing pH 11, he would have found that the "acceptable" pH of 7.2 was used (Fox, 1981g) (Experiments at pH 11 are simply some of those in which experimentalists "beat the bushes.")

3. In some cases, the printed criticisms are paraphrased in an attempt to reflect other critics as well.

REFERENCES

Allen, G. E. 1978. *Thomas Hunt Morgan* (Princeton University Press).

Ambrose, E. J. and Easty, D. M. 1970. *Cell Biology* (Addison-Wesley, Reading, Mass.), 479.

Asimov, I. 1967. *Is Anyone There?* (Avon Books, New York), 86.

Aw, S. E. 1976. *Chemical Evolution* (University Education Press, Singapore), 148.

Baker, J. J. W. and Allen, G. E. 1971. *The Study of Biology,* 2nd ed. (Addison-Wesley, Reading, Mass.), 792–95.

Baltscheffsky, H. 1981. Stepwise molecular evolution of bacterial photosynthetic energy conversion. *BioSystems 14,* 49–56.

Biological Sciences Curriculum Study. 1963. *Biological Science, Molecules to Man* (Houghton Mifflin, Boston).

Broda, E. 1975. *The Evolution of the Bioenergetic Processes* (Pergamon, Oxford).

Brooks, J. and Shaw, C. 1973. *Origin and Development of Living Systems* (Academic Press, London).

Calvin, M. 1969. *Chemical Evolution* (Oxford University Press).

Cloud, P. and Morrison, K. 1979. On microbial contaminants, micropseudofossils, and the oldest records of life. *Precambrian Res. 9,* 81–91.

Copeland, J. J. 1936. Yellowstone thermal myxophyceae. *Ann. N.Y. Acad. Sci. 36,* 1–232.

Corliss, J. B., Baross, J. A., and Hoffman, S. E. 1981. An hypothesis concerning the relationship between submarine hot springs and the origin of life on Earth. *Oceanologica Acta N°SP,* 59–69.

Crick, F. H. C. 1968. The origin of the genetic code. *J. Mol. Biol. 38,* 367–79.

Crick, F. H. C., Brenner, S., Klug, A., and Pieczenik, G. 1976. A speculation on the origin of protein synthesis. *Origins Life 7,* 389–97.

CRM. 1972. *Biology, An Appreciation of Life* (Del Mar, Calif.) 51–52.

Darnbrough, C., Goddard, J., and Stevely, W. S. 1981a. American creation. *Nature 292* (5818), 95–96.

Darnbrough, C., Goddard, J., and Stevely, W. S. 1981b. Creationism. *Nature 294* (5839), 302.

Darwin, C. Undated. *The Origins of Species by Means of Natural Selection or The Preservation of Favored Races in the Struggle for Life* (Random House, New York).

Dickerson, R. E. and Geis, I. 1976. *Chemistry, Matter and the Universe* (W. A. Benjamin, Menlo Park, Calif.), 641–42.

Dickerson, R. E. 1978. Chemical evolution and the origin of life. *Scientific American* 239 (3), 70–108.

Dose, K. and Zaki, L. 1971. The peroxidatic and catalatic activity of hemoproteinoids. *Z. Naturforsch. 26b,* 144–48.

Dowben, R. M. 1971. *Cell Biology* (Harper & Row, New York) 530.

Easton, T. A. 1973. A note on the mathematics of microsphere division. *Bull. Math. Biol.* 35, 259–62.

Eden, M. 1967. Inadequacies of neo-Darwinian evolution as a scientific theory. In Moorhead, P. S., & Kaplan, M. M. (eds.) *Mathematical Challenges to the Neo-Darwinian Interpretation of Evolution* (Wistar Institute Press, Philadelphia), 5–19.

Eigen, M. 1971. Selforganization of matter and the evolution of biological macromolecules. *Naturwissenschaften* 58, 465–523.

Etkin, W., Devlin, R. M., and Boufford, T. G. 1972. *A Biology of Human Concern* (J. B. Lippincott, Philadelphia).

Florkin, M. 1975. *Comprehensive Biochemistry* Vol. 29B (Elsevier, Amsterdam), 231–260.

Follmann, H. 1982. Deoxyribonucleotide synthesis and the emergence of DNA in molecular evolution. *Naturwissenschaften* 69, 75–81.

Fox, S. W. 1957. The chemical problem of spontaneous generation. *J. Chem. Ed.* 34, 472–79.

Fox, S. W. 1959. Review of book *The Biological Replication of Macromolecules. J. Chem. Ed.* 36, 706A.

Fox, S. W. 1964. Thermal polymerization of amino acids and production of formed microparticles on lava. *Nature* 201, 336–37.

Fox, S. W. 1965. Experiments suggesting evolution to protein. In Bryson, V. and Vogel, H. J. (eds.) *Evolving Genes and Proteins* (Academic Press, New York), 359–69.

Fox, S. W. 1968. Spontaneous generation, the origin of life, and self assembly. *Curr. Mod. Biol.* (Now *BioSystems*), 2, 235–40.

Fox, S. W. 1969. Self-ordered polymers and propagative cell-like systems. *Naturwissenschaften* 56, 1–9.

Fox, S. W. 1973a. The rapid evolution of complex systems from simple beginnings. In Marois, M. (ed.) *From Theoretical Physics to Biology* (S. Karger, Basel), 133–44.

Fox, S. W. 1973b. Molecular evolution to the first cells. *Pure Appld. Chem.* 34, 641–69.

Fox, S. W. 1974. The proteinoid theory of the origin of life and competing ideas. *Amer. Biol. Teacher 36,* 161–72, 181.

Fox, S. W. 1976a. Response to comments on thermal polypeptides. *J. Mol. Evol. 8,* 301–04.

Fox, S. W. 1976b. The evolutionary significance of phase-separated microsystems. *Origins Life 7,* 49–68.

Fox, S. W. 1976c. Contribution of experimental protobiogenesis to the theory of evolution. In Novak, J. A. and Pacltova, B. (eds.) *Evolutionary Biology* (Czech. Biol. Soc., Prague), 35–44.

Fox, S. W. 1977. Bioorganic chemistry and the emergence of the first cell. In van Tamelen, E. E. (ed.) *Bioorganic Chemistry* Vol. III, *Macro- and Multimolecular Systems* (Academic Press, New York), 21–32.

Fox, S. W. 1978. The origin and nature of protolife. In Heidcamp, W. H. (University Park Press, Baltimore), 23–92.

Fox, S. W. 1979. More on origin of life. *Chem. Eng. News* 57 (8) 4.

Fox, S. W. 1980a. Introductory remarks to the special issue on "assembly mechanisms." *BioSystems 12,* 131.

Fox, S. W. 1980b. Life from an orderly Cosmos. *Naturwissenschaften 67,* 576–81.

Fox, S. W. 1980c. Response to repeated statements on temperatures required for polycondensation of amino acids. *J. Mol. Evol. 15,* 539.

Fox, S. W. 1980d. The origins of behavior in macromolecules and protocells. *Comp. Biochem. Physiol. 67B,* 423–36.

Fox, S. W. 1980e. Metabolic microspheres. *Naturwissenschaften* 67, 378–83.

Fox, S. W. 1981a. From inanimate matter to living systems. *Amer. Biol. Teacher* 43 (3), 127–135, 140.

Fox, S. W. 1981b. Creation "copout." *Nature 292,* 490.

Fox, S. W. 1981c. Creationism, the random hypothesis, and experiments. *Science 213,* 290.

Fox, S. W. 1981d. A model for protocellular coordination of nucleic acid and protein syntheses. In Kageyama, M., Nakamura, K., Oshima, T., and Uchida, T. (eds.) *Science and Scientists* (Japan Sc. Soc. Press, Tokyo), 39–45.

Fox, S. W. 1981e. How many theories for the origin of (proto)life? In Srinivasan, R. (ed.) *Biomolecular Structure, Conformation, Function and Evolution* (Pergamon Press, Oxford), 643–46.

Fox, S. W. 1981f. Origins of the protein synthesis cycle. *Intl. J. Quantum Chem. QBS8,* 441–454.

Fox, S. W. 1981g. An acid test. *Nature 294,* 688.

Fox, S. W. and Dose, K. 1977. *Molecular Evolution and the Origin of Life,* rev. ed. (Marcel Dekker, New York).

Fox, S. W. and Harada, K. 1958. Thermal copolymerization of amino acids to a product resembling protein. *Science 128,* 1214.

Fox, S. W. and Krampitz, G. 1964. The catalytic decomposition of glucose in aqueous solution by thermal proteinoids. *Nature 203,* 1362–64.

Fox, S. W. and Nakashima, T. 1980. The assembly and properties of protobiological structures: the beginnings of cellular peptide synthesis. *BioSystems 12,* 155–66.

Fox, S. W. and Wang, C.-T. 1968. Melanocyte-stimulating hormone activity in thermal polymers of α-amino acids. *Science 160,* 547–48.

Fox, S. W. and Yuyama, S. 1963a. Effects of the Gram stain on microspheres from thermal polyamino acids. *J. Bacteriol. 85,* 279–83.

Fox, S. W. and Yuyama, S. 1963b. Abiotic production of primitive protein and formed microparticles. *Ann. N.Y. Acad. Sci. 108,* 487–94.

Fox, S. W., Harada, K., and Rohlfing, D. L. 1962. The thermal copolymerization of α-amino acids. In Stahmann, M. A. (ed.) *Polyamino Acids, Polypeptides, and Proteins* (University of Wisconsin Press, Madison), 47–54.

Fox, S. W., Joseph, D., McCauley, R. J., Windsor, C. R., and Yuyama, S. 1966. Simulation of organismic morphology and behavior by synthetic poly-α-amino acids. In Brown, A. H. and Florkin, M. (eds.) *Life Sciences and Space Research* Vol. IV (Spartan Books, Washington), 111–20.

Fox, S. W., McCauley, R. J., and Wood, A. 1967. A model of primitive heterotrophic proliferation. *Comp. Biochem. Physiol. 20,* 773–78.

Fox, S. W., McCauley, R. J., Montgomery, P. O'B., Fukushima, T., Harada, K., and Windsor, C. R. 1969. Membrane-like properties in microsystems assembled from synthetic protein-like polymer. In Snell, F., Wolken, J., Iverson, G. J., and Lam, J. (eds.) *Physical Principles of Biological Membranes* (Gordon & Breach, New York), 417–30.

Fox, S. W., Jungck, J. R., and Nakashima, T. 1974. From proteinoid microsphere to contemporary cell: formation of internucleotide and peptide bonds by proteinoid particles. *Origins Life 5,* 227–37.

Fox, S. W., Adachi, T., and Stillwell, W. 1980. A quinone-assisted photoformation of energy-rich chemical bonds. In Veziroglu, T. N. (ed.) *Solar Energy: International Progress* Vol. 2 (Pergamon, New York), 1056–74.

Fox, S. W., Harada, K., and Hare, P. E. 1981. Amino acids from the Moon: Notes on meteorites. *Subcell. Biochem. 8,* 357–73.

Fox, S. W., Nakashima, T., Przybylski, A., and Syren, R. M. 1982. The updated experimental proteinoid model. *Int. J. Quantum in Chem. QBS9*, 195–204.

Francis, S., Margulis, L., and Barghoorn, E. S. 1978. On the experimental silicification of microorganisms II. On the time of appearance of eukaryotic organisms in the fossil record. *Precambrian Res. 6*, 65–100.

Fuller, E. C. 1974. *Chemistry and Man's Environment* (Houghton Mifflin, Boston).

Gamow, G., Rich, A., and Yčas, M. 1956. Problem of information transfer from nucleic acids to proteins. *Advances Biol. Med. Phys. 4*, 23–68.

Gillespie, N. C. 1979. *Charles Darwin and the Problem of Creation* (University of Chicago Press).

Glasstone, S. 1968. *The Book of Mars.* (NASA, Washington), 193–94.

Grabiner, J. V. and Miller, P. D. 1974. Effects of the Scopes trial. *Science 185*, 832–37.

Grote, J. R., Syren, R. M., and Fox, S. W. 1978. Effect of product from heated amino acids on conductance in lipid bilayer membranes and nonaqueous solvents. *BioSystems 10*, 287–292.

Harada, K. and Fox, S. W. 1965. Thermal polycondensation of free amino acids with polyphosphoric acid. In Fox, S. W. (ed.) *The Origins of Prebiological Systems* (Academic Press, New York), 289–98.

Hardebeck, H. G., Krampitz, G., and Wulf, L. 1968. Decarboxylation of pyruvic acid in aqueous solution by thermal proteinoids. *Arch. Biochem. Biophys. 123*, 72–81.

Hartl, D. L. 1977. *Our Uncertain Heritage, Genetics and Human Diversity* (J. B. Lippincott Co., Philadelphia), 448–50.

Hayakawa, T., Windsor, C. R., and Fox, S. W. 1967. Copolymerization of the Leuchs anhydrides of the eighteen amino acids common to protein. *Arch. Biochem. Biophys. 118*, 265–272.

Heinz, B. and Ried, W. 1981. The formation of chromophores through amino acid thermolysis and their possible role as prebiotic photoreceptors. *BioSystems 14*, 33–40.

Hennon, G., Plaquet, R., and Biserte, G. 1975. The synthesis of amino acid polymers by thermal condensation at 105° without a catalyst. *Biochimie 57*, 1395–96.

Herrera, A. L. 1942. A new theory of the origin and nature of life. *Science 96*, 2497.

Hsu, L. L. 1975. Unpublished experiments.

Hsu, L. L., Brooke, S., and Fox, S. W. 1971. Conjugation of proteinoid microspheres: a model of primordial communication. *Curr. Mod. Biol.* (Now *BioSystems*) *4,* 12–25.

Huffman, H. M. 1942. Thermal data XV. The heats of combustion and free energies of some compounds containing the peptide bond. *J. Phys. Chem. 46,* 885–891.

Iben, I., Jr. 1973. *Molecules in the Galactic Environment* (John Wiley, New York), 19.

Ishima, Y., Przybylski, A., and Fox, S. W. 1981. Electrical membrane phenomena in spherules from proteinoid and lecithin. *BioSystems 13,* 243–251.

Jungck, J. R. and Fox, S. W. 1973. Synthesis of oligonucleotides by proteinoid microspheres acting on ATP. *Naturwissenschaften 60,* 425–27.

Kenyon, D. H. 1973. A theory of biogenesis. *Science 179,* 789.

Kenyon, D. H. 1974. Prefigured ordering and protoselection in the origin of life. In Dose, K., Fox, S. W., Deborin, G. A., and Pavlovskaya, T. E. (eds.) *The Origins of Life and Evolutionary Biochemistry* (Plenum Press, New York), 207–20.

Keosian, J. 1974. Life's beginnings—origin or evolution? In Dose, K., Fox, S. W., Deborin, G. A., & Pavlovskaya, T. E. (eds.) *The Origin of Life and Evolutionary Biochemistry* (Plenum Press, New York), 221–31.

Knight, C. A. 1974. *Molecular Virology* (McGraw-Hill, New York).

Korn, R. W. and Korn, E. J. 1971. *Contemporary Perspectives of Biology* (John Wiley, New York), 457–58.

Krampitz, G., Baars-Diehl, S., Haas, W., and Nakashima, T. 1968. Aminotransferase activity of thermal polylysine. *Experientia 24,* 140–42.

Krampitz, G., Haas, W., and Baars-Diehl, S. 1968. Glutaminsäure-Oxydoreductase-Aktivität von Polyanhydro-α-Aminosäuren (Proteinoiden). *Naturwissenschaften 55,* 345–46.

Kuhn, H. 1976. Model considerations for the origin of life. *Naturwissenschaften 63,* 68–80.

Kuhn, H. 1981. Molecular self-organization and the origin of life. *Angew. Chem. Internat. Ed. Engl. 20,* 500–20.

Lehninger, A. L. 1970. *Biochemistry* (Worth and Co., New York), p. 769.

Lehninger, A. L. 1975. *Biochemistry,* 2nd ed. (Worth and Co., New York).

Mader, S. S. 1976. *Inquiry into Life* (Wm. C. Brown Co., Dubuque, Iowa).

Marquis Who's Who Inc. 1976. *Who's Who in the World,* 3rd ed., p. 270.

Matsuno, K. 1981. Material self-assembly as a physicochemical process. *BioSystems 13,* 237–41.

Mayr, E. 1972. The nature of the Darwinian revolution. Acceptance of evolution by natural selection required the rejection of many previously held concepts. *Science 176,* 981–89.

Melius, P. and Sheng, Y. Y.-P. 1975. Thermal condensation of a mixture of six amino acids. *Bioorg. Chem.* 4, 385–91.

Miller, S. L. and Orgel, L. E. 1974. *The Origins of Life on the Earth* (Prentice-Hall, Englewood Cliffs, N.J.), 145.

Monod, J. 1971. *Chance and Necessity.* Translated by A. Wainhouse. (A. A. Knopf, New York).

Morgan, T. H. 1932. *The Scientific Basis of Evolution* (W. W. Norton Co., New York).

Morris, H. M. (ed.) 1974, second printing 1978. *Scientific Creationism,* Public School Edition (Creation-Life Publishers, San Diego).

Mueller, G. 1972. Organic microspheres from the Precambrian of South-West Africa. *Nature 235,* 90–95.

Nakashima, T. and Fox, S. W. 1980. Synthesis of peptides from amino acids and ATP with lysine-rich proteinoid. *J. Mol. Evol. 15,* 161–68.

Nakashima, T. and Fox, S. W. 1981. Formation of peptides by single or multiple additions of ATP to suspensions of nucleoproteinoid microparticles. *BioSystems 14,* 151–61.

Nakashima, T., Jungck, J. R., Fox, S. W., Lederer, E., and Das, B. C. 1977. A test for randomness in peptides isolated from a thermal polyamino acid. *Intl. J. Quantum Chem. QBS4,* 65–72.

Oparin, A. I. 1957. *The Origin of Life on the Earth* (Academic Press, New York).

Oparin, A. I. 1968. *Genesis and Evolutionary Development of Life* (Academic Press, New York).

Oparin, A. I. 1971. Routes for the origin of the first forms of life. *Sub-cell. Biochem. 1,* 75–81.

Oshima, T. 1968. The catalytic hydrolysis of phosphate ester bonds by thermal polymers of amino acids. *Arch. Biochem. Biophys. 126,* 478–485.

Osterberg, R. and Orgel, L. E. 1972. Polyphosphate and trimetaphosphate formation under potentially prebiotic conditions. *J. Mol. Evol. 1,* 241–48.

Pivcova, H., Saudek, V., Drobnik, J., and Vlasak, J. 1981. NMR study of poly (aspartic acid) I. α- and β-peptide bonds in poly (aspartic acid) prepared by thermal polycondensation. *Biopolymers 20*, 1605–14.

Popper, K. 1974. Scientific reduction and the essential incompleteness of all science. In Ayala, F. J. and Dobzhansky, T. (eds.) *Studies in the Philosophy of Biology* (University of California Press, Berkeley), 259–84.

Price, C. C. 1974. *Synthesis of Life* (Dowden, Hutchinson & Ross, Stroudsburg, Pa.).

Rohlfing, D. L. 1967a. Thermal poly-α-amino acids containing low proportions of aspartic acid. *Nature, 216*, 657–59.

Rohlfing, D. L. 1967b. The catalytic decarboxylation of oxaloacetic acid by thermally prepared poly-α-aminoacids. *Arch. Biochem. Biophys. 118*, 468–74.

Rohlfing, D. L. 1976. Thermal polyamino acids: synthesis at less than 100°C. *Science 193*, 68–70.

Rohlfing, D. L. and Fox, S. W. 1967. The catalytic activity of thermal polyanhydro-α-amino acids for the hydrolysis of *p*-nitrophenyl acetate. *Arch. Biochem. Biophys. 118*, 127–32.

Rohlfing, D. L. and Fox, S. W. 1969. Catalytic activities of thermal polyanhydro-α-amino acids. *Advances Catal. 20*, 373–418.

Ross, H. H. 1962. *A Synthesis of Evolutionary Theory* (Prentice-Hall, Englewood Cliffs, N.J.), 45–48.

Salisbury, F. B. 1971. Doubts about the modern synthetic theory of evolution. *Amer. Biol. Teacher 33*, 335–38.

Salthe, S. 1982. Original Life. *Nature 295*, 452.

Sanger, F. and Thompson, E. O. P. 1953. Amino acids in the glycyl chain of insulin. II Peptides from enzymic hydrolyzates. *Biochem. J. 53*, 366–74.

Saunders, P. T. and Ho, M. W. 1976. On the increase in complexity in evolution. *J. Theor. Biol. 63*, 375–384.

Sherman, I. W. and Sherman, V. G. 1975. *Biology. A Human Approach* (Oxford University Press, New York), 4.

Snyder, W. D. and Fox, S. W. 1975. A model for the origin of stable protocells in a primitive alkaline ocean. *BioSystems 7*, 222–29.

Temussi, P. A., Paolillo, L., Ferrara, L., Benedetti, E., and Andini, S. 1976. Structural characterization of thermal prebiotic polypeptides. *J. Mol. Evol. 7*, 105–10.

Turcotte, D. L., Nordmann, J. C., and Cisne, J. L. 1974. Evolution of the Moon's orbit and the origin of life. *Nature 251*, 124–25.

Usdin, V. R., Mitz, M. A., and Killos, P. J. 1967. Inhibition and reactivation of the catalytic activity of a thermal α-amino acid copolymer. *Arch. Biochem. Biophys. 122*, 258–61.

Vallery-Radot, P. 1922. *Fermentations et Générations dites Spontanées* Tome II (Masson et Cie, Paris), 328.

Vegotsky, A. and Fox, S. W. 1962. Protein molecules: intraspecific and interspecific variations. In Florkin, M. and Mason, H. S. (eds.) *Comparative Biochemistry* Vol. IV (Academic Press, New York), 185–244.

Volpe, E. P. 1977. *Understanding Evolution,* 3rd ed. (Wm. C. Brown Co., Dubuque, Iowa).

Weinberg, S. L. 1977. *Biology,* 4th ed. (Allyn & Bacon, Boston).

Wigner, E. 1961. The probability of the existence of a self-reproducing unit. In *The Logic of Personal Knowledge* (The Free Press, Glencoe, Ill., 231–38.

Wilhelm, R. D. 1978. A chronology and analysis of regulatory actions relating to the teaching of evolution in public schools. Ph.D. dissertation, University of Texas, Austin.

Wilson, E. O. 1978. *On Human Nature* (Bantam Books and Harvard University Press, Cambridge, Mass.).

Winchester, A. M. 1976. *Heredity, Evolution, and Humankind* (West Publishing Co., St. Paul).

Winchester, A. M. 1977. *Genetics,* 5th ed. (Houghton Mifflin Co., Boston), 410–11.

Wolfe, S. L. 1977. *Biology* (Wadsworth Publishing Co., Belmont, Calif), 366–68.

Wolken, J. J. 1975. *Photoprocesses, Photoreceptors and Evolution* (Academic Press, New York).

Wood, A. and Hardebeck, H. G. 1972. Light enhanced decarboxylations by proteinoids. In Rohlfing, D. L. and Oparin, A. I. (eds.) *Molecular Evolution* (Plenum, New York), 233–45.

Yockey, H. P. 1981. Self organization origin of life scenarios and information theory. *J. Theor. Biol. 91,* 13–31.

Young, R. S. 1966. *Extraterrestrial Biology* (Holt, Rinehart & Winston, New York).

L. BEVERLY HALSTEAD

EVOLUTION—THE FOSSILS SAY YES!

The poster advertising the oft-given lecture (and sometimes debate) by Dr. Duane T. Gish of the Institute for Creation Research, San Diego, carries the title "Evolution: Scholarship *versus* Dogmatism." In this article I shall endeavour to present the case for scholarship.

It is necessary first to admit of a major and fundamental area of agreement with the creationists. I personally do not see how the concept of evolution can be made consistent with that of creation by a personal god, or indeed any sort of god. I agree with Duane T. Gish in his book *Evolution? The Fossils Say No!*

> The reason that most scientists accept the theory of evolution is that most scientists are unbelievers, and unbelieving, materialistic men are forced to accept a materialistic, naturalistic explanation for the origin of all living things.
>
> Sir Julian Huxley, British evolutionist and grandson of Thomas Huxley, one of Darwin's strongest supporters when he first published his theory, has said that "Gods are peripheral phenomena produced by evolution." What Huxley meant was that the idea of God merely evolved as man evolved from lower animals. . . . Huxley believes that man is just as much a natural phenomenon as an animal or plant; that his body, mind and soul were not supernaturally created but are products of evolution, and that he is not under the control or guidance of any supernatural being or beings, but has to rely on himself and his own powers.

This is one area in which Gish and I agree.

The late Professor J. B. S. Haldane put the issue more bluntly:

> My practise as a scientist is atheistic. That is to say, when I set up an experiment I assume that no god, angel, or devil is going to interfere with its course; and this assumption has been justified by such success as I have achieved in my professional career. I should therefore be intellectually dishonest if I were not also atheistic in the affairs of the world. And I should be a coward if I did not state my theoretical views in public.

This is also where I stand. It is very important to illustrate the differences in approach between the scientific and religious standpoints. One of the best examples to illustrate this comes from the Genesis account of the fall. To me there are some inspiring passages in this book. But first I want to set the scene (Gen. 2:15–17).

> And the LORD GOD took the man, and put him into the garden of Eden to dress it and to keep it.
>
> And the LORD GOD commanded the man, saying, Of every tree of the garden thou mayest freely eat:
>
> But of the tree of the knowledge of good and evil, thou shalt not eat of it: for in the day that thou eatest thereof thou shalt surely die.

Here we have man being given an instruction by the supreme Authority, and he was expected to accept this quite uncritically—he was not expected to question it, he was certainly not expected to defy it, he was expected to obey it. Let us consider what this means. Here is a situation where you are placed in an environment where you have everything, all you must not do is think.

Samuel Butler in the last century wrote "The Kingdom of Heaven is the being like a good dog."

A good dog does what he is told, gets a pat on the head, and that is all. This is a prospect that no real human being should ever stand for. But we are very fortunate in this story—we have the hero of this entire episode, the serpent, and he gave very good advice (Gen. 3:5–7).

> For GOD doth know that in the day ye eat thereof, then your eyes shall be opened, and ye shall be as gods, knowing good and evil.

> And when the woman saw that the tree was good for food, and that it was pleasant to the eyes, and a tree to be desired to make one wise, she took of the fruit thereof, and did eat, and gave also unto her husband with her; and he did eat.
>
> And the eyes of them both were opened.

That, to my mind, is the most inspiring passage in this entire volume.

That was original sin, the defiance of the Lord God was original sin, and this sin is the one which every scientist worthy of the name is dedicated to uphold.

Let us go to the New Testament, the Gospel According to St. John (20:25–29) and note the approach of Thomas.

> The other disciples therefore said unto him, We have seen the Lord. But he said unto them, Except I shall see in his hands the print of the nails, and put my finger into the print of the nails, and thrust my hand into his side, I will not believe.

That is where the scientist stands—you should not believe something without evidence. Thomas was one of the early men of a scientific frame of mind.

Evidence, according to this story, was provided.

> And after eight days again his disciples were within, and Thomas with them, then came Jesus, the doors being shut, and stood in the midst and said, Peace be unto you.
>
> Then saith he to Thomas, Reach hither thy finger, and behold my hands; and reach hither thy hand, and thrust it into my side and be not faithless, but believing.
>
> And Thomas answered and said unto him, My Lord and my God.

He said that because he had the evidence. Thomas is still the hero, but this is where we come to the serious part.

> Jesus saith unto him, Thomas, because thou hast seen me, thou hast believed; blessed are they that have not seen, and yet have believed.

That is by far the most subversive statement in this book. And this is something that has characterised Western Civilisation for hundreds of years. You accept what authority says, do not question. And this is something of which I don't think we can be proud. Of course, people have defied authority throughout the millenia and will continue to do so. The motto of the Royal Society of London is *Nullius in Verba,* which, freely translated, means "We take no man's word for anything." This is a very proud motto for scientists to have. There is no question that any scientist or any man, for that matter, should not ask; there are no areas in which we should not question.

In striking contrast Gish writes, "It is apparent that acceptance of creation requires an important element of faith. Yes, it is true, creationists do have faith, and that faith is vitally important. In *Hebrews* 11:6 we read, 'But without faith it is impossible to please Him, for he that cometh unto God must believe that He is, and that He is a rewarder of them that diligently seek Him.' This faith is an intelligent faith, supported both by Biblical revelation and the revelation found in nature."

This attitude, I claim, seeks to close the mind. Gish and I stand on opposite sides of a divide which will not be crossed; he will not cross to my side and I will not cross to his. There is a fundamental division in the attitude of a believer and the attitude of a scientist. However, let me say a word or two about religion. There are two aspects. One is the social aspect, and this is quite an important one because it provides a framework of behaviour, a code of practice that can be codified under the umbrella of religion. I will not be discussing this aspect. The other aspect is the explanatory one. If you are faced with ignorance, it is very difficult to always say, "I do not know." It is much easier to simply invoke the deity to explain the unknown. That is a very comfortable stand to take, but that of science is to say, "I do not know, I cannot explain." It does not mean that the scientist doesn't believe that one day he may be able to explain. One of the techniques of Gish and his colleagues is to ask scientists to explain certain things regarding evolution and the fossil record, and the simple answer, and I will give it now, is that I cannot: I will not be able to explain them although I may be able to in the future. This is not what science is about. Science does not claim to have all the answers. Science is a way of looking at things, and the first approach of a scientist is to doubt and not believe.

For the sake of argument we can also accept, as did Gish, the definition of science in the Oxford English Dictionary.

> A branch of study which is concerned either with a connected body of *demonstrated truths* or with *observed facts* systematically classified and more or less colligated by being brought under general laws, and which includes trustworthy methods for the discovery of new truth within its own domain.

As Gish comments, "Thus, for a theory to qualify as a scientific theory, it must be supported by events, processes, or properties which can be observed, and the theory must be useful in predicting the outcome of future natural phenomena or laboratory experiments."

Just a word about the ideas of evolution—they did not begin in 1859. Neither did the idea of creation begin 8000 years ago. Western Civilisation has had these two contrasting basic ideas, which arose from two geographically different places. One can be traced back to the Fertile Crescent of the Tigris-Euphrates valley in Asia Minor and the other to the valley of the Nile. In the former there grew up the notion of sudden, violent, inexplicable events occurring through the agency of a capricious and violent supernatural power. The Babylonian (and biblical) Flood is a good example of this type of thing. Violence and sudden catastrophes characterised the climatic regime of the Fertile Crescent. In marked contrast, the annual cycle of the Nile overflowing its banks and bringing fertile silts to the land had a kind of timelessness about it. Changes that occurred were always gradual. So there arose a belief in gradual change and continuity on the one hand and another which involved not only the creation of the world but also a continual series of sudden interferences by the Creator in the affairs of the world. The Bible gives a clear idea of a vengeful, cruel god, which is nowhere more clearly illustrated than in the Book of Job. The philosophy that developed in Asia Minor has become entrenched in the culture of Western/Christian civilisation. The notion of gradual change through time (i.e., evolution) has been considered anathema and indeed still is by some fundamentalist Christian sects. The majority of the Christian churches, however, now accept the fact of evolution, just as they now also accept the fact that the earth revolves around the sun. The universe is no longer geocentric.

There is no problem with regard to establishing that evolution has occurred. The resistance to accepting the fact of evolution probably stems from the realisation that it does not leave the deity very much to do. If the process were directed it suggests that God was continually learning from his mistakes—a view held by Professor H. S. Lipson, a creationist, but hardly in accord with the orthodox creationist standpoint. Indeed, as J. S. Clarke stated in 1919, "no-one carefully studying the subsequent evolution of the biological, sociological and historical worlds can reach any other conclusion than that such evolution was the inspiration of an exaggerated and glorified criminal lunatic."

With regard to the concept of creation, we can do no better than accept Gish's definition.

> Creation. By creation we mean the bringing into being of the basic kinds of plants and animals by the process of sudden, or fiat, creation described in the first two chapters of Genesis. Here we find the creation by God of the plants and animals, each commanded to reproduce after its own kind using processes which were essentially instantaneous.
>
> We do not know how God created, what processes He used, *for God used processes which are not now operating anywhere in the natural universe.* This is why we refer to divine creation as special creation. We cannot discover by scientific investigations anything about the creative processes used by God.
>
> During the creation week God created all of these basic animal and plant kinds, and since then no new kinds have come into being, for the Bible speaks of a *finished* creation (Gen. 2:2).
>
> The first two chapters of Genesis were not written in the form of parables or poetry but present the broad outlines of creation in the form of simple historical facts. These facts directly contradict evolution theory. The Bible tells us that at one time in history there was a single human being upon the earth—a male by the name of Adam. This is in basic contradiction to evolution theory because, according to that theory, populations evolve, not individuals. After God had formed Adam from the dust of the ground, the Bible tells us that He used some portion from Adam's side (in the King James version this is translated as "rib") to form Eve. This, of course, cannot be reconciled with any possible evolutionary theory concerning the origin of man.
>
> The New Testament Scriptures fully support this Genesis ac-

> count. For example, in I Corinthians 11:8 we read, "Man is not of the woman, but the woman of the man." By any natural reproductive process, man is always born of a woman. We all have mothers. This Biblical account can, therefore, be referring only to that unique time in history when God created woman from man, just as described in Genesis 2:21, 22.

Dr. H. M. Morris, Director of the Institute for Creation Research, has listed the following 15 contradictions between evolution and creation:

1. Matter created in the beginning, versus matter created by God in the beginning.
2. Sun and stars before the earth, versus earth before the sun and stars.
3. Land before the oceans, versus oceans before the land.
4. Sun the earth's first light, versus light before the sun.
5. Contiguous atmosphere and hydrosphere, versus atmosphere between two hydrospheres.
6. Marine organisms first forms of life, versus land plants first life forms created.
7. Fishes before fruit trees, versus fruit trees before fishes.
8. Insects before birds, versus birds before insects ("creeping things").
9. Sun before land plants, versus land vegetation before the sun.
10. Reptiles before birds, versus birds before reptiles ("creeping things").
11. Reptiles before whales, versus whales before reptiles.
12. Woman before man (by genetics), versus man before woman (by creation).
13. Rain before man, versus man before rain.
14. "Creative" processes still continuing, versus creation completed.
15. Struggle and death necessary antecedents of man, versus man the cause of sin and death.

It is important to emphasize that this list was not compiled by me in order to ridicule creationists but was drawn up by the creationists themselves.

With regard to evolution, we can again accept Gish's definition with the proviso that it is equally conceivable that all living things could have arisen from multiple sources.

> *EVOLUTION.* When we use the term *Evolution* we are using it in the sense defined by the general theory of evolution. According to the *General Theory of Evolution,* all living things have arisen by a naturalistic, mechanistic evolutionary process from a single living source which itself arose by a similar process from a dead, inanimate world.
>
> According to this theory, all living things are interrelated. Man and ape, for example, are believed to have shared a common ancestor.

I think we can agree on our terms of reference.

There are two aspects of the theory of evolution which are all too frequently, and one suspects deliberately, confused: evidence of the processes by which it took place, the *how* of evolution, and the historical evidence of it actually having taken place. With regard to the process of evolution, Gish concedes the fundamental principle:

> Equally important to our discussion is an understanding of just what we are not talking about when we use the term evolution. We are not referring to the limited variations that can be seen to occur, or which can be inferred to have occurred in the past, but which do not give rise to a new basic kind.
>
> As noted earlier, then, the concept of special creation does not exclude the origin of varieties and species from an original created kind. It is believed that each kind was created with sufficient genetic potential, or gene pool, to give rise to all of the varieties within that kind that have existed in the past and those that are yet in existence today.

This, I would maintain, concedes the fundamental principle of the *how* of evolution. It depends naturally enough on what is meant by a Genesis "kind." The definition turns out to be rather elastic, and it is not possible to pin creationists down on this. For example, Gish writes:

> When we attempt to make fine divisions within groups of plants and animals where distinguishing features are subtle, only God can draw the line. Many taxonomic distinctions established by man are uncertain and must remain tentative."

There is no answer to that. It is perhaps, nevertheless, worth pointing out that the line is not drawn by God but by the animals themselves. Species reproduce to give fertile offspring, but if Gish concedes the origin of species, I am content.

It is important to consider the evidence not of the *how,* but whether or not evolution actually has taken place. We can argue interminably about the mechanisms, the processes involved, and scientists do this all the time. But what is the evidence that it has taken place? Dr. H. M. Morris, in the preface to Gish's book, correctly notes, "The fossil record must provide the critical evidence for or against evolution, since no other scientific evidence can possibly throw light on the actual history of living things."

With regard to determining whether or not evolution has taken place, I again agree with Gish that

> in the final analysis, however, whether evolution actually did happen or not can only be decided, scientifically, established by the discovery of the fossilized remains of representative samples of those intermediate types which have been postulated on the basis of the indirect evidence. In other words, the really crucial evidence for evolution must be provided by the palaeontologist whose business it is to study the evidence of the fossil record. As a matter of fact, the discovery of only five or six of the transitional forms scattered through time would be sufficient to document evolution.

I accept the challenge of Gish's book. I am a palaeontologist and the study of fossils is my profession.

I want to say a word or two about time—the geological record. Geologists consider that the Earth is very ancient and that the rocks that one sees represent a long period of time. Rocks are interpreted in terms of processes that can be observed in the present day—the theory of uniformitarianism. We see rivers depositing silts and sands at their mouths in lakes and on the seacoasts. We draw the conclusions that sedimentary rocks were first laid down as sediments, that those at the bottom of the pile are likely to have been the oldest, and that as one progresses upwards they become younger. We have what is termed the law of superposition. The one deposited on the top is younger than the one that has already been deposited—and we can see the same processes going on today.

There is also a method of radiometric dating. This involves studying the rate of decay of radioactive elements and gives absolute dates in millions of years. The working geologist, in fact, uses the relative dating based on which strata occur on the top of which others. I have done this myself, and I will not accept that it cannot be done.

Gish does not accept this. He believes in the account in Genesis, but he has said in his book that even when these assumptions (of the theory of uniformitarianism) are accepted, however, the data from the fossil record does not agree with the predictions of the evolution model.

> Therefore, whether or not the earth is ten thousand, ten million or ten billion years old, the fossil record does not support the general theory of evolution.

Let us now deal with some of the specific points on which Gish challenges the fossil record. Gish states:

> What do we find in rocks older than the Cambrian? Not a single, indisputable, multicellular fossil has ever been found in Precambrian rocks! Certainly it can be said without fear of contradiction that the evolutionary ancestors of the Cambrian fauna, if they ever existed, have never been found.

This statement is entirely incorrect. The stromatolites formed by ancient blue-green algae of the early Precambrian are well known throughout many sequences of Precambrian rocks from Australia, Canada, China, and many other regions. The later Precambrian is represented by a whole suite of multicellular animals, best known the Ediacara Formation near Adelaide in Australia. There are numerous types of jellyfish, sea pens, and *Spriggina* in an intermediate form between an arthropod and annelid worm. Precambrian fossils most definitely do exist.

It is true that there is a sudden and great explosion of animal life, at least in the fossil record, at the beginning of the Cambrian, and the feature which distinguishes the animals of that period is the possession of hard parts that make them thus easy to preserve. One must have very special conditions to preserve soft-bodied animals. We are extraordinarily lucky to have soft-bodied Precambrian fossils preserved.

There is one point that Gish emphasizes again and again—the

rarity of organisms that link major groups or phyla. There are two explanations of this. One is that they have not yet been found; the other is that they never existed. I think that it is likely they never existed. In fact, if one looks at the major divisions such as annelid worms, molluscs and arthropods, the early developmental stages are virtually indistinguishable. I agree with Professor Ralph Nursall, who postulated that from the one-celled stage, evolution proceeded along different pathways. This would account for the absence of intermediates between the major phyla, as these would have existed only in the early stages of embryological development, before the different pathways diverged.

One of the characteristic features of the fossil record which Gish and his colleagues continually draw to our attention and which I also accept—and I quote myself here—is that "the fossil record tells us that evolution always takes place somewhere else." What I meant by this is that in any succession of rocks containing fossils of, say, fishes, it will be found that as one follows the succession, the fossils will become more advanced, but there will be no evidence directly linking the stages. There is a break in continuity. Geologists know this from their experience in the field.

I want to tell you about my experience. When I graduated I researched for about eleven years on the very earliest fossil vertebrates, and I was fortunate because, as a full-time research worker, I was able to examine every single specimen that had been found of that group throughout the entire world. This involved quite a lot of travelling. What I discovered was that gradual evolution was taking place in one area, in Western Russia, and one could trace out a succession of migrations from that evolutionary centre to distant regions. That was the explanation of the apparent numerous gaps in the fossil record of these particular fishes. There were waves of migration from one major centre. I am not quoting out of context from someone else's books; I am speaking of the research I did myself.

When we come to the placoderms, early jawed vertebrates, Gish quotes the late Professor A. S. Romer as follows:

> "*No well-known placoderms can be identified as the actual ancestors of the Chondrichthyes,* but we have noted that some of the

> peculiar petalichthyids appear to show morphologically intermediate stages in skeletal reduction. Increasing knowledge of early Devonian placoderms, may some day bridge the gap." Earlier, with reference to the placoderms, Romer had said, ". . . we must consider seriously that at least the sharks and chimaeras may have descended from such impossible ancestors." Romer insists that special creation is not admissible as a scientific explanation for origins, but he is willing to appeal to "impossible ancestors" to support his own sagging theory! A consideration of the creation model certainly seems more reasonable than an appeal to "impossible ancestors."

In 1961 Dr. T. Ørvig described an unusual placoderm which he named *Ctenurella,* which he interpreted as the placoderm ancestor of the chimaeras. This is one piece of evidence where a gap in the fossil record is now filled. Furthermore the acanthodians are now recognised as the ancestors of the modern bony fishes.

When it comes to the transition from water to land, the detailed structure of the fins of the rhipidistian *Eusthenopteron*—the bones and associated musculature—have been shown by Professor T. S. Westoll and Dr. Mahala Andrews to be virtually identical to the fundamental arrangements of the limbs of the first land vertebrates.

The first amphibian *Ichthyostega* from Greenland has its tail composed of fish fin rays; the skull has fish opercular bones. I object to the sketch of *Ichthyostega* in Gish's book (1973 ed.) because, as every palaeontologist knows, that picture is nothing like *Ichthyostega.* The only explanations are not very charitable. Either this is a deliberate misrepresentation—and I am sure that is not the case—or it is a result of the author's ignorance.

Regarding the transition from reptile to mammal, Gish writes:

> The two most easily distinguishable osteological differences between reptiles and mammals, however, have never been bridged by transitional series. All mammals, living or fossil, have a single bone, the dentary, on each side of the lower jaw, and all mammals, living or fossil, have three auditory ossicles or ear bones, the malleus, incus, and stapes. In some fossil reptiles the number and size of the bones of the lower jaw are reduced compared to living reptiles. Every reptile, living or fossil, however, has at least four bones in the lower jaw and only one auditory ossicle, the stapes.

> There are no transitional forms showing, for instance, three or two jaw bones, or two ear bones. No-one has explained yet, for that matter, how the transitional form would have managed to chew while his jaw was being unhinged and rearticulated, or how he would hear while dragging two of his jaw bones up into his ear.

That is very good, quite amusing, but this transition is in fact very well known; the functional reasons for it were worked out in the early 1960s by Professor A. W. Crompton of Yale University. The splitting of the main jaw-closing muscles into two blocks, the masseter and temporalis, had the effect of reducing the pressure on the jaw joint, and in consequence the bones became reduced in size. In the early growth stages these bones were loose in the region of the ear apparatus, and Dr. J. Hopson in 1966 showed exactly how this process of the jaw bones becoming incorporated as sound-amplifying bones in the middle ear took place.

This same process can actually be seen today taking place in kangaroos and hedgehogs during their embryological development. The exact transition between the reptilian and mammalian jaw joint was described some years ago by Professor Romer from new fossils discovered in the Argentine. The new jaw articulation was already functioning while the other was becoming defunct.

Now a word about the other famous fossil which it is claimed links reptiles and birds, *Archaeopteryx*. It is considered to be intermediate between reptiles and birds. It has feathers, so it is classified as a bird; it also has a lot of reptilian features such as teeth and a long bony tail. Here we come to a problem of semantics. Gish quotes du Nouy as follows:

> Concerning the status of *Archaeopteryx*, du Nouy, an evolutionist, has stated: "Unfortunately the greater part of the fundamental types in the animal realm are disconnected from a palaeontological point of view. In spite of the fact that it is undeniably related to the two classes of reptiles and birds (a relation which the anatomy and physiology of actually living specimens demonstrates), we are not even authorised to consider the exceptional case of the *Archaeopteryx* as a true link. By link, we mean a necessary stage of transition between classes such as reptiles and birds, or between smaller groups.
>
> "An animal displaying characters belonging to two different

> groups cannot be treated as a true link as long as the intermediary stages have not been found, and as long as the mechanisms of transition remain unknown."

With many groups we have what we term mosaic evolution, beginning with animals that are completely reptilian one ends up with forms that are completely avian or mammalian. The whole picture gradually changes over as more birdlike or mammalian features develop. What we have to do, because the transition is so gradual, is draw an arbitrary line; if it has character X we will call it A, if not we will call it B. Hence, by definition there can never be an intermediate, because we have drawn arbitrary lines in such a way that an animal is forced to be either one thing or the other. The claim that there are no intermediates in these cases is a semantic trick that Gish is playing. I hope no one will be taken in by it.

Finally, a word or two about man. I will not go into great detail about man, but if man is considered a special creation then you will discover that his design is quite imperfect. To give one example, the viscera of human beings are supported by a musculature, which from the design point of view is only suitable for an animal going on all fours. When standing upright the weight is not properly supported and any undue exertion can lead to the tissues giving way, resulting in a hernia. This is bad design, and many other examples can be given.

One has to decide if the Creator was incompetent or had a strange sense of humour. And the curious thing about Genesis is that if you consider that man is perfect, then why on earth did God, when he instituted the covenant with Abraham, require the removal of bits of this perfection? Why did He insist that the male should have the foreskin cut off?

I think we could spend a long time on the issue of man, but I would like to conclude by quoting Gish with approval:

> Dr. George Gaylord Simpson, Professor of Vertebrate Palaeontology at Harvard University until his retirement and one of the world's best known evolutionists, has said that the Christian faith, which he calls the "higher superstition" (in contrast to the "lower superstition" of pagan tribes of South America and Africa), is in-

tellectually unacceptable. Simpson concludes his book, *Life of the Past,* with what Sir Julian Huxley has called "a splendid assertion of the evolutionist view of man" [which I personally echo. I only wish I could have expressed these sentiments so well. L.B.H.]. "Man," Simpson writes, "stands alone in the universe, a unique product of a long, unconscious, impersonal, material process with unique understanding and potentialities. These he owes to no-one but himself, and it is to himself that he is responsible. He is not the creature of uncontrollable and undeterminable forces, but his own master. He can and must decide and manage his own destiny."

ROGER J. CUFFEY

PALEONTOLOGIC EVIDENCE AND ORGANIC EVOLUTION

Introduction

Practicing paleontologists today, regardless of personal philosophical outlook, unanimously agree that the varied organisms inhabiting the earth originated by a process of gradual, continuous development or evolution over long periods of prehistoric time. Because the case for organic evolution had been adequately demonstrated in the late 1800's (principally by paleontologic evidence), scientists in this century turned their attention to many other important subjects. Consequently, most have been surprised by (Lewontin, 1971) and also ill-prepared to cope with the recent reappearance of anti-evolutionary ideas (such as Morris, 1963; Moore, 1970a, 1970b, 1971a, 1971b; Moore & Slusher, 1971). Therefore, presenting the paleontologic evidence relevant to the concept of evolution is most timely, particularly for an audience like that of the *Journal ASA.*

The participants in the current controversy about evolution seemingly agree that fossils (the study of which comprises the science of paleontology) are the remains (or direct traces) of formerly living organisms, preserved in the earth's crust since prehistoric times. This conclusion is incontrovertibly supported by the complete spectrum observable within the earth's crust between recently dead organisms and highly altered fossils.

In addition to the morphology of fossils, a paleontologist studies

From the *Journal of the American Scientific Affiliation,* vol. 24, no. 4, December 1972. Reprinted by permission from the *Journal of the American Scientific Affiliation.* Copyright held by the ASA, an affiliation of Christian men and women of science, Box J, Ipswich, MA 01938.

also various aspects of their distribution within the earth's crust. As Van de Fliert (1969) has ably discussed, the rock layers comprising that crust reveal a chronological framework (usually stated succinctly as the standard geological time scale) for the earth's history. This basic framework, founded upon repeatable observations of the succession of rock strata, is quite independent of any concept of organic evolution (Van de Fliert, 1969, pp. 75, 77); in fact, the standard time scale historically was worked out half a century before evolution was proposed and demonstrated.

Fossil Sequences

As a consequence, we can examine the fossils entombed in chronologically successive rock layers, and thereby learn what organisms inhabited this planet during successive intervals of past geologic time. When we do this, we find that the fossils naturally form sequences showing gradual and continuous morphologic changes from earlier forms to later forms of life, sequences which make evolutionary interpretations ultimately inescapable.

As working paleontologists interested in the history of particular organisms, we locate for detailed study a relatively thick succession of fossil-bearing rock layers whose observable physical features indicate continuous and uninterrupted deposition over a comparatively long time interval. We next examine those layers for the fossils in which we are interested. We initially find a few fossils, scattered widely among the different layers. Studying these specimens usually shows noticeable morphological differences between ones from various geologic ages, differences which we recognize formally in progress reports by referring the specimens to different species, genera, etc., depending upon the magnitude of those differences. Continued field collecting from the rock strata intervening between any two successive forms thus described frequently produces a series of fossils which begin with the earlier form, change in morphology gradually and continuously as we proceed upward, and end up with the later form. Because these new fossils demonstrate a morphological and parallel chronological transition from the earlier form to the later form, they are termed "transitional fossils."

Examples of Transitional Fossils

If we read the paleontologic literature (especially if with the background of professional paleontologic training and experience; Cuffey, 1970, p. 93), we find that the fossil record contains many examples of such transitional fossils. These connect both low-rank taxa (like different species) and high-rank taxa (like different classes), in spite of the record's imperfections and in spite of the relatively small total number of practicing paleontologists. Because of the critical role which transitional fossils played in convincing scientists of the occurrence of organic evolution, paleontologists have been appalled that many otherwise well-informed persons have repeated the grossly misinformed assertion that transitional fossils do not exist. Consequently, after a relatively brief and non-exhaustive search of the literature immediately available to me, I compiled the examples of transitional fossils presented here. At least enough of these can be readily examined by anyone seriously interested in this topic that he can be convinced of their implications, I believe; collectively, they (and the many other similar ones which more extended search would find) comprise a massive body of evidence which cannot be ignored or explained away.

Although the broad patterns and many details in the history of life are well known, many other details remain to be learned. Because of the unevenness of our knowledge, therefore, we can conveniently distinguish several different types of transitional-fossil situations. Let us consider these now, starting with that situation where our knowledge is most complete, and proceeding through situations in which knowledge is progressively less complete.

First, some groups have been so thoroughly studied that we know sequences of transitional fossils which grade continuously from one species to another without break (Table 1), sometimes linking several successive species which cross from one higher taxon into another (Table 2). We can say that situations of this kind display *transitional individuals.* Among the many available examples of transitional individuals, some particularly convincing examples can be noted. These involve:

corals (Carruthers, 1910, p. 529, 538; Easton, 1960, p. 175; Moore, Lalicker, & Fischer, 1952, p. 140; Weller, 1969, p. 123),

Table 1. Examples of transitional individuals grading continuously between successive species within the same higher taxon (genus).

Algae: Gartner, 1971.

Angiosperms: Chandler, 1923, p. 124, 132–133; Chaney, 1949, p. 197–198; Stebbins, 1949, p. 230–231.

Foraminiferans: Barnard, 1963, p. 82, 90; Rauzer-Chernousova, 1963, p. 48.

Corals: Carruthers, 1910, p. 529, 538; Cocke, 1970, p. 13; Easton, 1960, p. 175; Moore, Lalicker, & Fischer, 1952, p. 140; Ross & Ross, 1962, p. 1182–1184; Weller, 1969, p. 123.

Bryozoans: Cuffey, 1967, p. 38–39; Cuffey, 1971a, p. 158; Cuffey, 1971b, p. 38; Elias, 1937, p. 311, 317.

Brachiopods: Ziegler, 1966, p. 532.

Gastropods: Fisher, Rodda, & Dietrich, 1964; Lull, 1940, p. 19; Sohl, 1967, p. B12–13, B15–16; Thomson, 1925, p. 96.

Pelecypods: Charles, 1949; Charles & Maubeuge, 1952, 1953a, 1953b; Heaslip, 1968, p. 58, 69, 77–79; Imlay, 1959; Kauffman, 1965, p. 8–21; Kauffman, 1967; Kauffman, 1969, p. N198–200; Kauffman, 1970, p. 633; Kay & Colbert, 1965, p. 325; Lerman, 1965, p. 416, 431–432; MacNeil, 1965, p. G35–36, G42; Raup & Stanley, 1971, p. 191, 257; Stenzel, 1971, p. N1077; Waller, 1969, p. 26.

Ammonoids: Cobban, 1958, p. 114; Cobban, 1962a, 1962b; Cobban, 1969, p. 6; Cobban & Reeside, 1952, p. 1020–1022; Easton, 1960, p. 456.

Trilobites: Brouwer, 1967, p. 152–155; Kaufmann, 1933, 1935; Raup & Stanley, 1971, p. 292; Simpson, 1953, p. 250.

Echinoids: Beerbower, 1968, p. 136, 138; Durham, 1971, p. 1126–1127; Hall, 1962; Kermack, 1954; Nichols, 1959a, 1959b; Olson, 1965, p. 98; Rowe, 1899.

Conodonts: Clark, 1968, p. 21–23; Scott & Collinson, 1959, p. 562.

Mammals: Osborn, 1929, p. 20–21; Simpson, 1953, p. 387–388; Teilhard de Chardin, 1950; Trevisan, 1949; Watson, 1949, p. 47; Wood, 1949, p. 188–189.

Table 2. Examples of transitional individuals grading continuously between successive species, and crossing from one higher taxon into another.

Ginkgophytes: Andrews, 1961, p. 337–339; Brown, 1943, p. 863; Franz, 1943, p. 323; Scagel *et al,* 1965, p. 484; Seward, 1938; Weller, 1969, p. 66.

Angiosperms: Chaney, 1949, p. 193–199; Elias, 1942, p. 70–71, 88–89, 109–122; Stebbins, 1949, p. 230.

Foraminiferans: Banner & Blow, 1959, p. 21; Barnard, 1963, p. 86, 88–89; Gimbrede, 1962, p. 1121–1123; Jones, 1956, p. 274; Papp, 1963, p. 352–353; Woodland, 1958, p. 803–808; Zeller, 1950, p. 19.

Brachiopods: Boucot & Ehlers, 1963, p. 48–51.

Pelecypods: Newell, 1942, p. 21, 59.

Ammonoids: Arkell, Kummel, & Wright, 1957, p. L113–119; Brinkmann, 1929, 1937; Brouwer, 1967, p. 156–158; Cobban, 1951, p. 5–11; Cobban, 1964, p. I10–14; Easton, 1960, p. 455; Erben, 1966; Krumbein & Sloss, 1963, p. 369; Olson, 1965, p. 105–107; Raup & Stanley, 1971, p. 264, 306–307; Spath, 1938; Wenger, 1957.

Conodonts: Rexroad, 1958, p. 1158.

Mammals: Hanson, 1961, p. 50–51; Scott, 1937, p. 417; Simpson, 1951, p. 114–121, 148, 217–228, 232, 236, 257, 265, 282, pls. 20, 31; Wood, 1949, p. 186.

Hominids: Coon, 1962; Howells, 1967; Kummel, 1970, p. 578–583; Le Gros Clark, 1964; Uzzell & Pilbeam, 1971, p. 615.

gastropods (Fisher, Rodda, & Dietrich, 1964),
pelecypods (Kauffman, 1967; Kauffman, 1969, p. N198–200; Kauffman, 1970, p. 633),
echinoids (Beerbower, 1968, p. 136, 138; Kermack, 1954; Nichols, 1959a, 1959b; Olson, 1965, p. 98; Rowe, 1899).

Second, other fossil groups have been well enough studied that we know sequences of transitional fossils comprising a series of chronologically successive species grading from an early form to a later form (Table 3), again sometimes crossing boundaries separating different higher taxa (Table 4). This type of situation can be termed *successive species.* Published descriptions of successive

Table 3. Examples of successive species within the same higher taxon (genus).

Angiosperms: Chandler, 1923; Chaney, 1949, p. 197–199; Elias, 1942; Stebbins, 1949, p. 230–231.
Foraminiferans: Barnard, 1963, p. 82; Bronnimann, 1950, p. 406; Cita-Sironi, 1963, p. 119–121; Hottinger, 1963, p. 306–307; Schaub, 1963, p. 288–290, 292–294; Wilde, 1971, p. 376.
Brachiopods: Berry & Boucot, 1970, p. 30–31; Dunbar & Waage, 1969, p. 113; Greiner, 1957; Raup & Stanley, 1971, p. 124.
Gastropods: Franz, 1932; Franz, 1943, p. 272; Sohl, 1960, p. 100.
Pelecypods: Dechaseaux, 1934; Easton, 1960, p. 348; Heaslip, 1968, p. 74–77, 79–81; Kay & Colbert, 1965, p. 327; Lerman, 1965, p. 416; Moore, Lalicker, & Fischer, 1952, p. 447; Newell, 1937, p. 40, 80; Newell, 1942, p. 21, 42, 47–48, 51–52, 60, 63, 65; Olson, 1965, p. 97; Schafle, 1929, p. 79; Stenzel, 1949; Stenzel, 1971, p. N1056–1057, N1077, N1079–1080; Weller, 1969, p. 209; Zeuner, 1933, p. 317.
Trilobites: Grant, 1962, p. 983–998.
Crustaceans: Guber, 1971, p. 15–16; Sohn, 1962, p. 1207; Swartz, 1945; Weller, 1969, p. 267.
Carpoids: Barrande, 1887; Weller, 1969, p. 297.
Blastoids: Beaver, 1967, p. S303–305.
Graptolites: Berry, 1960, p. 9.
Fishes: Boreske, 1972, p. 3–4.
Amphibians: Olsen, 1965, p. 45–48.
Mammals: Lull, 1940, p. 189; McGrew, 1937, p. 448; Tedford, 1970, p. 671, 694.

species lack explicit discussion of individuals transitional between the species, although frequently such exist in the author's collection but are not discussed because they are not directly pertinent to his purposes. Again, some especially persuasive examples of successive species can be seen, among:

foraminiferans (Wilde, 1971, p. 376),
brachiopods (Greiner, 1957; Raup & Stanley, 1971, p. 124),
pelecypods (Easton, 1960, p. 348; Kay & Colbert, 1965, p. 327; Moore, Lalicker, & Fischer, 1952, p. 447; Newell, 1942, p. 21, 42, 47–48, 51–52, 60, 63, 65; Olson, 1965, p. 97; Stenzel, 1949; Stenzel, 1971, p. N1079–1080; Weller, 1969, p. 209),
ammonoids (Cobban, 1961, p. 740–741).

Table 4. Examples of successive species crossing from one higher taxon into another.

Ginkgophytes: Andrews, 1961, p. 337–339; Brown, 1943, p. 863; Franz, 1943, p. 323; Scagel *et al.*, 1965, p. 484; Seward, 1938; Weller, 1969, p. 66.
Foraminiferans: Berggren, 1962, p. 109, 116–126.
Bryozoans: Lang, 1921–1922; Easton, 1960, p. 268.
Gastropods: Fisher, Rodda, & Dietrich, 1964.
Pelecypods: Stenzel, 1971, p. N1057, 1078.
Nautiloids: Easton, 1960, p. 425; Flower, 1941, p. 526; Moore, Lalicker, & Fischer, 1952, p. 351.
Ammonoids: Arkell, Kummel, & Wright, 1957, p. L116; Cobban, 1961, p. 740–741; Easton, 1960, p. 446; House, 1970, p. 666–674; Miller, Furnish, & Schindewolf, 1957, p. L22; Wright & Wright, 1949.
Crustaceans: Glaessner, 1960, p. 40–41; Glaessner, 1969, p. R410–411.
Crinoids: Moore, Lalicker, & Fischer, 1952, p. 629.
Echinoids: Jackson, 1912, p. 231; Weller, 1969, p. 355.
Reptiles: Lull, 1940, p. 296; Olson, 1965, p. 99–101.
Reptile-Mammal Transition: Olson, 1965, p. 202.
Mammals: Kummel, 1970, p. 514; Lull, 1940, p. 524; Matthew, 1910; Nelson & Semken, 1970, p. 3734; Osborn, 1929, p. 35–37, 724, 761, 773, 784, 791, 801, pl. 48; Patterson, 1949, p. 243–244, 246, 263, 268; Scott, 1937, p. 429; Simpson, 1951, p. 148, 245; Wood, 1949, p. 188–189.

In many fossil groups, our understanding is relatively less complete, thus giving rise to a third type of situation which we can label *successive higher taxa.* Here, we may not have complete series of transitional individuals or successive species, but the genera (or other higher taxa) represented in our collections form a continuous series grading from an earlier to a later form, sometimes crossing from one higher-rank taxon into another (Table 5). Because genera are relatively restricted in scope, many series of successive genera have been published. However, families and higher-rank higher taxa are so broad in concept that they are not usually used to construct transitional-fossil sequences, although occasionally they are (Bulman, 1970, p. V103–104; Easton, 1960, p. 436; Flower & Kummel, 1950, p. 607).

Table 5. Examples of successive higher taxa (genera).

Coniferophytes: Florin, 1951; Scagel *et al.*, 1965, p. 491–492, 520–522, 596–597.
Foraminiferans: Dunbar, 1963, p. 42; Pokorny, 1963, p. 155, 192.
Corals: Wells, 1956, p. F364.
Brachiopods: Dunbar & Rodgers, 1957, p. 280; Shrock & Twenhofel, 1953, p. 346.
Nautiloids: Teichert, 1964a, p. K200–201; Teichert, 1964b, p. K325.
Ammonoids: Miller, Furnish, & Schindewolf, 1957, p. L23.
Coleoids: Easton, 1960, p. 476; Weller, 1969, p. 233.
Blastoids: Fay, 1967, p. S394–395; Tappan, 1971, p. 1087.
Crinoids: Moore, Lalicker, & Fischer, 1952, p. 631.
Echinoids: Kier, 1965; Tappan, 1971, p. 1088.
Graptolites: Moore, Lalicker, & Fischer, 1952, p. 726.
Fish-Tetrapod (Crossopterygian-Amphibian) Transition: Colbert, 1969, p. 71–78; Romer, 1966, p. 72–74, 86–88, 90; Romer, 1968, p. 71–72.
Amphibian-Reptile Transition: Colbert, 1969, p. 111–114; Romer, 1966, p. 94–96, 102–103; Romer, 1968, p. 86–87, 96.
Reptiles: Colbert, 1948, p. 153; Colbert, 1965, p. 170–171; Romer, 1968, p. 131, 137, 138.
Reptile-Mammal Transition: Beerbower, 1968, p. 477–480; Colbert, 1969, p. 130–144, 250, 254; Cuffey, 1971a, p. 159; Olson, 1965, p. 40–44, 193–209; Olson, 1971, p. 671–731; Romer, 1966, p. 173–174, 178, 186; Romer, 1968, p. 159, 163–164.
Mammals: Colbert, 1969, p. 368–369, 454, 457; Dunbar and Waage, 1969, p. 464; Lull, 1908, p. 180; Lull, 1940, p. 569, 615; McGrew, 1937, p. 448; Osborn, 1929, p. 759, 831; Scott, 1937, p. 335, 476; Stirton, 1959, p. 48; Thomson, 1925, p. 60.

Finally, in some fossil groups, our knowledge is quite fragmentary and sparse. We then may know of particular fossils which are strikingly intermediate between two relatively high-rank higher taxa, but which are not yet connected to either by a more continuous series of successive species or transitional individuals. We can refer to these as *isolated intermediates,* a fourth type of situation involving transitional fossils, a type which represents our least-complete state of knowledge.

Isolated intermediates include some of the most famous and spectacular transitional fossils known, such as *Archaeopteryx* (Colbert, 1969, p. 186–189; Romer, 1966, p. 166–167). This form is almost exactly intermediate between the classes Reptilia and Aves (Cuffey, 1971a, p. 159; Cuffey, 1972, p. 36), so much so that "the question of whether *Archaeopteryx* is a bird or a reptile is unimportant. Both viewpoints can be defended with equal justification" (Brouwer, 1967, p. 161). The fossil onychophorans (Moore, 1959, p. O19; Olson, 1965, p. 190) and the fossil monoplacophorans (Knight & Yochelson, 1960, p. I77–83; Raup & Stanley, 1971, pp. 308–309) have been regarded as annelid-arthropod and annelid-mollusk inter-phylum intermediates, respectively. Moreover, although invertebrate phylum origins tend to be obscure for several reasons (Olson, 1965, p. 209–211), recently discovered, Late Precambrian, soft-bodied invertebrate fossils may well alter that situation, particularly after certain peculiar forms are studied and compared with Early Cambrian forms (Kay & Colbert, 1965, p. 99, 103; Weller, 1969, p. 247).

Mention of this last prompts me to point out parenthetically that the appearance of shelled invertebrates at the beginning of the Cambrian has been widely misunderstood. The assertion is frequently made that all the major types of animals appeared suddenly and in abundance then. In actual fact, collecting in successive strata representing continuous sedimentation from Late Precambrian into Early Cambrian time reveals a progressive increase upward in abundance of individuals. Moreover, the various higher taxa—particularly the various classes and orders reflecting adaptation to different modes of life—appear at different times spread over the long interval between the Early Cambrian and the Middle Ordovician.

Finally, because of widespread interest in questions of man's origins, it is well worth emphasizing that a rather complete series of transitional fossils links modern man continuously and gradationally back to mid-Cenozoic, generalized pongids (see references in Table 2).

> In spite of statements to the contrary . . . , the fossil record of the Hominoidea, the superfamily containing man and the apes, is quite well known, and it is therefore possible to outline a tentative evolutionary scheme for this group (Uzzell & Pilbeam, 1971, p. 615).

Potential Complications of the Paleontologic Literature

Non-paleontologist readers examining examples of transitional fossils mentioned above should be aware of several common occurrences within the professional paleontologic literature which could conceivably be confusing.

Historically, continued paleontologic research on any particular fossil group tends to move our understanding of its fossil record from the least-complete to the most-complete type of transitional-fossil situation. For example, early paleontologists recognized that the goniatite ammonoids gave rise to the ceratite ammonoids (successive higher taxa, in this case superorders or infraclasses; Easton, 1960, p. 436); later work indicated the successive species by which this transition was accomplished (Easton 1960, p. 446; Miller, Furnish, & Schindewolf, 1957, p. L22). Other examples can also be cited (Simpson, 1953, p. 361–364; Cuffey, 1967, p. 38–39). Also, our ideas about particular lineages may sometimes change as more specimens are brought to light (Stenzel, 1971, p. N1068–1070, 1077).

Frequently, secondary references portray evolutionary lineages much more vividly than does the original paper reporting them. For instance, contrast the original presentation of one coral sequence (Carruthers, 1910, p. 529, 538) with several later presentations (Easton, 1960, p. 175; Moore, Lalicker, & Fischer, 1952, p. 140; Weller, 1969, p. 123).

Sequences of transitional individuals or successive species are often, especially for teaching purposes, presented instead as more generalized sequences of successive genera. One ammonite lineage including transitional individuals between families (Spath, 1938; Arkell, Kummel, and Wright, 1957, p. L113–116) appears elsewhere as merely successive genera (Olson, 1965, p. 105–107). The various successive species of the horse lineages (Simpson, 1951, p. 114–121, 217–228, 282) are often summarized as successive genera (Hanson, 1961, p. 50–51; Scott, 1937, p. 417).

Similarly, for instructional purposes, some authors illustrate a series of fossils which show a progression in morphology, but which are not chronologically successive. These therefore are not evolutionary sequences, even though they resemble such. Two examples of such morphological series involve foraminiferans (Pokorny, 1963, p. 312) and nautiloids (Easton, 1960, p. 426).

In many instances, transitional individuals exist but are not reported explicitly as evolutionary lineages, for several reasons. Fully documenting such complete sequences is rather expensive in both research effort and publication cost; thus, many remain unpublished (Berry & Boucot, 1970, p. 30–31). Moreover, the practicing paleontologist sees little need to repeatedly reprove well-established concepts, especially when his primary concern is with other matters such as biostratigraphic dating (Berry, 1960, p. 9).

Effect of Transitional Fossils on Taxonomic Practises

Still further, because the Linnean system of taxonomic nomenclature has been very useful historically, we tend to refer transitional individuals to that species which they resemble most, rather than calling attention nomenclaturally to their intermediate status (Bird, 1971; Crusafont-Pairo & Reguant, 1970). As a result, a casual reader might conclude erroneously that we see no evolutionary variations within species. However, the true situation is that paleontologists frequently ignore such variation because it is not pertinent to their immediate goals (Williams, 1953, p. 29), but that such variation is present as transitional individuals within the species (Anderson, 1971; Cuffey, 1967, p. 41, 85–86; Klapper & Ziegler, 1967; Scott & Collinson, 1959; Williams, 1951, p. 87).

Similarly, we also tend to refer transitional fossils to that higher taxon which they most resemble or to which their final representatives belong. Consequently, the fact that we are dealing with continuously gradational sequences may be obscured by our conventional practise of superimposing artificially discontinuous, higher-rank taxonomic boundaries across such lineages (Olson, 1965, p. 100–101, 202–203; Van Morkhoven, 1962, p. 105, 153; Williams, 1953, p. 29; Cuffey, 1967, p. 38–39). As a result, for example, in the middle of sequences of transitional fossils bridging the conceptual gaps between the various vertebrate classes, we find forms which sit squarely on the dividing line between these high-rank taxa and which can be referred to either of two. In addition to *Archaeopteryx* between reptiles and birds (discussed previously), we can also note *Diarthrognathus* between reptiles and mammals, the seymouriamorphs between amphibians and reptiles, and *Elpistostege* between fishes and amphibians (see references in Table 5).

Higher taxa—from genera on up through phyla—are useful concepts in handling data concerning organisms (in fact, they constitute what the layman terms "major kinds" of organisms); however, they are artificial mental constructs rather than "basic facts of nature" (Brouwer, 1967, p. 161; Olson, 1965, p. 100–101, 201–203). Moreover, although there are reasons why transitional sequences between higher taxa are not as frequent as we would like (Brouwer, 1967, p. 160–169; Olson, 1965, p. 118, 184–211; Simpson, 1953, p. 366–376; Simpson, 1960, p. 159–161), nevertheless we can cite some particularly impressive transitional fossils between higher taxa of various ranks. In addition to those mentioned previously as inter-phylum and inter-class transitions, others involve higher taxa of class-group rank (Erben, 1966; Raup & Stanley, 1971, p. 306–307), orders (Easton, 1960, p. 446; Miller, Furnish, & Schindewolf, 1957, p. L22; Teichert, 1964, p. K325), families (Arkell, Kummel, & Wright, 1957, p. L117–119; Brinkmann, 1937; Easton, 1960, p. 425; Flower, 1941, p. 526; Moore, Lalicker, & Fischer, 1952, p. 351), and genera (Arkell, Kummel, & Wright, 1957, p. L116–118; Brinkmann, 1929; Brouwer, 1967, p. 158; Gimbrede, 1962; Newell, 1942, p. 21, 59; Raup & Stanley, 1971, p. 264).

Evolutionary Implications of Transitional Fossils

Let us consider the implications of an observable sequence of transitional fossils, such as those examples cited above, linking an earlier form (A, in Figure 1) with a later form (I). At a preliminary stage of knowledge, when only the relatively distinct forms A and I are known, it could be thought (as was actually done in the early 1800's) that the earlier form (A) had been instantly created, lived for a time, was then eliminated by some catastrophic environmental event, and after extinction was replaced by special creation of the somewhat similar later form (I). As our knowledge of the paleontologic record begins to increase, we find a third form (such as E, in Figure 1) which is morphologically and chronologically

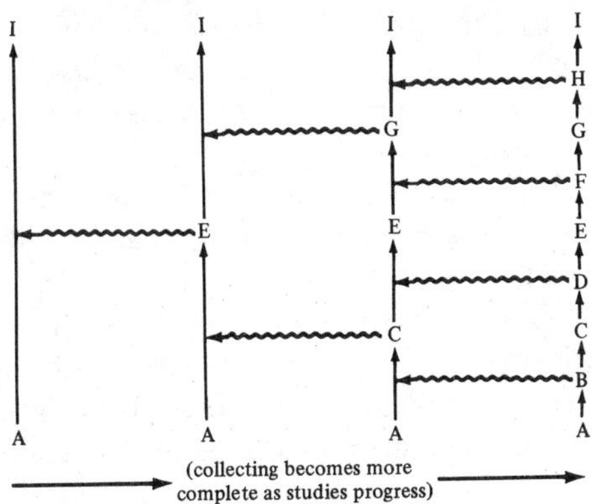

Fig. 1. The implications of an observable sequence of transitional fossils, linking an earlier form (A) to a later form (I).

intermediate between A and I. The gap between A and I is thus partly filled and replaced by two narrower gaps, and we must invoke an additional special creation and catastrophic extinction to explain the observed record. Continued collecting uncovers more morphologically and chronologically intermediate specimens (say C and G, and later also B, D, F, and H, in Figure 1); at each step, the new gaps we produce by partly filling existing ones are progressively smaller, and we must invoke ever more instantaneous creations and catastrophic extinctions. It is evident that, when we have accumulated a very large series of transitional fossils grading continuously from A to I (as we often now have in the course of population-oriented paleontologic studies), we must envision a very large number of creations and catastrophes—approaching, in fact, the probable number of reproductive generations involved in the sequence, allowing for the vagaries of the processes of fossilization and study. Invoking progressively more special creations until each generation is interpreted as the result of special creation becomes clearly implausible. Instead, noting that many fossils preserve ordinary reproductive structures, and also that the differences between successive fossil assemblages are of magnitude comparable to those observable between consecutive ancestor-descendent populations in nature today, we are forced to conclude that the entire series represents a chain of reproductive generations, descending one from the other by the usual natural reproductive processes, uninterrupted by any special creative acts from without.

As emphasized above, transitional fossils are known between groups of organisms classified at both low and high taxonomic ranks; i.e., between both low- and high-rank taxa.

Low-rank taxa—the many species known to us—have a real existence in nature, in that they consist of populations or morphologically similar, actually or potentially interbreeding individuals which live during a continuous segment of geologic time. Transitional fossils between morphologically distinct, chronologically successive species require us thus to conclude that a new species results from the operation of natural reproductive processes upon successive generations of a population without the intervention of special creative acts; i.e., through what the scientist terms "evolutionary processes."

On the other hand, higher taxa—those above species-rank, from genera up through phyla—do not have a real existence in nature in quite the same sense that species do. Instead, higher taxa of various ranks are simply the scientist's mental abstractions by which the many species comprising the organic world are grouped according to the various degrees of over-all morphologic similarity displayed. Species which are very similar may be grouped within one genus, while species which have only a little in common may be grouped together only in the same class or phylum. Since higher taxa are no more than aggregations of species, transitional fossils between higher taxa indicate simply that, in time, the same natural ancestor-descendent process producing new species eventually produces a chain of successive and progressively more different species, whose final member will be drastically different in morphology from its initial member and will therefore be classified by taxonomists in a different high-rank taxon. Consequently, the practice has developed among modern taxonomists that higher-rank classifications, which are based initially upon observable degrees of morphologic similarity among species, also should reflect evolutionary ancestor-descendent relationships among those species as much as possible. Moreover, it also is apparent that the amount of morphologic change producible by evolutionary processes is essentially unlimited, given the context of vast eons of geologic time.

As a still broader implication of these considerations, we can define "evolution" as the gradual and permanent change in the form and function of adult living organisms, of successive generations, over a long period of geologic time. Paleontologic evidence (discussed here) has played the critical role in developing this concept, but numerous other lines of evidence also suggest it. The interested reader can explore these in other excellent sources (especially Lull, 1940; Olson, 1965; Simpson, 1953), where he also can learn that the process termed "natural selection"—far from being carelessly equated to evolution as some anti-evolutionists assert—is an important part of the *method* by which evolution is accomplished. Moreover, the range in taxonomic ranks over which transitional fossils are observed (as described above) shows that what some anti-evolutionists label "general" and "special" evolution are merely extreme end-members in the scale of a single natural phenomenon, evolution, and thus usually do not warrant separate consideration.

As defined above, evolution is a scientific (rather than, say, philosophical) concept, and so comments about the nature of science are relevant here.

Using actual practice as the basis for definition, we can define "science" simply as the attempt to understand natural phenomena more completely by means of repeatable or verifiable observations of natural phenomena. (This is broader than the rigid, prediction- or experiment-oriented definitions developed by some philosophers not actively engaged in scientific work.) Also, unlike mathematics or logic, science does not deal in formally rigorous certainties, but instead strives for conclusions which are at best highly probable. Failure to understand this has made extensive, philosophically based discussions—by anti-evolutionists, among others—irrelevant. Moreover, while the search for ultimate or first causes moves into the realm of metaphysics, discussion of possible proximate or intermediate causes which might be implied by observational evidence clearly falls well within the scope of science.

Still further, we need to realize that there is no fundamental difference between what has been termed "historical science" and "empirical science." The scholar can be relatively certain of only what he is experiencing at the present moment, not of what the objects he is examining imply to him about the past. This is as true for the chemist reading his notebook describing yesterday's experiments and for the historian examining ancient Egyptian records, as it is for the paleontologist viewing the fossils and rock strata which form the pages of a natural textbook. None of these three can be rigorously certain that their world was not instantaneously created minutes ago with all its evidences of apparently longer history (Olson, 1965, p. 49); however, for each, his scholarly interpretations about events before the present moment are much more probable than would be purely conjectural imaginings.

Paleontologists studying sequences of transitional fossils are clearly operating in a scientific manner, because their data can be regenerated by anyone willing to examine the earth's crust independently. As more and more such sequences come to light, considering the processes which formed them becomes essential if we are to understand nature more thoroughly (i.e., still within the scope of science). As discussed above, interpreting these sequences as proximately due to evolutionary processes becomes ever more probable (in fact, overwhelmingly so, agree all who have been directly involved with the evidence), while a *fiat*-creationist interpretation becomes ever less likely. Because of the long time spans involved, we will never be rigorously certain that our view is a wholly accurate reflection of natural reality, but the many transitional fossils known render evolution already so highly probable that presentation of it as scientific fact is quite justified. Finally, as is generally true in the development of science, once a concept has been well documented, it can in turn provide a basis for further work; the concept of evolution has done just this most fruitfully for many areas within the earth and life sciences over the past years.

A few remarks are also appropriate about the theological implications of evolution as demonstrated by sequences of transitional fossils. As the reader may have noted, theological considerations do not enter at all into our demonstration of evolution as a very highly probable scientific conclusion. Consequently, like other scientific conclusions, this one cannot be viewed as inherently either pro- or anti-Christian. However, of course, Christians—especially theologians—will need to integrate evolutionary process into their views as being the proximate means which God uses to create various forms of life, just as He uses other scientifically demonstrable processes to maintain the natural universe.

Conclusion

In summary, the paleontologic record displays numerous sequences of transitional fossils, oriented appropriately within the independently derivable geochronologic time framework, and morphologically and chronologically connecting earlier species with later species (often so different that the end-members are classified in different high-rank taxa). These sequences quite overwhelmingly support an evolutionary, rather than a *fiat*-creationist, view of the history of life. Consequently, after carefully considering the implications of the fossil record, we must conclude that that record represents the remains of gradually and continuously evolving, ancestor-descendent lineages, uninterrupted by special creative acts, and producing successive different species which eventually become so divergent from the initial form that they constitute new major kinds of organisms.

REFERENCES

Anderson, E. J., 1971, Discriminant function analysis of variation among populations of the brachiopod *Gypidula coeymanensis:* Geol. Soc. Amer., Abs. Prog., v. 3, no. 1, p. 14–15.

Andrews, H. N., Jr., 1961, Studies in Paleobotany; Wiley, New York; 487 p.

Arkell, W. J., Kummel, B., & Wright, C. W., 1957, Mesozoic Ammonoidea: p. L80–L465 of Moore, R. C., ed., Treatise on Invertebrate Paleontology, pt. L (Mollusca 4, Ammonoidea), p. L1–L490.

Banner, F. T., & Blow, W. H., 1959, The classification and stratigraphical distribution of the Globigerinaceae: Palaeontology, v. 2, p. 1–27.

Barnard, T., 1963, Evolution in certain biocharacters of selected Jurassic Lagenidae: p. 79–92 of von Koenigswald, G. H. R., ed., Evolutionary Trends in Foraminifera; Elsevier, Amsterdam; 355 p.

Barrande, J., 1887, Système Silurien du Centre de la Bohème—Recherches Paléontologiques; Praha; v. 7 (Echinodermes), pt. 1 (Cystidées), 233 p.

Beaver, H. H., 1967, Morphology: p. S300–S344 of Moore, R. C., ed., Treatise on Invertebrate Paleontology, pt. S (Echinodermata 1), v. 2 (Blastoids), p. S297–S650.

Beerbower, J. R., 1968, Search for the Past—An Introduction to Paleontology, 2nd ed.; Prentice-Hall, Englewood Cliffs; 512 p.

Berggren, W. A., 1962, Stratigraphic and taxonomic-phylogenetic studies of Upper Cretaceous and Paleogene planktonic Foraminifera: Stockh. Contr. Geol., v. 9, p. 107–129.

Berry, W. B. N., 1960, Graptolite faunas of the Marathon region, west Texas: Univ. Tex. Bur. Econ. Geol., Pub. 6005, p. 1–179.

Berry, W. B. N., and Boucot, A. J., 1970, Correlation of the North American Silurian rocks: Geol. Soc. Amer., Spec. Pap. 102, p. 1–289.

Bird, S. O., 1971, On interpolative open nomenclature: Syst. Zool., v. 20, p. 469.

Boreske, J. R. A., Jr., 1972, Taxonomy and taphonomy of the North American amiid fishes (abs.): Geol. Soc. Amer., Abs. Prog., v. 4, no. 1, p. 3–4.

Boucot, A. J., & Ehlers, G. M., 1963, Two new genera of stricklandid brachiopods: Univ. Mich. Mus. Paleont. Contr., v. 18, p. 47–66.

Brinkmann, R., 1929, Statistischbiostratigraphische Untersuchungen an mitteljurassischen Ammoniten über Artbegriff und Stammesentwicklung: Gesell. Wiss. Göttingen, Abh., math.-phys. Kl., n. ser., v. 13, no. 3, p. 1–249.

Brinkmann, R., 1937, Biostratigraphie des Leymeriellenstammes nebst Bemerkungen zur Paläogeographie des Nord-westdeutschen Alb: Geol. Staatsinst. Hamburg, Mitt., v. 16, p. 1–18.

Bronnimann, P.., 1950, The genus *Hantkenina* Cushman in Trinidad and Barbados, B. W. I.: Jour. Paleont., v. 24, p. 397–420.

Brouwer, A., 1967, General Paleontology; Univ. Chicago Press, Chicago; 216 p.

Brown, R. W., 1943, Some prehistoric trees of the United States: Jour. Forestry, v. 41, p. 861–868.

Bulman, O. M. B., 1970, Graptolithina, 2nd ed.: p. V1–V163 of Teichert, C., ed., Treatise on Invertebrate Paleontology, 2nd ed., pt. V, p. V1–V163.

Carruthers, R. G., 1910, On the evolution of *Zaphrentis delanouei* in Lower Carboniferous times: Geol. Soc. Lond., Quart. Jour., v. 66, p. 523–538.

Chandler, M. E. J., 1923, Geological history of the genus *Stratiotes:* Geol. Soc. Lond., Quart. Jour., v. 79, p. 117–138.

Chaney, R. W., 1949, Evolutionary trends in the angiosperms: p. 190–201 of Jepsen, G. L., Simpson, G. G., & Mayr, E., eds., Genetics, Paleontology and Evolution; Princeton Univ. Press, Princeton; 474 p.

Charles, R. P., 1949, Essai d'étude phylogénique des gryphées liasiques: Soc. Geol. France, Bull., ser. 5, v. 19, pt. 1–3, p. 31–41.

Charles, R. P., & Maubeuge, P.-L., 1952, Les liogryphées du jurassique inférieur de l'est du bassin parisien: Soc. Geol. France, Bull., ser. 6, v. 1, pt. 4–6, p. 333–350.

Charles, R. P. & Maubeuge, P.-L., 1953a, Les liogryphées jurassiques de l'est du bassin parisien, II, Liogryphées du Bajocien: Soc. Geol. France, Bull., ser. 6, v. 2, pt. 4–6, p. 191–195.

Charles, R. P., & Maubeuge, P.-L., 1953b, Révision des liogryphées du Musée d'Histoire Naturelle de Luxembourg: Inst. Grand-Ducal de Luxemb., sec. sci. nat. phys. math., Arch., n. ser., v. 20, p. 183–186.

Cita-Sironi, M. B., 1963, Tendances évolutives des foraminifères planctiques (Globotruncanae) du Crétacé supérieur; p. 112–138 of von Koenigswald, G. H. R., ed., Evolutionary Trends in Foraminifera; Elsevier, Amsterdam; 355 p.

Clark, D. L., 1968, Fossils, Paleontology, and Evolution; Brown, Dubuque; 130 p.

Cobban, W. A., 1951, Scaphitoid cephalopods of the Colorado Group: U. S. Geol. Surv., Prof. Pap. 239, p. 1–42.

Cobban, W. A., 1958, Late Cretaceous fossil zones of the Powder River Basin, Wyoming and Montana: Wyo. Geol. Assoc., 13th Ann. Fld. Conf. Gdbk., p. 114–119.

Cobban, W. A., 1961, The ammonite family Binneyitidae Reeside in the western interior of the United States: Jour. Paleont., v. 35, p. 737–758.

Cobban, W. A., 1962a, *Baculites* from the lower part of the Pierre Shale and equivalent rocks in the western interior: Jour. Paleont., v. 36, p. 704–718.

Cobban, W. A., 1962b, New *Baculites* from the Bearpaw Shale and equivalent rocks of the western interior: Jour. Paleont., v. 36, p. 126–135.

Cobban, W. A., 1964, The Late Cretaceous cephalopod *Haresiceras* Reeside and its possible origin: U.S. Geol. Surv., Prof. Pap. 454–I, p. I1–I21.

Cobban, W. A., 1969, The Late Cretaceous ammonites *Scaphites leei* Reeside and *Scaphites hippocrepis* (DeKay) in the Western Interior of the United States: U.S. Geol. Surv., Prof. Pap. 619, p. 1–29.

Cobban, W. A., & Reeside, J. B., Jr., 1952, Correlation of the Cretaceous formations of the western interior of the United States: Geol. Soc. Amer., Bull., v. 63, p. 1011–1044.

Cocke, J. M., 1970, Dissepimental rugose corals of Upper Pennsylvanian (Missourian) rocks of Kansas: Univ. Kan. Paleont. Contr., art. 54, p. 1–67.

Colbert, E. H., 1948, Evolution of the horned dinosaurs: Evolution, v. 2, p. 145–163.

Colbert, E. H., 1965, The Age of Reptiles; Norton, New York; 228 p.

Colbert, E. H., 1969, Evolution of the Vertebrates, 2nd ed.; Wiley, New York; 535 p.

Coon, C. S., 1962, The Origin of Races; Knopf, New York; 724 p.

Crusafont-Pairo, M., & Reguant, S., 1970, The nomenclature of intermediate forms: Syst. Zool., v. 19, p. 254–257.

Cuffey, R. J., 1967, Bryozoan *Tabulipora carbonaria* in Wreford Megacyclothem (Lower Permian) of Kansas: Univ. Kan. Paleont. Contrib., Bryoz. art. 1, p. 1–96.

Cuffey, R. J., 1970, Critique of "The Dying of the Giants": Jour. Amer. Sci. Affil., v. 22, p. 93–96.

Cuffey, R. J., 1971a, Evidence for evolution from the fossil record: Jour. Amer. Sci. Affil., v. 23, p. 158–159.

Cuffey, R. J., 1971b, Transitional fossils well known: Jour. Amer. Sci. Affil., v. 23, p. 38.

Cuffey, R. J., 1972, More on *Archaeopteryx:* Jour. Amer. Sci. Affil., v. 24, p. 36.

Dechaseaux, C., 1934, Principales espèces de Liogryphées liasiques, valeur stratigraphique et remarques sur quelques formes mutantes: Soc. Geol. France, Bull., ser. 5, v. 4. no. 1–3, p. 201–212.

Dunbar, C. O., 1963, Trends of evolution in American fusulines: p. 25–44 of von Koenigswald, G. H. R., ed., Evolutionary Trends in Foraminifera; Elsevier, Amsterdam; 355 p.

Dunbar, C. O., & Rodgers, J. W., 1957, Principles of Stratigraphy; Wiley, New York; 356 p.

Dunbar, C. O., & Waage, K. M., 1969, Historical Geology, 3rd ed.; Wiley, New York; 556 p.

Durham, J. W., 1971, The fossil record and the origin of the Deuterostomata: N. Amer. Paleont. Conv., Proc., pt. H, p. 1104–1132.

Easton, W. H., 1960, Invertebrate Paleontology; Harper, New York; 701 p.

Elias, M. K., 1937, Stratigraphic significance of some Late Paleozoic fenestrate bryozoans: Jour. Paleont., v. 11, p. 306–334.

Elias, M. K., 1942, Tertiary prairie grasses and other herbs from the High Plains: Geol. Soc. Amer., Spec. Pap. 41, p. 1–176.

Erben, H. K., 1966, Über den Ursprung der Ammonoidea: Biol. Rev., v. 41, p. 641–658.

Fay, R. O., 1967, Phylogeny and Evolution: p. S392–S396 of Moore, R. C., ed., Treatise on Invertebrate Paleontology, pt. S (Eichinodermata 1), v. 2 (Blastoids), p. S297–S650.

Fisher, W. L., Rodda, P. U., & Dietrich, J. W., 1964, Evolution of *Athleta petrosa* stock (Eocene, Gastropoda) of Texas: Univ. Tex. Bur. Econ. Geol., Pub. 6413, p. 1–117.

Florin, R., 1951, Evolution in cordaites and conifers: Acta Hort. Bergiani, v. 15, p. 285–388.

Flower, R. H., 1941, Development of the Mixochoanites: Jour. Paleont., v. 15, p. 523–548.

Flower, R. H., & Kummel, B., Jr., 1950, A classification of Nautiloidea: Jour. Paleont., v. 24, p. 604–616.

Franz, V., 1932, *Viviparus;* Morphometrie, Phylogenie und Geographie der europäischen, fossilen und rezenten Paludinen: Med.-Naturw. Ges. Jena, Denkschr., v. 18.

Franz, V., 1943, Die Geschichte der Tiere: p. 219–296 of Heberer, G., ed., Die Evolution der Organismen; Fischer, Jena; 774 p.

Gartner, S., Jr., 1971, Phylogenetic lineages in the Lower Tertiary coccolith genus *Chiasmolithus:* N. Amer. Paleont. Conv., Proc., pt. G, p. 930–957.

Gimbrede, L. de A., 1962, Evolution of the Cretaceous foraminifer *Kyphopyxa christneri* (Carsey): Jour. Paleont., v. 36, p. 1121–1123.

Glaessner, M. F., 1960, The fossil decapod Crustacea of New Zealand and the evolution of the order Decapoda: N.Z. Geol. Surv., Paleont. Bull., v. 31, p. 1–63.

Glaessner, M. F., 1969, Decapoda: p. R399–R533 of Moore, R. C., ed., Treatise on Invertebrate Paleontology, pt. R (Athropoda 4), v. 2, p. R399–R651.

Grant, R. E., 1962, Trilobite distribution, upper Franconia Formation (Upper Cambrian), southeastern Minnesota: Jour. Paleont., v. 36, p. 965–998.

Greiner, H., 1957, "*Spirifer disjunctus*"—its evolution and paleoecology in the Catskill Delta: Yale Univ. Peabody Mus. Nat. Hist., Bull., v. 11, p. 1–75.

Guber, A. L., 1971, Problems of sexual dimorphism, population structure and taxonomy of the Ordovician genus *Tetradella* (Ostracoda): Jour. Paleont. v. 45, p. 6–22.

Hall, C. A., Jr., 1962, Evolution of the echinoid genus *Astrodapsis:* Univ. Cal. Pub. Geol. Sci., v. 40, p. 47–180.

Hanson, E. D., 1961, Animal Diversity; Prentice-Hall, Englewood Cliffs; 116 p.

Heaslip, W. G., 1968, Cenozoic evolution of the alticostate venericards in Gulf and East Coastal North America: Palaeontogr. Amer., v. 6, p. 55–135.

Hottinger, L., 1963, Les alvéolines paléogènes, exemple d'un genre polyphylétique: p. 298–314 of von Koenigswald, G. H. R., ed., Evolutionary Trends in Foraminifera; Elsevier, Amsterdam; 355 p.

House, M. R., 1970, On the origin of the clymenid ammonoids: Palaeont., v. 13, p. 664–676.

Howells, W., 1967, Mankind in the Making, rev. ed.; Doubleday, Garden City; 384 p.

Imlay, R. W., 1959, Succession and speciation of the pelecypod *Aucella:* U.S. Geol. Surv., Prof. Pap. 314-G, p. 155–169.

Jackson, R. T., 1912, Phylogeny of the Echini, with a revision of Paleozoic species: Boston Soc. Nat. Hist., Mem., v. 7, p. 1–491.

Jones, D. J., 1956. Introduction to Microfossils; Harper, New York; 406 p.

Kauffman, E. G., 1965, Middle and late Turonian oysters of the *Lopha lugubris* group: Smithson. Misc. Coll., v. 148, no. 6, p. 1–92.

Kauffman, E. G., 1967, Cretaceous *Thyasira* from the western interior of North America: Smithson. Misc. Coll., v. 152, no. 1, p. 1–159.

Kauffman, E. G., 1969, Form, function, and evolution: p. N129–N205 of Moore, R. C., ed., Treatise on Invertebrate Paleontology, pt. N (Mollusca 6, Bivalvia), v. 1, p. N1–N489.

Kauffman, E. G., 1970, Population systematics, radiometrics and zonation—a new biostratigraphy: N. Amer. Paleont. Conv., Proc., pt. F, p. 612–666.

Kaufmann, R., 1933, Variations-statistische Untersuchungen über die "Artabwandlung" und "Artumbildung" an der oberkambrischen Trilobitengattung *Olenus* Dalm: Geol. Pal. Inst. Univ. Greifswald, Abh., v. 10, p. 1–54.

Kaufmann, R., 1935, Exakt-statistische Biostratigraphie der *Olenus*—Arten von Sudöland: Geol. Foren. Stockholm Förhandl., v. 1935, p. 19–28.

Kay, M., & Colbert, E. H., 1965, Stratigraphy and Life History; Wiley, New York: 736 p.

Kermack, K. A., 1954, A biometrical study of *Micraster coranguinum* and *M.* (*Isomicraster*) *senonensis:* Roy. Soc. Lond., Philos. Trans., ser. B, v. 237, p. 375–428.

Kier, P. M., 1965, Evolutionary trends in Paleozoic echinoids: Jour. Paleont., v. 39, p. 436–465.

Klapper, G., & Ziegler, W., 1967, Evolutionary development of the *Icriodus latericrescens* group (Conodonta) in the Devonian of Europe and North America: Palaeontographica, ser. A., v. 127, p. 68–83.

Knight, J. B., & Yochelson, E. L., 1960, Monoplacophora: p. I77–I84 of Moore, R. C., ed., Treatise on Invertebrate Paleontology, p. I (Mollusca 1), p. I1–I351.

Krumbein, W. C., & Sloss, L. L., 1963, Stratigraphy and Sedimentation; Freeman, San Francisco; 660 p.

Kummel, B., 1970, History of the Earth, 2nd ed.; Freeman, San Francisco; 707 p.

Lang, W. D., 1921–1922, Catalogue of the Fossil Bryozoa (Polyzoa)—The Cretaceous Bryozoa (Polyzoa); British Museum (Natural History), London; v. 3 and 4.

Le Gros Clark, W. E., 1964, The Fossil Evidence for Human Evolution, 2nd ed.; Univ. Chicago Press, Chicago; 201 p.

Lerman, A., 1965, Evolution of *Exogyra* in the Late Cretaceous of the southeastern United States: Jour. Paleont., v. 39, p. 414–435.

Lewontin, R. C., 1971, The yahoos ride again: Evolution, v. 25, p. 442.

Lull, R. S., 1908, The evolution of the elephant: Amer. Jour. Sci., ser. 4, v. 25, p. 169–212.

Lull, R. S., 1940, Organic Evolution, rev. ed.; Macmillan, New York, 743 p.

MacNeil, F. S., 1965, Evolution of the genus *Mya,* and Tertiary migrations of Mollusca: U.S. Geol. Surv., Prof. Pap. 483-G, p. G1-G51.

Matthew, W. D., 1910, The phylogeny of the Felidae: Amer. Mus. Nat. Hist., Bull., v. 28, p. 289–316.

McGrew, P. O., 1937, The genus *Cynarctus:* Jour. Paleont., v. 11, p. 444–449.

Miller, A. K., Furnish, W. M., & Schindewolf, O. H., 1957, Paleozoic Ammonodoidea: p. L11–L79 of Moore, R. C., ed., Treatise on Invertebrate Paleontology, pt. L (Mollusca 4, Ammonoidea), p. L1–L490.

Moore, J. N., 1970a, Evolution—required or optional in a science course?: Jour. Amer. Sci. Affil., v. 22, p. 82–87.

Moore, J. N., 1970b, Should Evolution Be Taught?; privately published, East Lansing; 28 p.

Moore, J. N., 1971a, On chromosomes, mutations, and phylogeny: Amer. Assoc. Adv. Sci., 138th Ann. Mtg., paper, 16 p. (mimeogr.).

Moore, J. N., 1971b, Retrieval system problems with articles in *Evolution:* Amer. Inst. Biol. Sci., 22nd Ann. Mtg., Paper 279, 13 p. (mimeogr.).

Moore, J. N., & Slusher, H. S., 1971, Biology, A Search for Order in Complexity; Zondervan, Grand Rapids.

Moore, R. C., 1959, Protarthropoda: p. O16–O20 of Moore, R. C., ed., Treatise on Invertebrate Paleontology, pt. O (Arthropoda 1), p. O1–O560.

Moore, R. C., Lalicker, C. G., & Fischer, A. G., 1952, Invertebrate Fossils; McGraw-Hill, New York; 766 p.

Morris, H. M., 1963. The Twilight of Evolution; Baker, Grand Rapids; 103 p.

Nelson, R. S., & Semken, H. A. 1970, Paleoecological and stratigraphic significance of the muskrat in Pleistocene deposits: Geol. Soc. Amer., Bull., v. 81, p. 3733–3738.

Newell, N. D., 1937, Late Paleozoic Pelecypods–Pectinacea: Kan. Geol. Surv., (Publ.) v. 10, pt. 1, p. 1–123.

Newell, N. D., 1942, Late Paleozoic Pelecypods–Mytilacea: Kan. Geol. Surv., (Publ.) v. 10, pt. 2, p. 1–115.

Nichols, D., 1959a, Changes in the Chalk heart-urchin *Micraster* interpreted in relation to living forms: Roy. Soc. Lond., Philos. Trans., ser. B, v. 242, p. 347–437.

Nichols, D., 1959b, Mode of life and taxonomy in irregular sea-urchins: Syst. Assoc., v. 3, p. 61–80.

Olson, E. C., 1965, The Evolution of Life; Mentor, New York; 302 p.

Olson, E. C., 1971, Vertebrate Paleozoology; Wiley-Interscience, New York; 839 p.

Osborn, H. F., 1929, The titanotheres of ancient Wyoming, Dakota, and Nebraska: U.S. Geol. Surv., Mon. 55, p. 1–953.

Papp, A., 1963. Über die Entwicklung von Heterosteginen: p. 350–355 of von Koenigswald, G. H. R., ed., Evolutionary Trends in Foraminifera; Elsevier, Amsterdam; 355 p.

Patterson, B., 1949, Rates of evolution in taeniodonts: p. 243–278 of Jepsen, G. L., Simpson, G. G., & Mayr, E., eds., Genetics, Paleontology and Evolution; Princeton Univ. Press, Princeton; 474 p.

Pokorny, V., (Transl. Allen, K. A.), 1963, Principles of Zoological Micropalaeontology; Macmillan, New York; 652 p.

Raup, D. M., & Stanley, S. M., 1971, Principles of Paleontology; Freeman, San Francisco; 388 p.

Rauzer-Chernousova, D. M., 1963, Einige Fragen zur Evolution der Fusulinideen: p. 45–65 of von Koenigswald, G. H. R., ed., Evolutionary Trends in Foraminifera; Elsevier, Amsterdam; 355 p.

Rexroad, C. B., 1958, The conodont homeomorphs *Taphrognathus* and *Streptognathodus:* Jour. Paleont., v. 32, p. 1158–1159.

Romer, A. S., 1966, Vertebrate Paleontology, 3rd ed., Univ. Chicago Press, Chicago; 468 p.

Romer, A. S., 1968, Notes and Comments on Vertebrate Paleontology; Univ. Chicago Press, Chicago; 304 p.

Ross, C. A., & Ross, J. P., 1962, Pennsylvanian, Permian rugose corals, Glass Mountains, Texas: Jour. Paleont., v. 36, p. 1163–1188.

Rowe, A. W., 1899, An analysis of the genus *Micraster,* as determined by rigid zonal collecting from the zone of *Rhynchonella cuvieri* to that of *Micraster cor-anguinum:* Geol. Soc. Lond., Quart. Jour., v. 55, p. 494–547.

Scagel, R. F., *et al.*, 1965, An Evolutionary Survey of the Plant Kingdom; Wadsworth, Belmont; 658 p.

Schäfle, L., 1929, Ueber Lias und Doggeraüstern: Geol. u. Palaeont. Abh., n. ser., v. 17, no. 2, p. 1–88.

Schaub, H., 1963, Über einige Entwicklungsreihen von *Nummulities* und *Assilina* und ihre stratigraphische Bedeutung: p. 282–297 of von Koenigswald, G. H. R., ed., Evolutionary Trends in Foraminifera; Elsevier, Amsterdam; 355 p.

Scott, A. J., & Collinson, C., 1959, Intraspecific variability in conodonts—*Palmatolepis glabra* Ulrich & Bassler: Jour. Paleont., v. 33, p. 550–565.

Scott, W. B., 1937, A History of Land Mammals in the Western Hemisphere, rev. ed., American Philosophical Society (repr. Hafner, New York) ; 786 p.

Seward, A. C., 1938, The story of the maindenhair tree: Sci. Progr., v. 32, p. 420–440.

Shrock, R. R., & Twenhofel, W. H., 1953, Principles of Invertebrate Paleontology, 2nd ed.; McGraw-Hill, New York; 816 p.

Simpson, G. G., 1951, Horses; Oxford Univ. Press, Oxford; 323 p.

Simpson, G. G., 1953, The Major Features of Evolution; Columbia Univ. Press, New York; 434 p.

Sohl, N. F., 1960, Archeogastropoda, Mesogastropoda, and stratigraphy of the Ripley, Owl Creek, and Prairie Bluff Formations: U.S. Geol. Surv., Prof. Pap. 331-A, p. 1–151.

Sohl, N. F., 1967, Upper Cretaceous gastropods from the Pierre Shale at Red Bird, Wyoming: U.S. Geol. Surv., Prof. Pap. 393-B, B1–B46.

Sohn, I. G., 1962, Stratigraphic significance of the Paleozoic ostracode genus *Coryellina* Bradfield, 1935: Jour. Paleont., v. 36, p. 1201–1213.

Spath, L. F., 1938, A Catalogue of the Ammonites of the Liassic Family Liparoceratidae; British Museum (Natural History), London; 191 p.

Stebbins, G. L., Jr., 1949, Rates of evolution in plants: p. 229–242 of Jepsen, G. L., Simpson, G. G., & Mayr, E., eds., Genetics, Paleontology and Evolution; Princeton Univ. Press, Princeton; 474 p.

Stenzel, H. B., 1949, Successional speciation in paleontology—the case of the oysters of the sellaeformis stock: Evolution, v. 3, p. 34–50.

Stenzel, H. B., 1971, Oysters: p. N953–N1214 of Moore, R. C., ed., Treatise on Invertebrate Paleontology, pt. N (Bivalvia), v. 3 (oysters), p. N953–N1224.

Stirton, R. A., 1959, Time, Life, and Man—The Fossil Record; Wiley, New York; 558 p.

Swartz, F. M., 1945, Zonal Ostracoda of the Lower Devonian in New York and Pennsylvania (abs.): Geol. Soc. Amer., Bull., v. 56, p. 1204–1205.

Tappan, H., 1971, Microplankton, ecological succession and evolution: N. Amer. Paleont. Conv., Proc., pt. H, p. 1058–1103.

Tedford, R. H., 1970, Principles and practices of mammalian geochronology in North America: N. Amer. Paleont. Conv., Proc., pt. F, p. 666–703.

Teichert, C., 1964a, Actinoceratoidea: p. K190–K216 of Moore, R. C., ed., Treatise on Invertebrate Paleontology, pt. K (Mollusca 3, Nautiloidea), p. K1–K519.

Teichert, C., 1964b, Nautiloidea-Discosorida: p. K320–K342 of Moore, R. C., ed., Treatise on Invertebrate Paleontology, pt. K (Mollusca 3, Nautiloidea), p. K1–K519.

Teilhard de Chardin, P., 1950, Sur un cas remarquable d'orthogénèse de groupe—l'évolution des siphnéidés de Chine: Colloq. Internat. Centre Nat. Rech. Sci., v. 21, p. 169–173.

Thomson, J. A., 1925, Concerning Evolution; Yale Univ. Press, New Haven; 245 p.

Trevisan, L., 1949, Lineamenti dell'evoluzione del ceppo di elefanti eurasiatici nel Quarternario: La Ricerca Scientifica, v. 19 (suppl.), p. 105–111.

Uzzell, T., & Pilbeam, D., 1971, Phyletic divergence dates of hominoid primates—a comparison of fossil and molecular data: Evolution, v. 25, p. 615–635.

Van de Fliert, J. R., 1969, Fundamentalism and the fundamentals of geology: Jour. Amer. Sci. Affil., v. 21, p. 69–81.

Van Morkhoven, F. P. C. M., 1962, Post-Paleozoic Ostracoda; Elsevier, Amsterdam; 204 p.

Waller, T. R., 1969, The evolution of the *Argopecten gibbus* stock (Mollusca: Bivalvia), with emphasis on the Tertiary and Quaternary species of eastern North America: Paleont. Soc. Mem. 3, p. 1–125.

Watson, D. M. S., 1949, The evidence afforded by fossil vertebrates on the nature of evolution: p. 45–63 of Jepsen, G. L., Simpson, G. G., & Mayr, E., eds., Genetics, Paleontology and Evolution; Princeton Univ. Press, Princeton; 474 p.

Weller, J. M., 1969, The Course of Evolution; McGraw-Hill, New York; 696 p.

Wells, J. W., 1956, Scleractinia: p. F328–F444 of Moore, R. C., ed.,

Treatise on Invertebrate Paleontology, pt. F (Coelenterata), p. F1–F498.

Wenger, R., 1957, Die germanischen Ceratiten: Palaeontogr., ser. A, v. 108, p. 57–129.

Wilde, G. L., 1971, Phylogeny of *Pseudofusulinella* and its bearing on Early Permian stratigraphy: Smithsonian Contr. Paleobiol., no. 3, p. 363–379.

Williams, A., 1951, Llandovery brachiopods from Wales with special reference to the Llandovery district: Geol. Soc. Lond., Quart. Jour., v. 107, p. 85–136.

Williams, A., 1953, North American and European stropheodontids—their morphology and systematics: Geol. Soc. Amer., Mem. 56, p. 1–67.

Wood, H. E., II, 1949, Evolutionary rates and trends in rhinoceroses: p. 185–189 of Jepsen, G. L., Simpson, G. G., & Mayr, E., eds., Genetics, Paleontology and Evolution; Princeton Univ. Press, Princeton; 474 p.

Woodland, R. B., 1958, Stratigraphic significance of Mississippian endothyroid Foraminifera in central Utah: Jour. Paleont., v. 32, p. 791–814.

Wright, C. W., & Wright, E .V., 1949, The Cretaceous ammonite genera *Discohoplites* and *Hyphoplites* Spath: Geol. Soc. Lond., Quart. Jour., v. 104, p. 477–497.

Zeller, E. J., 1950, Stratigraphic significance of Mississippian endothyroid Foraminifera: Univ. Kan. Paleont. Contr., Protoz., art. 4, p. 1–23.

Zeuner, F., 1933, Die Lebensweise der Gryphäen: Palaeobiologica, v. 5, p. 307–320.

Ziegler, A. M., 1966, The Silurian brachiopod *Eocoelia hemisphaerica* (J. de C. Sowerby) and related species: Palaeontology, v. 9, p. 523–543.

ROY A. GALLANT

TO HELL WITH EVOLUTION

Was the Arkansas creationism law, Act 590, in violation of the First Amendment of the U.S. Constitution, which provides for the separation of church and state, as charged by the American Civil Liberties Union, May 27, 1980? Clearly it was, ruled Judge William Overton on January 5, twelve days after the close of what has been called the "Scopes-2 Monkey Trial," which took place in Little Rock and lasted from the 14th to the 24th of December, 1981. Overton declared that Act 590 "was simply and purely an effort to introduce the Biblical version of creation into the public school curricula."

On the eve of that historic decision the American Association for the Advancement of Science, convening in Washington, adopted a resolution branding legislation requiring the teaching of "creation science" in the public schools "a real and present threat to the integrity of education and the teaching of science."

How did Arkansas's legislature manage to push Act 590 through the legal process? It was easy. The Arkansas House reportedly had held fewer than 30 minutes of hearings on Act 590, and the Senate none at all. Governor Frank White conceded that he signed the Act before reading it.

Seven months later, Federal District Judge Adrian Duplantier granted a request for summary judgment to plaintiffs challenging Louisiana's creation-science law. The judge ruled that the First Amendment to the Constitution prohibits the government from dictating not only what subjects be taught but also how they will be taught. The Louisiana law, passed in 1981, would have required that schools teaching evolution must also teach "creation science"—which postulated that the universe and life within it were created "from nothing," that man and apes have separate ancestry, and that mutation alone cannot account for the variety of life forms on Earth.

Had this law not been struck down, the plaintiffs (including the Louisiana State Board of Elementary and Secondary Education and the American Civil Liberties Union) were prepared to argue that creation science as described in the law was a thinly veiled disguise for the fundamentalist Christian view of creation derived from the Bible. As such, they charge that teaching creationism in public schools would violate the separation of church and state guaranteed by the First Amendment.

While both the Louisiana and Arkansas decisions were cheered by the scientific community, it did little more than dampen the spirits of that sect of creationists to whom the idea of evolution is anathema. Evolution is that biological process of descent that has resulted in present-day species having risen from those of the past. Traditionally, the fundamentalists have singled out evolution as being responsible for periodic declines in traditional social values. According to the 1970-founded Creation-Science Research Center (CSRC) affiliated with the Christian Heritage College in San Diego, California, its research proved that evolution "fostered the moral decay of spiritual values, which contributes to the destruction of mental health and . . . [the prevalence of] divorce, abortion, and rampant venereal diseases."

This group of fundamentalists also abhors evolution because they see it as an attack on the Bible and their inflexible belief in the inerrancy of the Scriptures. Evolution is a lie, they maintain, it is nothing more than a religion based on atheism. They want evolution thrown out of the public school science curriculum or, failing that, they want divine creation to be given equal time in the biology classroom.

In 1976, two California housewives, Nell Segraves and Jean Sunrall, both associates of a California organization called The Bible-Science Association, learned of the U.S. Supreme Court's ruling in a case protecting atheist students from required prayers in public schools. It was the Madalyn Murray case. Murray's success in protecting her child "from religious exposure suggested to Segraves that parents like herself 'were entitled to protect our children from the influence of beliefs that would be offensive to our religious beliefs,'" according to a report in the magazine *Science* (5 Nov. 1982).

Battered by creationist demands for equal time over the years, in 1976 California's State Board of Education notified the state's teachers that whenever human origins were discussed in the context of evolution, "alternative theories" should be presented. Such alternative theories—whatever they might be—presumably could include the biblical account of creation in Genesis. In August 1980 President Ronald Reagan left no doubts about where he stands on the issue. During a Dallas, Texas, press conference when he was asked if evolution should be taught in our public schools, he replied, "If it is going to be taught, then I think that also the Biblical theory of creation, which is not a theory but the Biblical story of creation, should also be taught."

I wonder to what extent Reagan's attitude is being written into governmental department policy guidelines. When the Minnesota Association for Improvement of Science Education recently applied to the Internal Revenue Service for tax-exempt status, certain of its members were disturbed by the implications of the questions put to it by the IRS. As a sample of its literature distributed to the public, the Association, on request, submitted a copy of a brochure, "The Place of Evolution in Science Education." Of the 16 questions posed by the IRS, eight related to creationism. Among them: "What do you consider to be the pseudo-scientific versions of the origins of life on Earth? What gives you the standing or the prerogative to deem certain versions of the origin of life on Earth as pseudo-scientific? Why are you opposed to permitting the granting of equal time in school curricula to the teaching of the theory of creationism?"

In Georgia an equal-time bill was passed in the Senate but failed in the House, largely because the state's Department of Education had vigorously argued that curriculum details are the business of local school boards, not the state. But a new attempt to introduce a "Balanced Treatment for Creation-Science and Evolution-Science Act" is expected. Atlanta creationist lobbyist Judge Braswell Dean, who brands evolution an "animal fairy tale," accused Georgia's Department of Education of defending "the monkey mythology of Darwin." According to the National Association of Biology Teachers (NABT), biblical creationist lobbyists in nearly twenty other states have introduced, or are now introducing, legislation designed to make "scientific creationism" a required, or optional, subject in the biology curriculum. Among them: Illinois, New York, Iowa, Minnesota, Virginia, Ohio, Tennessee, Florida, Arkansas, and Louisiana.

EXCERPTS FROM ARKANSAS'S ACT 590

SECTION 1. Public schools within this State shall give balanced treatment to creation science and to evolution science. Balanced treatment to these two models shall be given in classroom lectures taken as a whole for each course.

SECTION 2. Treatment of either evolution science or creation science shall be limited to scientific evidences for each model and inferences from those scientific evidences, and must not include any religious instruction or reference to religious writings.

SECTION 4. (a) "Creation science" means the scientific evidences for creation and inferences from those scientific evidences. Creation science includes . . . evidences and related inferences that indicate: (1) sudden creation of the universe, energy, and life from nothing; (2) changes only within fixed limits of originally created kinds of plants and animals; (4) separate ancestry for man and apes; (5) explanation of the earth's geology by catastrophism, including the occurrence of a worldwide flood; and (6) a relatively recent inception of the earth and living kinds.

A revised draft creationist bill is presently circulating in legislatures throughout the country. "The new draft bill is very tight indeed," says Paul Ellwanger, the bill's architect. Its title has been changed to the "Unbiased Presentation of Creation-Science and Evolution-Science Bill." The words "from nothing" have been deleted from Section 4 (a) (1), as has reference to a worldwide flood. Also creationist dogma for a relatively young Earth has been considerably toned down. Ellwanger is head of an Anderson, South Carolina group called Citizens for Fairness in Education. "My group is not affiliated with any political or religious organization," he says.

Creationists are having much more success on local than on state levels. In areas including Dallas, Atlanta, and Chicago, school committees have granted the creationists equal time in science classrooms. Schools have been authorized to buy creationist-biased books and pamphlets, sometimes for required reading, other times for optional supplementary use. In Mississippi and Indiana, creationist text materials have made the approved list of books that can be selected by local school boards and paid for by the state.

Wherever an equal-time controversy flourishes, the students tend to be the losers. In 1979 in Cobb County, Georgia, biology courses were omitted from high school graduation requirements in an attempt to avoid an equal-time ruling passed by the local school board. "Where will we be if any pressure group can win, by legislative fiat, the ordered inclusion of its favorite doctrine into school curricula?" asks Harvard's Stephen Jay Gould, renowned paleobiologist. "Every conceivable position has some advocates," he adds.

It has been 58 years since the celebrated "Monkey Trial" jury found the young biology teacher John T. Scopes guilty of violating Tennessee's law forbidding the teaching of evolution. But once again the anti-evolutionists are abroad in the land. No longer do they openly advocate teaching the biblical Genesis account of the origin of life as an alternative to evolution, since it is clearly in violation of the separation of church and state provision in the U.S. Constitution. Try as they did in the Arkansas trial, the creationists consistently failed to prove their claim that so-called "creation science" is science. Through their own admissions under cross-examination, they revealed creation science as nothing more than religion incognito, and a very narrow sect of religion at that.

Here is what creation scientist Wayne Friar, a zoologist at The King's College, Briarcliff Manor, New York, had to say when cross-examined by Bruce Ennis, a counsel for the plaintiff. Friar had been asked to read from his book *The Case for Creation* a passage describing a separate ancestry for humans and apes, based solely on the scriptures.

> Ennis: You believe that the choice between evolution and creation is a matter of faith, don't you?
>
> Friar: There's certainly an element of faith in it.
>
> Ennis: Do you recall in your deposition my asking you the following question and your giving the following answer?

Ennis: You believe the choice between evolution and creation is a matter of faith, don't you?

Friar: Basically, yes.

Ennis: No further questions.

The creationists are serving up their old biblical wine in new nonsectarian bottles labeled "scientific creationism"—a semantic fallacy claiming that scientific data favor the notion that a divine creator relatively recently fashioned Earth and its varied living forms. It is that view they want pitted as a "viable alternative model" to evolution through natural selection. Evolution holds that over hundreds of millions of years living forms on Earth have originated by natural descent from simpler life forms, which themselves arose from prebiological aggregates of molecules some 3.5 billion years ago. The party-line tenet of the creationists is that scientific creationism is no more religious or less scientific than evolution. They at once attempt to reduce evolution to a religious belief and raise creationism to the level of science. They then claim that if the teaching of one violates the U.S. Constitution then so must the teaching of the other.

Today in more than half the states textbook selection committees are being pressured by creationists and their supporters to adopt books that teach divine creation, not as theology, but as science. Elementary and high school biology textbooks are this minute being changed despite creationist losses such as the Arkansas and Louisiana cases. Further, new texts, and magazines designed for the elementary school market, are being written and edited with the assurance that the word *evolution,* or any suggestion of evolution as an indisputable biological force, are being omitted.

By the early 1960s the creationist movement had achieved an impressive reincarnation of strength—notably in California and Texas. A major coup was winning control of California's State Board of Education in the early sixties. From 1966 on, creationism and the California Board of Education were almost synonymous. In 1969 the board revised the "Science Framework for California Public Schools" in such a way as to distort its scientific content.

In 1971 John Ford was vice-president of California's Curriculum Commission to advise on the choice of textbooks. Ford is an M.D., a Seventh-Day Adventist, and an avid creationist. He fired off a memo to the other 15 members of the commission reminding them that "no textbook should be considered for adoption . . . that has not clearly discussed at least two major contrasting theories of origin." What biology textbooks would California adopt? The state has about one million school children and buys 10 percent of the nation's textbooks. Any publisher winning a California adoption makes a bundle, and all are eager to have their elementary and high school science series accepted in California.

In June of 1972 Vernon Grose, a member of the Board of Education Curriculum Commission, was responsible for discussing with textbook publishers how they planned to include divine creation in light of the then new guideline. Grose is a Los Angeles aerospace engineer who belongs to the Assemblies of God, a Pentecostal denomination. He is a "commission sitter," having served on at least 14 state commissions, turns out numerous archconservative essays on the decline of morality in America, and is an enthusiastic creationist. In reply to his challenge, one publisher eagerly offered his fourth-grade science text, pointing out that it contained an "investigation" of the biblical account of creation. Another suggested deleting Richard Leakey's discoveries about early man and substituting a reproduction of Michelangelo's Sistine Chapel painting of the creation, and possibly tossing in a bonus drawing of Moses. This book subsequently appeared in two editions, one for California, with Michelangelo, and another for the rest of the U.S., with Leakey. In December, to further meet their new Framework, the board—all appointees of former Governor Reagan—took it upon themselves to edit some evolution out of and some creationism into science books headed for California's elementary classrooms. The following year one of the nation's largest and most highly respected publishers of elementary school textbooks issued the following directive to the authors, editors, and consultants of its new science series then in preparation: "Change the word 'evolution' to 'Biological Change' in the student texts, grades 1–8, where it is a heading and/or where it appears in indexes and glossaries, and throughout the story line of the text." If the creationists get their way, a whole generation of public school children will grow up ignorant of evolution, except as distorted and condemned by creationist dogma.

Those in the scientific community who cheer the Arkansas and

Louisiana decisions and bask in the deceptive suggestion that biology teachers have been let off the hook should be a bit more realistic. The anti-evolution movement flexes its muscles not in the courts, where it has every right to feel uncomfortable, but around the purse strings of the commercial press that feeds the nation's schools its millions of textbooks and supporting periodicals.

How can one organization—and the creationists are extremely well organized—wield enough power to make America the laughingstock of the educated world by turning the clock back on both biology and geology more than a century to the pre-Darwinian era? Says Wayne A. Moyer, executive director of NABT, "I'm convinced that if we fail to confront this issue squarely and publicly, we will have an American equivalent of the Lysenko affair." (Trofim Lysenko was the Russian biologist responsible for the Communist Party's 1948 outlawing of all work in Mendelian genetics on the supposition that genes do not exist! The affair virtually killed genetics research in Russia until only a few years ago; Lysenkoism also had disastrous effects on Soviet agriculture.

Numerous creationists are members or supporters of the Institute for Creation Research (ICR), a tax-free division of biblically-oriented Christian Heritage College. Says one of their spokesmen, Richard Bliss, who holds a doctoral degree in science education from the University of Sarasota, Florida: "I believe there's a ground swell moving across this country at the level of parents and teachers. I can cite hundreds of teachers using the two-model approach. It's catching on like wildfire, and when it happens down at the community level that's when it will work."

Earlier, in 1922, William Jennings Bryan, Presbyterian layman and Woodrow Wilson's Secretary of State, cheered a Kentucky effort to suppress the teaching of evolution in that state's public schools. His hopeful prediction, like that of Bliss: "The movement will sweep the country and we will drive Darwinism from our schools. . . . Commit your case to the people," Bryan urged. "Forget, if need be, the high-brows both in the political and college world, and carry this cause to the people. They are the final and efficiently corrective power.

Bliss supports his claim with "statistically sound surveys" taken in widely differing communities. He says that the surveys consistently show that only about 10 percent favor the teaching of evolution alone, 10 to 15 percent favor creation alone, and 75 to 80 percent favor a two-model approach. Polls conducted by *Time* magazine and others support those figures. The following excerpt is from a *New York Times* article of August 1982:

> The American public is almost evenly divided between those who believe that God created man in his present form at one time in the last 10,000 years and those who believe in evolution or an evolutionary process involving God, according to the Gallup Poll. George H. Gallup, Jr., said his organization had not previously polled Americans on the same questions regarding creation and so no comparisons could be made with beliefs in years past.
>
> The findings dismayed some prominent religious leaders, who said, among other things, that human existence on Earth is much older than 10,000 years, but the results came as no surprise to a leading anthropologist. Of the participants in the poll, 44 percent, nearly a quarter of whom were college graduates, said they accepted the statement that "God created man pretty much in his present form at one time within the last 10,000 years." Four statements were offered to respondents on a card and they were asked to select the one that came closest to describing their views "about the origin and development of man."
>
> Nine percent agreed with the statement: "Man has developed over millions of years from less advanced forms of life. God had no part in this process." Thirty-eight percent said they agreed with the suggestion that "man has developed over millions of years from less advanced forms of life, but God guided this process, including man's creation." Nine percent of those interviewed simply said they did not know.
>
> The views of Roman Catholics and Protestants were divergent, with Protestants more likely to believe in the biblical account of creation and Catholics more likely to believe in evolution guided by God. The Gallup organization reported that Southerners and Middle Westerners were slightly more likely to accept creationism than those living elsewhere.

In 1979 a public opinion poll published in *Christianity Today* reported that half the adults in America continue to believe that "God created Adam and Eve to start the human race."

What is this so-called "two-model" approach to teaching about the origins of life? Bliss travels extensively for ICR explaining the two-model approach to teacher groups and school administrators and exposing them to his student text, *Origins: Two Models—Evolution, Creation.* Since the two-model approach in this ICR text claims to examine scientific evidence to find out whether it better supports evolution or divine creation, teachers are encountering few objections to their use of the book as source material in science classes. Since the creationists know they will be in violation of the U.S. Constitution if they openly try to teach biblical creationism in public schools, they avoid mention of the Bible or Genesis and speak only of "scientific creationism."

Judge Overton came down hard on the two-model approach. In his decision, he wrote: "The two-model approach of the creationists is simply a contrived dualism which has no scientific factual basis or legitimate educational purpose. . . . Application of the two models . . . dictates that all scientific evidence which fails to support the theory of evolution is necessarily scientific evidence in support of creationism and is, therefore, creation science 'evidence'."

Overton also left no doubts about his attitude toward public opinion polls being used to alter school curricula:

> The application and content of First Amendment principles are not determined by public opinion polls or by a majority vote. Whether proponents of Act 590 constitute the majority or the minority is quite irrelevant under a constitutional system of government. No group, no matter how large or small, may use the organs of government, of which the public schools are the most conspicuous and influential, to foist its religious beliefs on others.

ICR scientists and educators—all with postgraduate degrees in science or education—turn out a prodigious number of technical monographs, filmstrips, cassette tapes, and books. Many of the books and pamphlets are written for children. The books presently number more than 40 and have been translated into a dozen languages. The most popular include *Scientific Creationism* and *Troubled Waters of Evolution,* both written by ICR Director Henry M. Morris, who holds a Ph.D. in hydraulic enginering from the University of Minnesota. He is past chairman of the Department of Civil Engineering at Virginia Polytechnic Institute, where his fundamentalist writings led to at least one unsuccessful attempt to get him ousted from the faculty. Another best-seller is *Evolution: The Fossils Say No!,* written by ICR Associate Director Duane T. Gish, a biochemist with a Ph.D. from the University of California, Berkeley, who has taught biochemistry at New York City's Cornell University Medical College. Gish's book has sold about 150,000 copies. ICR also broadcasts a weekly radio program, "Science, Scripture, and Salvation," over more than ninety stations in 35 states and overseas, and publishes a 12-page monthly newsletter, *Acts and Facts,* with a circulation of more than 75,000.

ICR has a sister institution, the Creation Research Society (CRS) of Ann Arbor, Michigan, of which Morris is past president. Both organizations claim a total of about 700 scientist-supporters, all with postgraduate degrees, and well over 100,000 direct financial supporters. Their general following must number hundreds of thousands more among the millions who make up the nation's membership in evangelical churches.

CRS requires its scientist-applicants to sign a form affirming that they believe in the following:

> 1. The Bible is the written word of God, and because we believe it to be inspired thruout, all of its assertions are historically and scientifically true in all the original autographs. To the students of nature, this means that the account of origins in Genesis is a factual presentation of simple historical truths.
> 2. All basic types of living things, including man, were made by direct creative acts of God during Creation Week as described in Genesis. Whatever biological changes have occurred since Creation have accomplished only changes within the original created kinds.
> 3. The great Flood described in Genesis, commonly referred to as the Noachian Deluge, was an historical event, worldwide in its extent and effect.
> 4. Finally, we are an organization of Christian men of science, who accept Jesus Christ as our Lord and Savior. The account of the special creation of Adam and Eve as one man and one woman, and their subsequent fall into sin, is the basis for our belief in the necessity of a Savior for all mankind. Therefore, salvation can come only thru accepting Jesus Christ as Savior.

Five of the State's witnesses defending Arkansas's Act 590 are members of CRS. The impossibility of such scientists conducting research objectively, rather than searching out data that support their biblically oriented hypothesis, was brought out when counsel Ennis cross-examined CRS scientist Harold Coffin, of the Geoscience Research Institute, Loma Linda University, California. The lack of scientific credibility of such scientists quickly became apparent:

> Ennis: You have had only two articles in standard scientific journals since getting your Ph.D. in 1955, haven't you?
> Coffin: That's correct.
> Ennis: The Burgess Shale is said to be 500 million years old, but you think it is only 5,000 years old, don't you?
> Coffin: Yes.
> Ennis: You say that because of information from the Scriptures, don't you?
> Coffin: Correct.
> Ennis: If you didn't have the Bible you could believe the age of Earth to be many millions of years, couldn't you?
> Coffin: Yes, without the Bible.
> Ennis: Creation science is not falsifiable, is it?
> Coffin: No, it is in the same category as evolution science.
> Ennis: No further questions.

Another creationist argument against evolution is that it is not falsifiable, as a body of knowledge must be in order to be considered a science. Counsel might have added that the discovery of a fossil mammal in rock strata 500 million years old would immediately falsify the principle of evolution.

When I first visited Christian Heritage College in 1980, its faculty members were looking forward to the day it has its planned scientific research center for its many scientists. Comments Preston Cloud, research biogeologist with the U.S. Geological Survey and professor emeritus of biogeology at the University of California, Santa Barbara, "Creation 'research' is a contradiction in terms. There's no research to be done if the task is complete, perfect, and fully described in the Bible." The expressions "creation research" and "scientific creationism" are at best muddled terminology that only reduces our ability to think rationally and leads to a hopelessly confused and confusing blend of religious beliefs and facts. The major task of education is to make distinctions. Says Gould, "[The expression] 'scientific creationism' is nonsense and an absurd misnomer."

Accusations that ICR is attempting to introduce religion into the classroom instill rage in Morris and Gish. Both brand such claims a lie. All they want is the opportunity to challenge evolution on purely scientific grounds, they say. To do so the creationists are continually on the lookout for ambiguities, conflicting scientific views, and any hint of uncertainty in papers published by evolutionary scientists. They are quick to broadcast any such conflicts they find and cite them as evidence against evolution. Says paleobiologist Niles Eldredge, of New York's American Museum of Natural History: "[The creationists] delight in finding dissenters, like myself, known to be dissatisfied with one or another aspect of current evolutionary science. They try to use internal disagreement among evolutionary biologists as evidence that somehow evolutionary biology isn't science after all. In doing so they mistake the nature of science." Science operates by offering hypotheses that can be tested and that always are vulnerable to being proven wrong. Eldredge teaches biology at the City University of New York and geology at Columbia University.

Dissenter Eldredge says that the fossil record is hard put to verify the evolutionary prediction that there should be gradual and progressive change leading from one species, or one genus, to the next. *Aha!* cry the creationists, *the evolutionists themselves admit that the fossil record contradicts evolution.* Not at all, Eldredge replies, pointing out that such smooth transitions do indeed exist. But he asks why we must suppose that evolutionary ancestry must demonstrate a gradual progressive pattern of improvement. Both Eldredge and Gould think it need not, and that the evolutionary sacred cow is in trouble, but not the general notion of evolution itself. "To progress, science needs conflicting views," Eldredge emphasizes. "The very process of science thrives on disagreement. Without it any science loses its vigor and dies."

In monographs and many debates with evolutionists the creationists cite the sudden appearance of numerous "basic kinds" of organisms in the fossil record of the early Cambrian period, some 600 million years ago. They use this sudden appearance of numerous and varied life forms as evidence to support divine creation and contradict evolution. The Cambrian period forms the base of that idealized geologic rock column extending from the present back to about 600 million years, back to the time when an explosive diversity of life indeed does appear suddenly in the fossil record. Where are the evolutionary ancestors of these organisms? the creationists demand to know. Not one can be found in the earlier Precambrian rock record going back some 3.8 billion years, they claim. Not so, answer the evolutionists. The Precambrian rock record contains many fossils. As early as 680 million years ago a limited variety of multicellular organisms had appeared. And fossils of primitive microorganisms have been dated back to 3.5 billion years. Then where are the transition forms linking basic kinds of life of the Precambrian with basic kinds of the Cambrian, and so on up to the present? the creationists again demand to know. They claim that the older rock strata that are supposed to contain the earlier forms of life should be followed by hundreds and thousands of transitional forms linking fishes with amphibians, and reptiles with birds, for example. New basic kinds should not appear suddenly in the fossil record, as they indisputably do, the creationists point out. "Our museums should be jam-packed with these transitional forms," argues Gish. "They should be coming out of our ears."

What Is a Basic Kind?

The term "basic kind," used as scientific terminology by the creationists, derives from Genesis. The term is virtually meaningless and never is found in scientific literature outside the creationist community. The creationists use it as a loose taxonomic category that can include a number of different species of the same genus—for instance, wolves, dogs, and coyotes. But the term's very looseness gives even the creationists problems in understanding what they are talking about. During the Arkansas trial, Ennis cross-examined CRS scientist Friar, who earlier had offered criticism of evolution by reading passages from texts published in 1929 and 1930. The question of "basic kinds" was the focus of this cross-examination:

> Ennis: How many originally created kinds were there?
> Friar: Let's say 10,000 plus or minus a few thousand.

Ennis: Some creationists believe kinds to be synonymous with species, some with genera, some with family, and some with order, don't they?

Friar: The scientists with whom I am working . . . well . . . it tends more toward the family. But it may go to order in some cases.

Ennis: You have been studying turtles for many years, haven't you?

Friar: Yes.

Ennis: Is a turtle an originally created kind?

Friar: I'm working on that.

Ennis: Are all turtles within the same created kind?

Friar: That's what I'm working on.

Like other creation scientists cross-examined, Friar was unable to define "kind" or explain how his research would ever lead to an explanation of this fuzzy concept.

Virtually all biologists scoff at Gish's extravagant claim that there are no transitional forms linking major groups of organisms, and further deny that such transitional forms should be coming out of our ears. Says Gould, "Transitional forms exist in abundance—especially forms linking reptiles and mammals, and reptiles and birds." *Archaeopteryx* has features both of birds (feathers and a birdlike pelvis) and reptiles (teeth). Gish maintains that *Archaeopteryx* was 100 percent bird since it had wings and flew; but in his logic he conveniently omits the fact that a good number of reptiles living at the same time as *Archaeopteryx* also had wings and flew. From his own work, Eldredge cites indisputable evidence of transitional forms—a series of four groups (genera) of trilobites some 350 million years old, "connected by a compelling array of intermediates."

"The creationists understand no more about the fossil record than they do about radioactive dating," says William V. Mayer, who teaches biology at the University of Colorado and is director of the Biological Sciences Curriculum Study (BSCS). In the early 1960s BSCS produced three high school biology textbooks that became widely adopted throughout the country. Mayer adds that "the wonder is not that there are gaps, but that the fossil record is as complete and compelling as it is. While there are gaps they are no more significant than occasional patches of unpaved road in a highway, the highway being sufficiently continuous to go where it is supposed to."

Dating Earth's Past with Atomic Clocks

Since the early 1900s geochemists have been using atomic clocks to date Earth's rocks. They also can date such organic materials as the charcoal remains of a prehistoric campfire, a mummy, or an ancient peat bed. In brief, an atomic clock works like this:

Through a process of radioactive decay, "parent" elements, such as uranium and potassium, gradually break down into the "daughter" elements lead and argon, respectively. Each radioactive element has its own known decay rate, ranging from microseconds to trillions of years. Since the rate of decay is constant under any circumstances, a given radioactive element incorporated into a newly crystallized mineral or rock can be used to date the rock. What the geochemist does is measure the ratio of parent product to daughter product. At the time of formation the ratio ideally is zero, then with time daughter product forms through radioactive decay and increases. The larger the amount of daughter product when the sample is measured, the older the sample is.

The geochemist makes two basic assumptions about a rock sample to be dated: (1) that there was no daughter product present when the rock was formed; and (2) that neither parent nor daughter product has been added to or removed since the rock was formed.

Gish maintains that there is no direct way of determining the age of any rock. While he admits that the presently used method of measuring the ratio of parent to daughter product in a given rock sample is accurate, he claims that the two basic assumptions are in error. Without knowing the original ratio when the rock was formed, he maintains, even the most accurate measurement of the ratio today can tell us nothing about the rock's age.

"Complete nonsense!" replies Gould. "We can know the original ratio of uranium to lead, for example, because dating can be done on crystals into which no lead atoms could be originally fitted when the crystals formed. Therefore, all lead now wedged into the crystals must be a product of radioactive decay."

Geochemists readily admit that in certain cases daughter product was present originally, but they have ways of reliably estimating the amount present and compensating for it. They also point out that the main source of inaccuracy in radioactive dating is loss or gain of parent or daughter product with time. But this possibility can be guarded against by cross-checking the results of one dating method with the results of other methods. If all are in good agreement then the age estimate is considered reliable.

The carbon-14 method of radioactive dating is used to measure the age of organic materials up to about 40,000 years old. It also relies on two basic assumptions, both of which Gish claims are in error. But in numerous instances the method has been cross-checked with objects of known historical dates—including wood, peat, manuscripts, clothes—and the method consistently has proven reliable.

Thousands of scientists have produced data about Earth's age, and the age of the Universe. Their motive has been a search for truth. If Earth's age were only about 10,000 years the data gathered by those thousands of scientists would be more likely to indicate that age than the much older age of about 4.5 billion years. As a result, the age of Earth accepted by virtually all scientists the world over, except the creationists, is the one based on radiometric dating.

The creationists and evolutionists alike agree that millions of years are needed for evolution to work. But the creationists continue to maintain that Earth is young. "Turn this argument around," challenges Mayer, "what evidence exists that Earth is *less* than 10,000 years old? To provide some validity to their young-Earth argument, the creationists have to disregard the fossil record, disregard all dating mechanisms, and dispute an entire cosmology. The creationists have an exceptionally hard time covering up, discrediting, or ignoring the wide variety of data that tell the age of Earth." Adds John W. Patterson, professor of materials science and engineering, Iowa State University: "The tragedy, in my view, is the unwillingness of enough qualified experts to brand the creationists for the incompetents they are—in biology, hydrology, paleontology, and many other areas, including biblical scholarship and exegesis."

Another argument of the creationists is that while gene alterations produce variations in a given species, they never permit one species to evolve into a new species. For instance, while the dog species *Canis familiaris* has the gene potential to produce variations such as collies, English setters, and Great Danes, dogs as a species do not have the gene potential to evolve into a new species, the creationists maintain. Says Gish: "No matter what combinations [of genes] may occur, the human kind always remains human, and the dog kind never ceases to be dog. The [genetic] transformations proposed by the theory of evolution never take place." Aerospace engineer and New York creationism lobbyist Luther Sunderland agrees. "A wing is a wing, a feather is a feather, an eyeball is an eyeball, a horse is a horse, and a man is a man," he reasons. He might have added that "a rose is a rose is a rose."

Some variations, say the creationists, were planned by God to enhance the survival of a basic kind. But other variations "are simply an expression of the Creator's desire to show as much beauty of flower, variety of song in birds, or interesting types of behavior in animals as possible," according to CRS's high school biology textbook *Biology: A Search for Order in Complexity*. The Dallas, Texas Independent School District approved this creation-biased text "as mandatory supplementary reading in high school biology courses."

Studies on DNA, the chemical substance of genes, provide molecular evidence for evolution that was unavailable only a few years ago. But creationists tend to ignore, or not believe in, DNA or in the fact that new genes are introduced into populations through a process called mutation. Mutations are known to be the source of variations, and it is variation that leads to evolution. Recent studies show that DNA molecules from closely related species resemble each other more closely than DNA from distantly related species. For instance, DNA from humans resembles that from apes and other primates more closely than DNA from dogs or cats. Molecular biologists interpret this fact to mean that humans resulted from mutations in the primate branch of the evolutionary tree.

The creationists invoke divine intervention in an attempt to explain away another biological fact—homologies as evidence of genetic relationships among different species. Homologies are structural similarities found among such major taxonomic groups as humans, birds, and horses, for instance, due to genetic (DNA) similarity. The creationists firmly deny that the homologous forelimbs shown in the diagram in any way provide evidence of genetic relationship among these animals. According to the CRS biology textbook, homologies were designed by God when "He used a single blueprint for the body plan but varied the plan so that each 'kind' would be perfectly equipped to take its place in the wonderful world He created for them." The evolutionists maintain there is only one logical interpretation of these homologous structures. They unequivocally demonstrate modifications of a common ancestral pattern, an anatomical pattern intelligible only through common inheritance, meaning evolution.

In any debate with ICR scientists, evolutionists can expect to hear the second law of thermodynamics cited in yet another attempt to discredit evolution. In essence the law says that an isolated natural system, such as a star, left to itself can go in one direction only—from a state of higher order of energy to one of lower order. In other words, it is destined to run down like a battery that is never recharged. And it can never go in the opposite direction without an input of energy from the outside. The Sun, as a star, is a closed energy system, and it eventually must run down and ultimately stop shining. ICR debaters pervert the second law of thermodynamics by treating Earth as a closed system. Clearly it is not. It receives energy from an outside source, the Sun, which drives Earth's many living and nonliving processes. Earth is a battery kept charged by the Sun. The distorted argument of the creationists is that evolution could not possibly have occurred on Earth because it requires a continuing process of increasing order—from simpler organisms and life processes of long ago to the more complex ones of the present. They further claim that no such energy source has been available, since, from the original divine creation of the Universe as a whole, the Universe has been running down.

"This silly argument about thermodynamics," says Gould, "shows that the creationists know as little physics as biology." Adds Mayer, "The creationists have deliberately distorted the laws of thermodynamics, either out of maliciousness or stupidity."

Periodically, ICR stages public debates with well-known evolutionary biologists, paleontologists, anthropologists, and geologists. The debates attract audiences that number about 30,000 a year. Until 1978 about 20 debates were held annually, most on university campuses. But, according to Morris, the number has declined

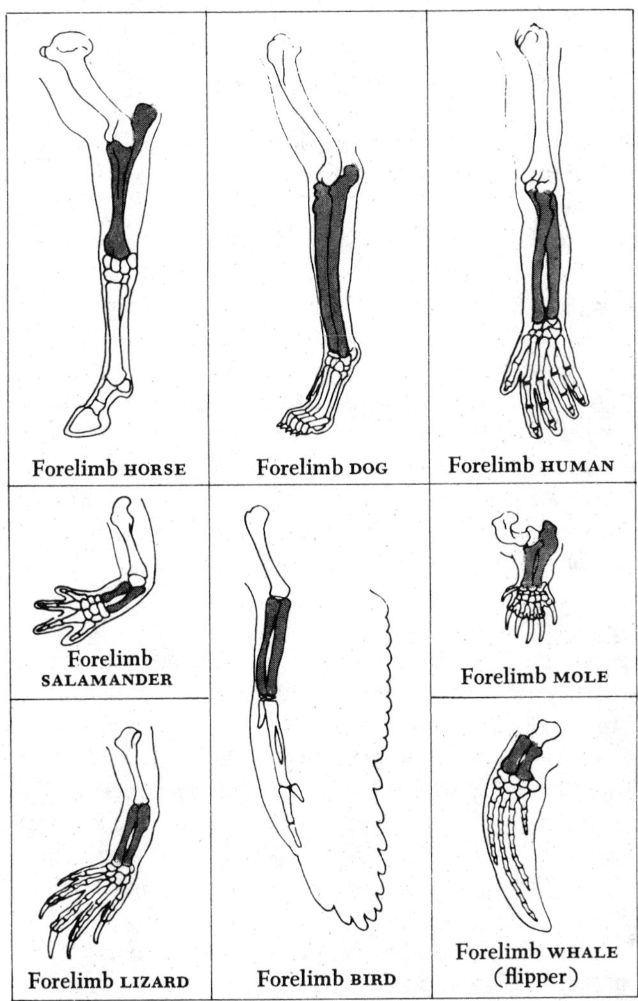

HOMOLOGIES

"The conclusion is inescapable that the limb bones of man, the bat, and the whale are modifications of a common ancestral pattern. The facts admit of no other logical interpretation . . . the forelimbs of all tetrapod vertebrates exhibit a unity of anatomical pattern intelligible only on the basis of common inheritance."—Peter Volpe, *Understanding Evolution* (William C. Brown Co.)

about 50 percent since then. Gish claims that the "evolutionists now realize that it's difficult to oppose a creationist on the basis of a scientific challenge." The bigger the name of the evolutionist, the better ICR debaters enjoy the show. In April, 1979 Gish debated world-renowned anthropologist Ashley Montagu in Princeton's gymnasium before a crowd of more than 2,000.

Afterward, Montagu said that he would never again engage in debate with any creationist. Why? "Gish had agreed to limit his discussion to the scientific 'facts.' Yet it was clear to everyone that what he was attempting was conversion to creationism—to prove his belief. This is neither honest nor scientific. A scientist is interested neither in proving nor disproving. What he is interested in is discovering what the facts are." Montagu added that he regretted not having had the opportunity to present technical arguments "that would have demolished Gish." He said that it was quite clear that "Gish had no understanding of the laws of thermodynamics and that he seemed to be quite unfamiliar with developments in modern physics, especially in thermodynamics. . . . My own answer to Gish was that the very process of evolution was an outstanding example of disorder leading to order, the maximization of the improbable, in that the vast majority of mutations are harmful, lethal, and yet out of the few that are not, the great variety of living forms have come into being."

University of California geneticist G. Ledyard Stebbins also says that he will never debate ICR people again, particularly Morris and Gish. One reason—they are highly skilled and practiced debaters, and Stebbins admits that he is not. Another is that the house usually is packed with fundamentalist supporters carrying Bibles and frequently shouting "Amen!" whenever ICR seems to score a point, an observation also made by Montagu. "They're brought in by the busloads," says Stebbins. He adds that his real revulsion to these public spectacles is that "they are not honest searches for truth, but tests of cleverness." Another evolutionist debater has said that an ICR team "has all of the persuasiveness of a snake-oil salesman, and a similar standard of honesty." Concludes Stebbins, "I'm convinced that these people are best answered not by filling the audience with a lot of facts that it doesn't understand, but by knowing Gish and company well enough to be ready for what they are going to say, and pointing out clearly and forcefully the obvious flaws in their tactics."

According to Mayer: "The debates I have had were not debates. The creationists come with a prepared script they present come hell or high water." That script invariably includes what ICR continues to regard as its most devastating arguments—irrelevant gaps in the fossil record, misinterpreted laws of thermodynamics, and erroneous statements about radiometric dating.

Creationism in the science classroom, on any level, is anathema to virtually all scientists. Says biologist Edward J. Kormondy, past president of NABT: "After reading ICR literature I must conclude that they are not only anti-evolution, but anti-science as well."

In science we propose a model and make predictions about how a star or a butterfly can be expected to behave on the basis of that model. Meanwhile we continue to make observations and examine the data we find. If the data support the model, we stick with the model, but as soon as the data begin to contradict the model then we either change the model accordingly or we may even have to throw it out and start all over again. Always we must emphasize the tentative nature of much of the knowledge of science, and its susceptibility to change. Now the creationists' model for the origin of life is the creation story in Genesis, an inflexible model. Like scientists, creationists examine the data to see if the data can support the model. But unlike scientists, if the data fail to support their model, instead of altering the model or throwing it out, which they can't do, the creationists throw out the data. They must be highly selective about what data they accept.

As creationists in virtually every state continue to lobby for ICR's two-model approach, biology teachers frequently find themselves in a bind: what to do if they are challenged for teaching evolution, or told to give creationism equal time in their science classrooms? To aid such teachers, the Iowa Academy of Science has set up a panel of members "willing to assist teachers who encounter difficulties because of teaching evolution." Teachers throughout the state have been informed of the panel's existence and given the names of its members. At least 42 states and two Canadian provinces have such committees. On the national level NABT has encouraged Committees of Correspondence on Evolution. According to Mayer, it is a national information network designed to provide biologists willing to speak on evolution, keep abreast of creationist activities, assist classroom teachers, and publish critiques and rebuttals of creationist writings.

The creationists are enjoying a new wave of publicity. Monthly they win new ground in local areas despite defeats on the state level. Creationist pressure groups have always been around, waiting for an opportunity to broadcast their fundamentalist religious views in public school classrooms. The last wave of creationist dogma swept the country in the 1920s when legislatures in 20 states heard proposals for laws requiring that the biblical account of creation be taught in secondary school biology courses. In response to that surge of creationist lobbying, an academic group called the American Association of University Professors was formed. In a position paper every bit as relevant today as it was some 50 years ago, the group reported that the fundamental issue really has nothing to do with whether evolution is true or false. "The real question is whether or not we wish to make an intellectual slave of every teacher in a state-supported institution and force him to square his teaching with the dogmas of any group which succeeds in getting legislative protection for its doctrines."

Sidney W. Fox, director of the University of Miami's Institute for Molecular and Cellular Evolution, points out that there are numerous citizens' groups organized to attack the teaching of evolution in the public schools. But until now there have been few, if any, groups organized to improve the teaching of evolution. Says Fox, "The way the principles of evolution are taught in the public schools—late introduction and often inadequate explanation—has been a serious national educational deficiency, and ICR can take some of the blame, although they regard it as credit." He adds that ICR also is partly responsible for the present uneven public understanding of evolution. Other biologists agree, saying that many high school biology teachers are poorly prepared in biology, especially in evolutionary biology, so concepts often are presented in an inflammatory way and, equally often, incorrectly.

How can we account for the explosive rebirth of the creationist movement? The reason that the creationists emerged from dormancy in the late 1960s, according to Mayer, "can be directly attributed to the reintroduction of the word 'evolution' [by BSCS] into high school materials." An accompanying opportunity was the wave of disenchantment with the "godless materialism of science," and what many regarded as its misuses, that swept into the 1970s and is still with us. The resulting intellectual vacuum was quickly filled by supernatural explanations of nature—astrology, alchemy, mystical morsels of Far Eastern religions, not to mention the pop cosmologies of Immanuel Velikovsky, Erich Von Daniken, and the UFO cultists. Gould points out that "creationism is just one manifestation of the right-wing revival now sweeping America." Also, there are demands for equality from minority groups and women, who feel left out of a society that promises equal opportunity for all. As a minority, the fundamentalists see bias in the science curriculum. They want creationism to have an equal voice along with evolution, and they want it where evolution is most effectively voiced—right up front in the science classroom.

ROBERT M. MAY

CREATION, EVOLUTION, AND HIGH SCHOOL TEXTS

Anyone who has seen Clarence Darrow—as played by Spencer Tracy—triumph over William Jennings Bryan in the powerful movie *Inherit The Wind* will be pardoned for thinking that the Scopes Trial helped establish evolution over creation in the United States in the 1920s. This impression is, indeed, fairly widely held, and is explicitly affirmed in at least one university-level biology text.

To the contrary, Grabiner and Miller[1] have shown that the pressures exerted by fundamentalists in the 1920s were successful in bringing about the systematic removal of references to evolution and Darwin in high school biology texts, particularly in the Southern States. Before the Scopes Trial in 1925, several of the widely used texts discussed evolution[1]: Hunter's *A Civic Biology* (American Book, 1914), from which Scopes himself taught, had a three page section on evolution together with other associated material elsewhere in the book; Gruenberg's *Elementary Biology* (Ginn, 1924) was outspokenly evolutionist, with several chapters devoted to the subject and classification discussed from an evolutionary standpoint; Moon's *Biology for Beginners* (Holt, 1921) had several chapters on evolution and a picture of Darwin as the frontispiece. Other successful texts were more reticent: Smallwood, Reveley and Bailey's *New Biology* (Allyn and Bacon, 1924), possibly the most widely adopted text of the time, gave about two pages to

From *Nature,* vol. 296, March 11, 1982. Reprinted by permission of *Nature.*

evolution and omitted any discussion of the origins of man; Clement's *Living Things* (Iroquois, 1924) gave evolution only brief mention; and Peabody and Hunt's *Biology and Human Welfare* (Macmillan, 1924) explicitly excluded evolution. Grabiner and Miller give a simple rule for identifying texts published in the decade following 1925: "Merely look up the word 'evolution' in the index or the glossary; you almost certainly will not find it." Concerned about the effect the Scopes Trial might have on sales of his text, Hunter rechristened the 1926 edition *New Civic Biology* and removed most of the explicit discussion of evolutionary ideas; the word "evolution" is no longer in the index. The 1926 edition of Moon's *Biology for Beginners* has the frontispiece portrait of Darwin replaced by a cartoon diagram of the digestive system, although the substantial treatment of evolution in the text is retained and "evolution" remains in the index (at last disappearing, however, in the 1933 edition). The treatment of evolution in Smallwood *et al.*'s *New General Biology* (1929) is even more perfunctory than in their earlier *New Biology,* and again the entry for "evolution" is removed from the index. The most widely used text in the 1930s was Baker and Mills' *Dynamic Biology* (Rand McNally, 1933), which discusses evolutionary ideas in the final chapter; the book[2], however, manages to avoid ever using the term 'evolution' and concludes with the extraordinary statement that "Darwin's theory, like that of Lamarck, is no longer generally accepted."

As emphasized by Simpson[3], it is not so much that the Scopes Trial precipitated these changes, but rather that both the trial and the changes were consequences of the upsurge of anti-intellectual conservatism and biblical literalism during the period in question. The essential point, however, is that evolutionary biologists were deluded in seeing the drama enacted in the Tennessee courthouse as a significant victory. The real battles were being fought over state and local decisions about which textbooks to adopt for high school biology courses, and here creationists advanced on a broad front throughout the 1920s and 1930s.

We would do well to keep these facts in mind today. Even as the community of professional biologists takes comfort from Judge Overton's incisive and unequivocal verdict against the state of

Arkansas' law mandating equal time for "creation science" in the biology classroom, there are signs that new editions of major high school biology texts are being eviscerated.

As stressed by Nelkin[4], a sociologist whose *Science Textbook Controversies and the Politics of Equal Time* (MIT Press, 1977) is the canonical book on this general subject, it is "too early to get a detailed reading on actual changes; they are just being implemented and publishers won't talk." But there are some clear signs, summarized recently by Zuidema[5]. The index to the 1973 edition of *Biology: Living Systems* (Charles Merrill) gave 17 lines of page references under the heading 'evolution'; in the 1979 edition, 'evolution' is indexed in 3 lines. Recent editions of three Harcourt Brace series texts omit all mention of Charles Darwin, and one excludes the index entry 'evolution' (though the subject is covered in the text). This is sad, as Simpson[3] has praised Harcourt Brace for "consistently and effectively oppos[ing] anti-evolutionist pressure" in the earlier period around 1930–1960. Another brief survey[6] shows the 1977 edition of an Otto and Towle high school biology text has diminished the coverage of evolution by one-third compared with the 1973 edition.

Taking a different tack, some texts now include material on the Genesis account of creation, or on creation myths generally[5,6]. Books in which special creation now appears include two editions of *Biology: An Inquiry Into the Nature of Life* published by Allyn and Bacon in 1974 and 1977, a 1974 Smallwood and Green text, and the 1980 Houghton Mifflin text *Biology: The Science of Life.* These books can be characterized as teaching about creationism, but not supporting it. More controversial is *Biology: A Search for Order in Complexity* (Zondervan), written by Moore who is a 'born again' professor of natural science at Michigan State University and a founder of the Creation Research Society. The book was chosen as one of seven officially approved biology texts by the Indiana State textbook commission in 1975, and in two of Indiana's districts was the only ninth-grade biology text available to students. Later, in 1977, the book was barred from use in public schools in Indiana as sectarian; it was relegated to library use as a reference work in Texas, but remains approved by state commissions in Alabama, Georgia, Oklahoma and Oregon[5].

Underlying much of this is the system whereby 19 of the 50 United States have prescribed selection systems under which textbooks are approved for adoption in schools. According to rules which vary from state to state among these 'adoption states,' one or several books may be approved for school use at a given level[7]. These texts may then be provided to schools, paid for with tax dollars; alternative books may be used but they must be bought by the individual schools or by individual students. These 19 'adoption states' are preponderantly southern, but importantly include California (which accounts for about 10 per cent of the national textbook market) and Texas (which currently has an annual budget of $45 million for high school textbooks)[7]. Given the highly competitive nature of the textbook market, the pressures that this system—with all its political and populist overtones—puts on publishers is obvious. Typical, and understandable, is the havering of one publisher about his company's text[8]: "evolution runs like a thread throughout, but is mentioned specifically only in the last chapter."

Why are creationists not a force in Europe? The reasons are many and varied, but one simple set of numbers helps to characterize the difference. Writing of "revived dogma and new cult," Martin[9] observes that in Northern Europe, around three to five per cent of those nominally Protestant are found in church on any given Sunday. In contrast, the latest Gallup polls show 51 per cent of all American teenagers in church on any given Sunday—and at least one-third of these are fundamentalists.

Grabiner and Miller's conclusions[1] about the Scopes Trial era have a message for today. "Readers may choose their own villain in the story we have told. Like us, some will find the greatest culpability in the scientific community itself, for the large-scale failure to pay attention to the teaching of science in the high schools. Others will blame the textbook authors and publishers for pursuing sales rather than quality. Some will attach blame to the politicians who exploited anti-evolution sentiment to get into, or remain, in office . . . But whatever the lesson one wishes to draw from the history of biology textbooks since the Scopes trial, we think the story itself is worth knowing. That the textbooks could have downgraded their treatment of evolution with almost nobody noticing is the greatest tragedy of all."

REFERENCES

1. Grabiner, J. V. & Miller, P. D. *Science* 185, 832 (1974).
2. Baker, A. O. & Mills, L. H. *Dynamic Biology,* 681 (Rand McNally, New York, 1933).
3. Simpson, G. G. *Science* 187, 389 (1975).
4. Nelkin, D. Personal communication (1982).
5. Zuidema, H. P. *Creation/Evolution* 2, 18 (1981).
6. *Science 81,* 58 (December 1981).
7. Dahlin, R. *Publishers Weekly* 220, 28 (1981).
8. Henig, R. M. *Bioscience* 29, 513 (1979).
9. Martin, D. *Daedalus* 111, 53 (1982).

MICHAEL RUSE

A PHILOSOPHER'S DAY IN COURT

On Sunday, December 6, 1981, I found myself on a flight south, from Toronto, Ontario, to Little Rock, Arkansas. That week I was to appear as an expert witness in a U.S. Federal Court case. What was the case? Why would I, a philosophy professor from a small town in Canada, be summoned? What happened? Where are we today? These are some of the questions I hope to answer.

Act 590

My story is about evolutionary biology, and as always when dealing with that topic the best place to begin is in 1859. It was that year which saw the publication of Charles Robert Darwin's major work, *On the Origin of Species by Means of Natural Selection, or the Preservation of Favoured Races in the Struggle for Life.*[1] In the course of some 450 pages, Darwin wrote vigourously in support of two major theses. First, he argued to the actual occurrence of *evolution,* that is, to the claim that all organisms (including ourselves) are descended by a slow natural process, gradually modifying from "one or a few" original forms. Second, Darwin supplied a *mechanism* for the evolutionary process: natural selection.

Darwin started with the potential Malthusian population explosion which exists everywhere in the living world. Given the obvious constraints due to limited supplies of food and space, we get a universal struggle for existence; more particularly, we get a struggle for reproduction. Drawing on the analogy of artificial selection as practiced by animal and plant breeders, Darwin then went on to argue that the struggle fuels a form of "natural" selection. Given enough time, this force leads to full-blown evolution.

> How will the struggle for existence . . . act in regard to variation? . . . Let it be borne in mind in what an endless number of strange peculiarities our domestic productions, and, in a lesser degree, those under nature, vary; and how strong the hereditary tendency is. . . . Can it, then, be thought improbable, seeing that variations useful to man have undoubtedly occurred, that other variations useful in some way to each being in the great and complex battle of life, should sometimes occur in the course of thousands of generations? If such do occur, can we doubt (remembering that many more individuals are born than can possibly survive) that individuals having any advantage, however slight, over others, would have the best chance of surviving and of procreating their kind? On the other hand, we may feel sure that any variation in the least degree injurious would be rigidly destroyed. This preservation of favourable variations and the rejection of injurious variations, I call Natural Selection (Darwin, 1859, pp. 80–81).[2]

In Victorian Britain, Darwin's ideas had a somewhat mixed reception. Many people recoiled from everything that he wanted to claim, refusing to have any truck whatsoever with filthy evolutionism. However, among the intelligentsia—scientists obviously, but extending right across the spectrum even to liberal clergymen—evolution per se was acceptable and accepted. Long before Darwin's *Origin* was published, the English had come to realize that a literal reading of Genesis—six days of creation, short time-span for Earth (about 6000 years, as calculated from the genealogies of the Bible), universal flood—was simply not tenable. The empirical facts speak against such a reading. Hence, the idea of evolution, binding and explaining so many different aspects of the organic world, was welcomed with enthusiasm.[3]

Where people had trouble was with Darwin's mechanism of selection. They simply could not see how "blind" law could lead to the intricate adaptations which we see around us in the world—the hand, the eye, the beautiful colours of the butterfly, and, above all, those qualities of intelligence and morality which raise us humans up above the apes. Here, critics felt, one simply has to add "something more." Thus we find Herbert Spencer relying on a full-blown Lamarckian inheritance of acquired characteristics; Darwin's "bulldog" Thomas Henry Huxley turning to large variations, "saltations"; and religious people of all stripes insisting that every now and then God gives His handiwork a little shove.

> We must suppose the idea of *Jumps* . . . as if for instance a wolf should at some epoch of lapine history take to occasionally littering a dog or a fox among her cubs. Through such a process we introduce *mind, plan, design,* and to the . . . obvious exclusion of the haphazard view of the subject and the casual concourse of atoms.[4]

This general sense of unease about Darwin's mechanism of selection is something which has persisted down through the years even to the present. Virtually no active scientist today wants God to intervene personally in the course of evolution (except, perhaps, when it comes to immortal souls), but still all sorts of different rival mechanisms are proposed. One young researcher has recently argued for a form of Lamarckism, and there is a very articulate group of paleontologists pushing a form of neo-Saltationism. Stephen Jay Gould and others argue that every now and then evolution takes a leap forward, followed by periods of relative unchange. (This is the theory of "punctuated equilibria.")[5]

But for all of the troubles of selection, in Darwin's homeland at least, evolution had a relatively smooth ride. With scientists, liberal clergymen, and other educated laypeople all early converts to evolution, no formally organized opposition was really able to make much progress. In fact, Darwin was very lucky, for it was shortly after publication of the *Origin* that universal school education became available in Britain. And those very people who were involved in setting up such education were often the *same* people most concerned to spread evolutionism! For instance, T. H. Huxley, the chief spokesman for descent with modification, was a founding member of the incredibly influential London School Board. One can well imagine what the little East Enders got in their classes!

> In particular he [Huxley] advocated the teaching of "the first elements of physical science"; "by which I do not mean teaching astronomy and the use of the globes, and the rest of the abomina-

> ble trash—but a little instruction of the child in what is the nature of common things about him; what their properties are, and in what relation this actual body of man stands to the universe outside of it." "There is no form of knowledge or instruction in which children take greater interest."[6]

In the U.S., particularly in the South, matters were otherwise. In times of stress and unhappiness, people frequently look to simplistic doctrines for support and comfort. Obviously, after the Civil War, in the South, one did indeed have such times of stress and unhappiness. Naturally, people turned to the most obvious place of consolation: the Holy Bible. In particular, Genesis gave people a sense of where they are and where they belong. Leviticus analogously gave people a guide for moral conduct. Thus, before long there was the flourishing of a peculiarly strong, indigenous brand of biblical literalism.

One consequence was that zealots started to monitor carefully the teaching and contents of school science classes. Woe and behold anyone who dared to step beyond the strictest bounds of the Old Testament. Those sufficiently foolish as to suggest that Adam may not have sprung up from mud soon found themselves in want of a job.

It was after the First World War that Fundamentalists (as biblical literalists were now called) scored some of their most striking successes. Several states of the Union, including Tennessee, passed laws prohibiting the teaching of evolutionism. As is well known, this led to a famous trial, when a young Dayton, Tennessee schoolteacher, John Thomas Scopes, let himself be prosecuted for teaching evolution. Matters soon took on a carnival air, attracting the attention of the whole nation. Three-time presidential candidate William Jennings Bryan led the prosecution, and noted freethinker and devastating advocate Clarence Darrow appeared for the defense. Refused permission to introduce experts on evolution, Darrow had the brilliant idea of cross-examining Bryan on the literal truth of the Bible. Before long, Darrow had Bryan tied into knots as Bryan tried to defend such an unexpected doctrine.

Thus, although Scopes was found guilty—he had, after all, confessed to teaching evolution—evolutionists rightly felt they had won a moral victory. Moreover, thanks to the savage pen of *Balti-*

more Sun reporter H. L. Mencken—who referred to the good people of Dayton as anthropoid rabble—many other states wisely and quietly shelved their proposed antievolutionary "monkey laws."[7]

Thus matters rested for long years, although it was not until the 1960's that the U.S. Supreme Court finally overturned the Tennessee law, ruling the teaching of evolution constitutional. I have a friend who grew up in Tennessee in the 1950's and 1960's. Her father worked at Oak Ridge National Laboratory. She tells me that in her high school—which may well not have been typical—they read Darwin. But in literature classes, not in science! At the local library, the copy of the *Origin* was missing from the Great Books Series. It was kept under the desk, along with all the other dirty books. Naturally enough, it was the most-read book of the whole series.

Fundamentalism lost some of its virulence after the 1920's, but in the 1960's it started to grow again. It was the Russians, of all people, who had a major part in this. In 1957 they put up Sputnik. This simply terrified America, which saw itself behind Russia, both in science and in technology. Typically, therefore, lots of money was thrown at the problem, and this (wisely) included money directed towards the improvement of American science education. As a consequence, high-powered committees were struck and fine new up-to-date textbooks produced.[8]

Naturally enough, the biology textbooks took evolution for granted. Unfortunately, when children started to bring these new books home, trouble developed all over again. Evolutionary biology was too much for biblical literalists, who felt that they simply had to do something about the situation.

But times had moved on since the days of the original Scopes Trial. By the late 1960's, evolution could no longer be the direct issue, given the Supreme Court's ruling on the constitutionality of its teaching. Furthermore, in the past half century, the Court has ruled with increasing force that the First Amendment's separation of church and state means precisely that. In particular, one may not teach religion as religion in schools. One certainly may not teach it in biology classes.

This puts modern-day Creationists in a bind. They would like to exclude evolution, but they cannot. They would like to include

Genesis, but they cannot. As a compromise, therefore, they try to slide Genesis into classrooms, sideways. They argue that—Surprise! Surprise!—all of the claims of Genesis can be supported by the best principles and premises of empirical science. In other words, *as scientists,* people can argue for instant creation of the universe, separate ancestry for man and apes, short time-span for the Earth (between approximately 6,000 and 20,000 years), and a universal flood over everything, at some later date.[9]

Hence, we have the growth of "Scientific Creationism" or "Creation-science." An institute has been set up; a college; a museum; and an organization of Creation-scientists. Full membership in the latter demands that one have a graduate degree in the sciences; however, most of the 500 full members have nonbiological degrees in such areas as mining. Additionally, many Creationist books, magazines, and films have been produced.[10] Most successfully, leading Creation-scientists—notably Henry M. Morris and Duane T. Gish—travel the campus circuit, debating with local evolutionists on "Creation *versus* Evolution." The orthodox scientists tend not to be that skilled in public debate, and sometimes have lost their tempers at the Creationists' lies and underhand tricks. Hence, the Creationists usually garner major propaganda value from these circuses.

The cry today is for "equal time" or "balanced treatment" in the schools between what the Creationists refer to as "Evolution-science" and their own "Creation-science." It is a cry which has a powerful appeal, for it seems that it is those who would deny the Creationists' demands who are the bigots. Why should Creationists not have the chance to make their case? Indeed, surely the best principles of education demand that children be exposed to all kinds of ideas, and not just to those of one group, however powerful. The denial of the rights of minorities to make their case smacks of fascism.

Many people outside of the Creation-science movement have responded favourably to this plea for equal time for Creationist ideas in the schoolroom. One such respondent is no less a person than the incumbent President of the United States. When he was on the campaign trail, he spoke as follows:

> Well, it is a theory, it is a scientific theory only, and it has in recent years been challenged in the world of science and is not yet

> believed in the scientific community to be as infallible as it once was believed. But if it was going to be taught in the schools, then I think that also the biblical theory of creation, which is not a theory but the biblical story of creation, should also be taught.

Finally, in 1981, the breakthrough hoped for by the Creationist movement became a reality. The state legislature of Arkansas considered a bill requiring of its teachers that if they talk at all of origins in the classroom, then they must talk of Creation-science as well as evolution. The pertinent parts of Act 590 (as it was called) were as follows.

> (a) "Creation-science" means the scientific evidences for creation and inferences from those scientific evidences. Creation-science includes the scientific evidences and related inferences that indicate: 1) Sudden creation of the Universe, energy and life from nothing, 2) The insufficiency of mutation and natural selection in bringing about development of all living kinds from a single organism, 3) Changes only within fixed limits of originally created kinds of plants and animals, 4) Separate ancestry for man and apes, 5) Explanation of the Earth's geology by catastrophism, including the occurrence of a world-wide flood, and 6) A relatively recent inception of the Earth and living kinds.
>
> (b) "Evolution-science" means the scientific evidences for evolution and inferences from those scientific evidences. Evolution-science includes the scientific evidences and related inferences that indicate: 1) Emergence by naturalistic processes of the Universe from disordered matter and emergence of life from non-life, 2) The sufficiency of mutation and natural selection in bringing about development of present living kinds from simple earlier kinds, 3) Emergency [*sic*] by mutation and natural selection of present living kinds from simple earlier kinds, 4) Emergence of man from a common ancestor with apes, 5) Explanation of the Earth's geology and the evolutionary sequence by uniformitarianism, and 6) An inception several billion years ago of the Earth and somewhat later of life.
>
> (c) "Public schools" mean public secondary and elementary schools.[11]

Act 590 did not originate in Arkansas. Indeed, it was a "model bill," drawn up by out-of-state Creationists. Their hope was that sympathetic politicians in several southern states would respond favourably to this model. In Arkansas, their hopes came to fruition, for a Fundamentalist senator steered it through both houses with no opposition whatsoever. Thus, the bill came before Governor Frank J. White. On March 19, 1981, apparently without taxing himself so far as to read it, White signed the bill, and so Act 590 became law.[12]

Preparation

"Hi! My name's David Klasfeld. I'm with the New York law firm of Skadden, Arps. I've been told that you might be able to help me." And so, I got personally involved with the fight against Act 590.

The American Civil Liberties Union is an organization dedicated to the preservation and support of the Constitution. As soon as Act 590 was passed, it sprang into active opposition, first at the state level and then at the national office in New York City. As noted, the First Amendment to the Constitution separates church and state. Fairly obviously, the ACLU saw Act 590 as a clear violation of this separation. Eventually, although separation of church and state was always the main thrust of the case, the ACLU added two other reasons why it thought Act 590 unconstitutional: that Act 590 infringes on the teacher's "academic freedom" to teach his/her subject properly, and that the Act is unconstitutionally vague. (The ACLU were not the plaintiffs as such. They were acting for Arkansas individuals, who claimed their civil liberties were being infringed. Many of the actual plaintiffs were clergymen.)

In fact, as the ACLU geared up for action, within the state of Arkansas itself there was growing opposition to Act 590. Of all groups, one that was most upset was the Junior Chamber of Commerce! Arkansas, like other sunny states in the South and Southwest, is busily trying to attract high-technology industry, such as computer firms. But the last thing that a bright young computer engineer wants is transference to a state where his kids have to learn Creationism as a matter of course. Hence, Arkansas businessmen began to sense that Act 590 might prove a disaster for the well-being of the state. They inquired whether the Act could be quietly shelved, but it was too late. Nothing could be done. Therefore, as the ACLU began to prepare its side of the case for the families, the office of the Attorney General of Arkansas began to prepare the defense. The case was to come to trial at the beginning of December before a Federal judge, William J. Overton, sitting on his own without a jury.

The ACLU does not have a large staff of its own. In a major case, such as this promised to be, it looks for help. This time, help came from an apparently unlikely source. The New York law firm of Skadden, Arps, Slate, Meagher, and Flom is a huge organization, with 250 lawyers and some 400 support staff. It specializes, very successfully, in aiding the biggest of American corporations in their aims of swallowing up all competitors, or in aiding the just-less-than-biggest companies in avoiding being swallowed up by predators. Just as the Arkansas case was coming to trial, it was helping Marathon Oil in its successful escape from the jaws of Mobil Oil.

For reasons which were never made absolutely clear, Skadden, Arps agreed to let about ten of its young lawyers go to work for the ACLU, *pro bonum* (that is, for nothing). I cannot believe that this was an act of purely disinterested altruism. Probably, the truth is that even the most successful of firms need to look to their image. Skadden, Arps has to go out with everyone else, recruiting the cream of the top law schools. Involvement with the ACLU case in Arkansas would certainly help to soften the picture of an enterprise concerned solely with money and power.

Whatever the reasons, the ACLU got together its men, one woman partner and a number of female legal assistants. And very bright people they were too. I have never encountered so sharp and enthusiastic and hardworking a group of young folk in my whole life. They could pick out the flaw in any argument instantly. They could start work after supper, build a case, take all of the components to pieces, put them back together again properly, and have everything typed, in multiple copies, on one's desk by nine o'clock the next morning. The ACLU looked for the best, and Skadden, Arps gave it that.

Since the ACLU was working with the plaintiffs—the ones objecting to the law—it had to make the positive case. It had to show why the law was unconstitutional. It was decided to make as broad a case as possible, because it was obvious that "Scopes II" would attract a great deal of media attention—newspapers, radio, and television. The more that Creation-science could be shown to be the travesty that it is, the more public opinion could be brought against it. Hence, the less success Creationists would be likely to have in the future.

To this end, the ACLU decided to divide its case into three parts. The first part of the case would be devoted to *religion.* Expert witnesses would be sought to prove that Creation-science is no more than Genesis by another name. Second, the case would move to *science.* It would be shown that Act 590 is thoroughly confused about evolution and that Creation-science has no right whatsoever to be called "science." Third, the case would conclude with *education,* and various witnesses would be called to show the difficulties of translating Act 590 into actual classroom practice.

Where did I come in? I am neither a theologian nor a scientist. I am not an expert on education. I am not an American.[13] And, as the state took some pains to point out, I have never taken a biology class in my life. This gap in my education, incidentally, was not due to a conscious act on my part. I grew up and was educated in England. It is, or at least was in the 1950's, virtually impossible for anyone to specialize in mathematics and physics—as I did—and yet take courses in biology. Regretfully, biology, like Spanish and geography, was for those who could not really handle the "hard" subjects.[14]

I am a historian and philosopher of science, and my claim to fame—more modestly, that which made the ACLU interested in me—was that I have written extensively on evolutionary theory. Moreover, when David Klasfeld contacted me in September 1981, I had just completed a manuscript which dealt in detail with evolutionary ideas.[15] Most pertinently, I had completed an in-depth, highly critical study of Creation-science. I was therefore already primed to fight for the ACLU.

But why a historian and philosopher at all? The ACLU decided that it wanted a historian to give the judge some of the general background to the story of man's quest for ultimate origins. Most particularly, they wanted to show that, contrary to Creationist complaints, evolutionists have not simply walked in and excluded all opposition. Creationism had a good run for its money—about 2000 years. As noted earlier, Creationism began to collapse of its own accord *before Darwin.* Hence, the ACLU wanted someone who could talk about men like the Reverend Adam Sedgwick, Professor of Geology at Cambridge and a deeply committed Christian. In 1831, the year in which Darwin graduated from Cambridge, Sedgwick publicly declared that one can no longer take Genesis as a literally true account of Earth's history. Most particularly, Sedgwick openly conceded that there is simply no evidence of a worldwide flood, and that he was working with a time frame far greater than that of the traditional 6000 years.[16]

An historian, therefore, could show the Court that juxtaposing Creationism against evolution sets up a false dichotomy. Creationism was simply judged not to work as science anyway. And a philosopher could then go on to hammer home the nonscientific nature of Creationism. It is the job of the philosopher to look critically at science from afar, asking such questions as: What is the nature of a scientific theory? Are all sciences similar in logic and structure? What is the relationship between theory and evidence, and why do scientists sometimes change their minds?

Obviously, therefore, a philosopher can ask many pertinent questions about Creation-science. Having given a general account of science, one can then move right into an analysis of Creation-science, showing where it fails as science. Of course, this in itself does not show Creation-science to be a religion. There are many things which are neither science nor religion. Philosophy, for instance. But demolition of Creation science as science would be an important first step. Furthermore, seeing where Creation science falls down as science can surely give important clues as to its true religious nature.

For reasons such as these, the ACLU sought out someone with historical and philosophical training. As I have explained, I myself attracted attention, because I had already worked on the very issues that the ACLU wanted to put before the judge. I readily agreed to lend a hand, although I did not then realise just how hard I was going to have to work.

I should say now, unequivocally, that I found it a terrifically exciting experience to work for the ACLU. Also, now that the fight is over in Arkansas, I look back with a great sense of personal satisfaction. Following Socrates, I believe that the unexamined life is not worth living. For me, philosophy is a consuming passion, and not just a job. However, one does not usually get the immediate, tangible sense of the worth of one's work as was afforded by the Arkansas trial. It was good to be able to take the theoretical training of years and apply it to an actual practical problem, fighting something which I believe to be a real intellectual and moral evil.

In the course of preparing for the trial, I made two trips to New York. The first was to meet the lawyers and to get acquainted with the general facts of the case. For two days my brains were picked constantly about the nature of science, until I was fairly reeling (a condition that was alleviated when I took time out to see a World Series game, including Billy Martin in action). Returning to Guelph, I continued to stay busy. In the next month I got three requests for position/discussion papers. These were to be brief essays touching on pertinent points, written clearly and directly so that the lawyers concerned in the case could grasp essential factors about the history and philosophy of evolution and Creation-science. In them I state that I believe the key distinguishing factor about science to be its appeal to and reliance on *law:* blind, natural regularity. Everything else follows from an unpacking of this notion: explanation, prediction, testing, confirmation, falsifiability, tentativeness. Judged by these various criteria, evolutionary theory is a genuine scientific theory, whereas Creation-science is just not science.

Although, as noted, evolution was not on trial, it was clearly at the back of everyone's mind. The lawyers, therefore, wanted to know as much as possible about it, and this was why I dealt so fully with it in my discussion papers. I feel no need at all to apologise for my discussion papers, but I would point out that, necessarily, they were written at great speed: two days for the first and a day each for the second and third. Therefore, they could not be researched, but had to be written "off the top of my head." This, of course, was what the lawyers wanted: elementary lectures, not contributions to the literature. I would like to pride myself that the points I made, particularly in the first paper, went straight through into Judge Overton's decision. (See especially, section IV[C].)[17]

Before I returned to New York, I went off to a previously arranged conference in East Germany. On the way, stopping over in England, I spoke to an elderly zoologist, L. Harrison Matthews, who wrote the introduction to Darwin's *Origin* in the Everyman Edition.[18] In phrases which have been seized on by Creationists, Matthews argues that belief in Darwinism is like a religious commitment.[19] This was going to be used by the State of Arkansas, who would argue that belief in Creation-science is logically identical to belief in evolution. Hence, since one can teach the latter, one should be allowed to teach the former. (A more rigorous conclusion would be that since both are religion, neither should be taught. But no matter.)

Would Matthews recant? He was happy to do so, and wrote me a strong letter about the misuse that he felt Creationists had made of his introduction. Reading between the lines, I got the strong impression that what motivated Matthews in his introduction was not the logic of evolutionary theory at all. He wanted to poke the late Sir Gavin de Beer in the eye. De Beer was a fanatical Darwinian, and Matthews was dressing him down for the undue strength of his feelings!

On to East Germany, where the scientists from the other side of the Iron Curtain talked nonstop about Creationism. Nothing I could say would prevent them from tying Creationism in directly with Western capitalism. Creationism is the paradigmatic example of what goes wrong if you are not Marxist-Leninist! Of course, the men arguing this way really were not that motivated by ideological purity. By tying in Creationism with capitalism, at the same time noting that they themselves were not Creationists, they hoped to curry favor with their own superiors. "Look at us," they were saying. "We and we alone do proper science. Therefore encourage us." Given the predicament of scientists in the Soviet block, it is hard to blame them.

Upon returning to New York, I met for the first time Jack Novik, the lawyer from the ACLU who was to lead me in direct examination. Tall, dark, heavily mustached, Novik is everything one expects of a bright, aggressive, young New York lawyer. He has a fantastic ability to cut through to the central point in a complex discussion and a total inability to relax. He thinks, eats, and breathes his cases. When you are in a jam, you could not have a better partner.

I was made to bone up on some twenty or so Creationist texts. Fortunately, there is not much variation on the same theme. Then, for one whole exhausting day I had my deposition taken by the assistant attorney general of Arkansas, David Williams. He was undoubtedly the toughest of the defense team, and by the time he had finished with me I was as limp as a rag. For the uninformed, having your deposition taken is pure reciprocal altruism in action. The lawyers from the other side are allowed to examine your witnesses in order to find out what these witnesses will say. You let this happen, because then you can have a crack at their witnesses. A witness obviously cannot lie, but he can be as unhelpful as possible, simply by not volunteering any information other than that directly demanded. Your own lawyer sits next to you (a court reporter is also present, taking everything down), and he can monitor for fairness. Like all professors I talk too much, and Jack Novik kept me quiet simply by making me more terrified of his wrath than of Williams's.

The deposition ranged widely. What were my religious beliefs? (Agnostic with flashes of deism.) Did I care about my children's religious beliefs? (I do, to the extent of sending my children to private Anglican schools.) Had I ever heard of Sir Karl Popper? (Yes.) Did I think him the greatest philosopher of science that had ever lived? (No.) Was I aware that some people so considered him? (Yes.) Did I think that one could have morality in a world of evolution? (Yes.) How do I regard morality? (I intuit moral values as objective realities. [Don't ask me what that means. Fortunately, Williams didn't ask me either, and so we moved on.])

There was one point at which I ran into rather heavy weather. In the first book I wrote, in defending the synthetic theory of evolution (i.e., the Darwin-Mendel synthesis), I asserted somewhat passionately that it "was established beyond all reasonable doubt."[20] Did this not show my own dogmatism? Did it not show that, contrary to my own claims about the tentative nature of science, I simply could not conceive of modern evolutionary theory being thrown over?

I sweated for a minute or two and then realised that the phrase could be turned to advantage. Scientists obviously don't keep questioning their theories every day of the week. When something is reasonably well established, they accept it. Similarly, in the realm of law one tries to reach a verdict which will be accepted without constant question: "it is beyond reasonable doubt." But because someone has been found guilty, it does not follow that the case can never be reopened. If important new evidence comes up, then there are ways of looking again at verdicts, and perhaps of overturning them. Similarly in science. Even if something is established "beyond all reasonable doubt," there is always the possibility that new evidence will make one reopen the case.

I think Williams realised what a tight spot he had me in. But when we came to trial, Novik and I brought out this analogy in direct examination, using it for our own purpose. We heard no more on the subject.

Finally my deposition came to an end, and with it nearly all of the pretrial work. There was only one more thing to be done: preparing what lawyers refer to as one's "Q's and A's." This is the script for the direct examination (thus "Questions and Answers"). Novik produced a 50-page script, and I was told to learn it. Since it was based on my own position papers, it was hardly new. But I confess that I did shiver rather at the thought of blanking out at some crucial moment. Suppose I forgot what a law is! I really pored over that script. It was made very clear to me by the ACLU that its witnesses were expected to perform without notes.

The Trial

I arrived in Little Rock at four o'clock in the afternoon of December 6 and was taken straight to the hotel: the Sam Peck, a comfortable but plain building in downtown Little Rock, directly opposite the huge greystone Federal building that contains the law courts. Since the religion experts were to appear first, it was agreed that I should arrive with them in order to see the kinds of points they would make. We wanted no gaffes, with me saying the very opposite to the previous witness.

The religion witnesses were Bruce Vawter, a Catholic priest and Old Testament expert; George Marsden, an historian who has specialised in the study of development of Fundamentalism; Dorothy

Nelkin, a sociologist of science, who has studied the ways in which Creationists operate and how they have kept evolutionary ideas at a minimum in school texts; and Langdon Gilkey, a rather trendy and superarticulate theologian from Chicago Divinity School. As a side business, Gilkey works for the IRS, flying off to places like California, seeing if weird new sects qualify as religions and thus merit tax benefits.

There were no lawyers to be seen. Later we found that they were all huddled in last-minute preparations, still squabbling about who was to do what. Frankly, the Arkansas trial meant a lot to those young people, and a fair number of selfish genes were working hard to ensure that their survival machines had maximum time in court. Adding complexity to complexity, we not only had the New York lawyers involved, but now we had two Little Rock law firms too.

Fortunately or foolishly, a hospitality room had been set up in the hotel (i.e., unlimited free liquor). Religion witnesses and lonely philosopher trucked in, and by the time that the lawyers came back to earth an hour or so later, no one was feeling much pain. To be honest, I think all of the witnesses were feeling a little scared. How would we do the next day? Would we be ripped apart on cross-examination, as in a Perry Mason show? Would the judge be hostile or difficult, as in the original Scopes Trial?

The lawyers were furious. They had hoped to work us hard all evening. As it was, we got through supper in a comfortable haze, and any discussion was at a pretty low level. Later in the evening, Novik, Klasfeld, and I tried running through my testimony. The rehearsal was an absolute disaster. I forgot points and—far worse—simply would not get to the crux of what I was supposed to be saying. I would wander on and on like the worst kind of teacher. Nevertheless, looking back, I don't feel all that sorry. All of the witnesses had been chosen because they would be able to perform well in open court, as well as for obvious expertise. I think people rather forgot how we too needed to relax and be reassured. Even those (especially those?) who appear most confident and self-assured in public have preperformance fears. I know that thanks to several large gins I myself settled in much more rapidly than I would have done otherwise.

On Monday we started off. The courtroom was a large oak-paneled hall on the fifth floor of the Federal Building. The judge, impressive in his simple black gown, sat high at a desk at one end. Before him was the court reporter, speaking nonstop into her machine. Then came the witness box, isolated before the two large tables used by the opposing lawyers. The plaintiffs (the ACLU's side) had at least three times as many people as the defense (the State of Arkansas). To one side were some twenty or so reporters, including several artists busily sketching faces for the evening newscasts. On the other side sat the citizens of Arkanas who had brought the original complaint against Act 590.

Then, in the main body of the hall one saw the various witnesses, together with the several experts who were helping the two sides. We had the support and advice of the distinguished evolutionists Niles Eldredge and Joel Cracraft. The attorney general was being aided by leading Creation scientist, Duane Gish. Finally, behind the direct participants in the trial was the audience of about two hundred: schoolchildren, bused in for lessons in "civics"; Fundamentalist ministers, with that overgroomed look which is their trademark; and any number of others.

Following brief opening statements in which both sides laid out their main claims (the ACLU was against Act 590, and Arkansas was for it!), we got down to business.[21] Our first witness was the Methodist bishop of Arkansas, who said he was all in favour of religion but not in schools. This was a theme to be repeated again and again. No one is more upset by the Creationist movement than orthodox American churchmen, Christians and Jews. They hold the First Amendment separation of church and state very dear. So many of their ancestors came to freedom in America, driven from Europe because of religious persecution.

Hence, orthodox churchmen do not want religion—any religion—taught in schools. In such a mingling of church and state, we see the road leading to such religiously torn countries as Northern Ireland. Moreover, the churchmen loathe Creationism. They do not view it as the only true form of religion. Rather they see it as a perverted blasphemy. God did not give us our reason just to have us hide our heads in the arid, comforting sands of Genesis. The Bible is not a work of science, and to pretend otherwise is to lose its true meaning. The Holy Writ is the story of God, man, and the relationship between the two. What does God expect of us? What promise does He hold out for us in the future? The Bible is a work of spiritual and moral significance.

After the bishop came the expert religion witnesses. They explained in very short order that the description of Creation-science in Act 590 is no more and no less than the story of Creation in Genesis. Indeed, they explained this in rather too short order, for whereas we had expected these witnesses to take two full days, it suddenly seemed that they might be through before the first day was over. I was packed off back to the hotel to put on a tie. Then, Jack Novik, David Klasfeld, and I went to a quiet room to run through my testimony. At long last things started to come right. Questions and answers flowed smoothly, one following on the next—a combination of theatre and philosophical symposium. At the end, David Klasfeld said quite simply and truthfully: "Mike! That was the first time I heard you do it better than I could have done it."

Tuesday morning came—rather early for me. I woke at three A.M. and was unable to get back to sleep. Finally, after *The Lone Ranger* and *Sunrise Semester,* it was seven o'clock and time for breakfast with Novik. I realized suddenly that, for all his outer confidence, he was as tense as I was. This was his big moment too. Nine o'clock finally arrived, we rose for the judge, I took the stand, and the show was on.

With cross-examination, I was up in front of everyone for three and a half hours in the morning and a little more in the afternoon. Direct examination went like a dream as Jack and I looked at each other, words coming straight to our lips, ideas bubbling out, and one point after another getting answered and hammered right home.

Q: What is science?
A: Science is an attempt to understand the physical world, primarily through law, that is, through unbroken natural regularity.
Q: Would you elaborate on that please?
A: Yes indeed. Understanding in science means explanation and prediction. Through this comes test, confirmation, and the potential to falsify. This means that a crucial mark of science is that it is tentative.

Q: Let's now stop and go through these various attributes of science, one by one, giving exmples to explain them to the Court.

And so we did explain and give examples, and the judge wrote everything down. He listened intently and would nod when he had grasped and noted a point. Then we would go on to the next topic.

We gave the judge a short history of evolutionary theory and of the growth of the idea of science separate from religion. We explained the essential attributes of science. Then we turned our attention to Act 590. Those crucial passages quoted earlier were examined word by word, line by line. Consider 4(a)1 "[Creation-science includes ideas implying] sudden creation of the universe, energy, and life from nothing." Why is there no description of how this creation will occur? In the corresponding phrase about evolution, 4(b)1, it is explicitly stated that origins are "naturalistic" that is, governed by normal law. Clearly, Creation-science implies that origins are nonnaturalistic, that is, miraculous. Origins come through some sort of supernatural forces. This force is not a scientific notion. It is, however, a religious notion. (Notice how here we were able to make the point that Creation-science is not merely not science, but is religion.)

Moving on through the various clauses, we noted the talk of originally created "kinds" in the description of Creation-science. "Kind" is not a regular taxonomic term at all. But it does occur in the Genesis description of Creation. Analogously, it is very odd that a worldwide flood is singled out for special attention. If this is not a direct reference to Noah's Flood, why not talk of other natural disasters also? Why not talk of the Chicago fire or the San Francisco earthquake? And most particularly, what on earth could one mean by "a relatively recent inception of the earth and living kinds?" As I noted, and as virtually every other ACLU witness noted, this really is a meaningless statement. Is "relatively recent" a million years, the 6000 years traditionally calculated from the genealogies of the Bible, or (as Christopher Robin would have it) a week ago Friday?

At the same time as Novik and I went over the clauses describing Creation-science, we also told the Court a little about the supposed rival, "evolution-science." First we noted that "evolution-science" is both a name and a concept unknown outside Act 590 and the Creationist literature. Biologists just don't roll everything together in one overall hybrid. The ultimate origins of life and the subsequent evolution of life were separated by Darwin (he never mentions the former in the *Origin*). They have been kept separate ever since. I am sure that there are many more questions about ultimate origins than about evolution. Even if one had no idea whatsoever about how life originated, it could still be reasonable to believe in evolution.[22]

Second, we noted that Act 590 gets evolutionary theory wrong. No one—certainly not Darwin—has ever thought that natural selection is a sufficient mechanism for evolution. Today, everyone allows some randomness in evolution—for instance, that which comes in genetic drift. Third, the very juxtaposition of evolution with Creation-science—what Creationists call the "two-model approach"—is itself fallacious. One cannot prove Creationism by disproving evolution. There are alternative positions, as the State was to find to its embarrassment. For instance, recently the physicists Fred Hoyle and N. C. Wickramasinghe have proposed an earth-picture supposing life to come here, through the ages, from outer space.[23] This is not conventional evolution, but neither is it Creation-science. Apart from anything else, Hoyle and Wickramasinghe believe the earth to be very old. The attorney general was to call Wickramasinghe as a witness. When he had finished, he had done more damage to the State's case than to the ACLU case. The judge was left mystified as to why Wickramasinghe was called.[24]

Novik and I were almost through. But, as in the marriage at Cana, the best was left until last. We turned finally to the voluminous Creation-science literature, and for some twenty-five wonderful minutes showed its total failure as science. With Gish in the audience, I delighted in reading the following passage from *Evolution? The Fossils Say No!* The ACLU team could not have made the point more nicely if it had written the passage itself.[25]

> CREATION. By creation we mean the bringing into being of the basic kinds of plants and animals by the process of sudden, or fiat, creation described in the first two chapters of Genesis. Here we

> find the creation by God of the plants and animals, each commanded to reproduce after its own kind using processes which were essentially instantaneous.
>
> We do not know how God created, what processes He used, *for God used processes which are not now operating anywhere in the natural universe.* This is why we refer to divine creation as special creation. We cannot discover by scientific investigations anything about the creative processes used by God.

Incidentally, many people have suggested that Steve Clark, the Arkansas attorney general, threw the State's case by not using as witnesses the major Creation-scientists, such as Gish and Morris. The simple fact of the matter is that he couldn't. Can you imagine what an ACLU lawyer would have done with Gish on the stand? How lingeringly and how lovingly they would have dealt with that statement!

We next looked at questions of explanation and prediction. We showed that in Creation-science these never occur in a genuine scientific way. I talked of the facts of homology—nonfunctional isomorphisms between organisms of different species. Evolution explains them as a result of common descent. Creation-science has no answer at all. Lamely, its supporters suggest that homology is irrelevant, because all classification is arbitrary anyway. Just imagine. We are supposed to accept that the classification of organisms into birds, whales, dogs, men, etc., is arbitrary.[26]

We emphasized the nontentative nature of Creation-science. Again and again, the Creationists assert dogmatically that one must accept their position. It is not open to doubt and debate, as all genuine science must be. I read from the manifesto that all Creation-scientists must sign when they join the Creation Research Society. They have to affirm their belief in the literal truth of every word of the Bible. Whatever this may be, this is not the way in which science is done. Can you imagine every evolutionist having to sign a statement accepting the literal truth of Darwin's *Origin?*[27]

Finally, we came to the question of honesty. Novik and I wanted to prove not merely that Creation-science is not science, but that it is a dishonest and thoroughly corrupt enterprise, violating every standard of intellectual integrity. Again and again, I was able to show that statements made by eminent evolutionists are lifted by Creation-scientists and quoted out of context. Evolutionists are made to say the very opposite to what they intended.

A classic case of such distortion occurs in a recent Creationist book, *Creation: The Facts of Life* by Gary E. Parker. Repeatedly, Parker refers to "noted Harvard geneticist [Richard] Lewontin's views" that things like the hand and the eye are the best evidence of God's design. Can this really be so? Has the distinguished author of *The Genetic Basis of Evolutionary Change*[28] really thrown over Darwin for Moses? When you actually look at the source of Lewontin's views, you find that you must wait awhile for Lewontin's full conversion.[29] What he says, in fact, is that *before Darwin* people believed that such phenomena as the hand and the eye were evidence of God's direct design. Now we accept evolution through natural selection.

> The theory about the history of life that is now generally accepted, the Darwinian theory of evolution by natural selection, is meant to explain two different aspects of the appearance of the living world: diversity and fitness. . . . It was the marvelous fit of organisms to the environment, much more than the great diversity of forms, that was the chief evidence of a Supreme Designer. Darwin realized that if a naturalistic theory of evolution was to be successful, it would have to explain the apparent perfection of organisms and not simply their variation. . . . The modern view of adaptation is that the external world sets certain "problems" that organisms need to "solve," and that evolution by means of natural selection is the mechanism for creating these solutions.

But the reader learns nothing of this. Parker makes Lewontin say the very opposite. And on this high note we ended my direct testimony. Cheerfully, I referred to the Creation-scientists as "sleazy." And that term certainly gave a keen edge to cross-examination, as Williams and I crossed wits for the rest of the morning and into the afternoon. He wanted to make me seem stupid. I dearly wanted to return the compliment. Actually, however, I never got a sense of personal bitterness. There were a lot of jokes cracked through the trial, and I confess I rather grew to like the members of the defense team. Certainly, Steve Clark was the epitome of the gracious southern gentleman, who always had a friendly word when you bumped into him.[30]

And so, Williams and I went over and over the territory again. "What about Popper?" "Well, what about him?" "Hasn't he said such and such?" "Yes he has, but since then he's changed his mind, and he didn't know what he was talking about in the first place." "But that's just your opinion, isn't it?" "Yes, but I think it's a good opinion. I'd be happy to go into matters in some detail, if you'd like, Mr. Williams." "No, I don't think we need bother about that just at the moment, thank you, Dr. Ruse."[31]

My religious beliefs gave us all several minutes' entertainment. What did I mean by "agnostic"? What did I mean by "deist"? Were these claims consistent with what I'd said in my deposition? Wasn't my equivocation just one more proof of the godless, atheistic nature of evolutionism? Finally, half in embarrassment, half defiantly, I blurted out: "I'm sorry, Mr. Williams! Surely you can see that I'm not an expert witness on my own religious beliefs!" The burst of laughter that greeted my confession squelched any further discussion, and so we moved on.

One thing which occupied us for quite some time was the manuscript of my book, *Darwinism Defended.* (The defense had been given copies of everything that I have written. The Xeroxing bills in the Arkansas trial were quite colossal.) In that book, not only do I examine Creation-science in some detail, but I explain why I do not think it should be taught in schools. I argue that free speech does not imply that crazy ideas belong in the biology classroom. Nevertheless, quoting John Stuart Mill, I defend the right of anyone to believe anything they like.

Q: Dr. Ruse. In your new book you quote John Stuart Mill. Would you please read to the Court the words that you quote.

A: "If all mankind minus one were of one opinion, and only one person were of the contrary opinion, mankind would be no more justified in silencing that one person, than he, if he had the power, would be justified in silencing mankind."[32]

Q: Don't you think you're being inconsistent in saying that? At the same time you want to ban Creation-science?

A: I don't want to ban Creation-science. Anyone who wants to can believe it and say it out loud.

Q: But not in the schools?
A: But not in the schools. Creation-science is religion and has no place in biology classrooms.

Finally, I was finished. We had covered just about everything under the sun, with the possible exception of L. Harrison Matthews's claims about the religious nature of Darwinism. When Williams saw the scathing letter that Matthews wrote to me about Creationism, he decided not to introduce Matthews into the testimony.

I was tired but satisfied. It was now the turns of others. And magnificent turns they were, too. Francisco Ayala gave a brilliant exposition of population genetics and of modern work on evolutionary mechanisms. Similarly, Stephen Jay Gould put us all right on the fossil record, showing the dishonest stupidity of those who think the record speaks against evolution. Ayala and Gould were nice complements, for the first is an ardent Darwinian and the second has led the attack on conventional Darwinism. By putting the two together, the ACLU neatly defused a major Creationist misrepresentation, namely that differences between evolutionists over mechanisms imply that evolution itself is in doubt. Both men stood strongly for one of the greatest of all ideas. To hear Ayala talking lovingly of his fruit flies and Gould of his fossils was to realise so vividly that it is those who deny evolution who are anti-God, not those who affirm it.

A major plank in the Creationist attack on science is based on thermodynamics. Creationists argue that life could not have originated and then evolve, because order can never come from disorder. Hence, the ACLU called Harold Morowitz of Yale, who talked about the second law of thermodynamics, showing why it does not disprove a possible natural origin of life or evolution. (Simply, the second law applies only to closed systems and this earth is an open system.)

Rounding out the science witnesses was G. Brent Dalrymple of the U.S. Geological Survey. He gave a quite brilliant disquisition on methods of dating the earth. One would not think that such a topic could be all that intrinsically interesting, but Dalrymple gave this assumption the total lie. He held us absolutely spellbound as he talked of various dating techniques and how geologists compensate for weaknesses in one direction by strengths from another. My sense was that Dalrymple was so good and so firm that he rather broke the back of the State's case. He had checked all of the Creationist arguments, and showed in devastating detail the trail of misquotations, computational errors, out-of-date references, and sheer blind stupidity which allows the Creationists to assign the earth an age of 6000 years. After Dalrymple, the State seemed far less ready to tangle with witnesses.

The science witnesses had done their job. Then came the most moving testimony of all, as ordinary schoolteachers from Arkansas explained how they simply could not teach the travesty of Creation-science. I shall never forget the man who cried out under cross-examination: "Look, sir! I'm not a martyr or anything! But I just can't teach that stuff. I'm not a scientist. I'm a science educator. I'm like a traffic cop, directing ideas down from scientists to schoolchildren. My pupils respect me. All teachers are like parents in a way. How can I go into my classroom, spreading ideas that I know to be wrong? My students will despise me, and I'll not be able to live with myself."

What a man! How one would have loved to have had him as a teacher. How one would love one's children to have him as a teacher.

The ACLU case drew to a close. Then it was time for the defense to make its case. Unfortunately, I had to leave. Family, students, and exams called me back to Guelph. Would that I could have been there as the ACLU lawyers took on the State's witnesses. I would love to have heard the expert who allowed that he believed in flying saucers because he had read about them in the *Reader's Digest.* His explanation was that they are emissaries from Satan. In case this sounds too ridiculous to be true, I should say that Morris himself is on record as believing that the canals of Mars are an aftereffect of the fight between Satan and the Archangel Michael. He thinks that evolution is undoubtedly a pit dug for the unwary by the Evil One.

Generally, none of the State's witnesses were able to make any progress in showing that Creation-science is not religion but genuine science. One after another, they admitted that they believe what they believe for religious reasons.[33] One witness was so intimidated by what he saw in Court before he himself was to appear that he left on the next plane. He did not wait to be demolished. The only defense witness of any scientific calibre was Wickramasinghe, and I have explained already why the State probably wished they had never called him in the first place.

All the Creationist arguments crumbled. "Were the Creationists excluded by the professional scientific community?" "Yes, they were." "Had any Creationist submitted any articles to an established scientific journal in the past twenty years." "No, they had not." "Did the Creationists agree that scientists should have open minds?" "Yes, they did." "Was there any evidence whatsoever that would make the witnesses alter their own opinions?" "No, there was not." "Had the witnesses done original research on their claims, rather than simply combining the evolutionists' literature?" "No, they had not." And so on, and so on, and so on.

Creationists have gotten a lot of mileage in recent years claiming that they have a valid case which should be heard. A case which merits "balanced treatment" in the classroom. When they had their day in Court—literally—they had nothing to offer.

Decision and Retrospect

Judge Overton handed down his decision on January 5, 1982.[34] In no uncertain terms, he ruled that Creation-science is religion and therefore is constitutionally barred from the classroom. Act 590 may not be enforced. Quoting Justice Felix Frankfurter, Overton concluded that "good fences make good neighbors."[35] The State of Arkansas decided not to appeal, and so a great victory has been won.

But as the months slip by, thoughts of the Arkansas trial stay with me. What is there to be said in retrospect, and also prospectively in looking forward?

First, as I have said, personally I found the Arkansas trial a very rewarding experience. To stand in defense of the nobility of science, along with Gilkey, Dalrymple, and the Arkansas teachers, was a once-in-a-lifetime privilege. Several professional colleagues have since criticised me for participating, arguing that one cannot really make philosophical points clearly in a courtroom: one has to ignore all sorts of subtleties and distinctions. But without in any way conceding that I had to compromise, let me say simply that when the discipline of Socrates, John Locke, John Stuart Mill, Bertrand Russell, and Jean-Paul Sartre can no longer get up and defend right from wrong, then indeed it has collapsed ignobly and become irrelevant.

Moreover, I took particular satisfaction in doing something for America. At a personal level, many of my best friends are Americans, I was a student for some years at a U.S. college, and American publishers and journals have taken up my ideas and put them before the public. At a general level, for all the faults (and they are there), only a fool or a knave would deny the worth of American democracy—for the whole world, as well as for America itself. Going to Arkansas was partial payment of a large debt.

Second, let me pay full compliment to the fine people of Arkansas: judge, lawyers, church people, teachers. They got up and fought. Act 590 was a wretchedly stupid law to have passed in the first place, but once the Act became law, good people came out and opposed it. One of the plaintiffs, a clergyman, said to me: "Everyone thinks we're red-necks, but we're not. We don't want this kind of thing in our schools any more than anyone else does. We're just ordinary people." There was nothing at all ordinary about the people I met.

Third, let me warn against complacency. We have won a great victory. We have not yet won the war. The State of Louisiana has also passed a Creation-science law, so the ACLU must do battle there. No doubt other states will pass similar laws. And informally, Creation-scientists have announced their intentions to fight at the levels of individual teachers and school boards. There will be a long, hard, unglamorous slog before evolution gets fair, unfettered treatment in the classrooms of the U.S. And the same, I am afraid, holds true of my own country, Canada, where Creationist ideas are already making some headway in the schools.

Those of us who know better must roll up our sleeves and fight for the truth and for our children's education. Remember, although we may be little higher than the apes, we are also little lower than the angels. What better way could there be of showing this than by opposing Creation-science? No brute ever played Beethoven. No brute ever inquired into his past, opening up the magnificent picture that Darwin bequeathed to us.

Epilogue

Let me end on a lighter note. By the evening of the third day of the trial, it was clear to all that the ACLU was building an absolutely devastating case. The attorney general's lawyers were simply being brushed aside by the might of our science forces. Indeed, I have mentioned how, after Dalrymple's testimony, the defense had virtually given up fighting. In actual order of presentation, Gould was the final science witness. I have never seen such a disappointed man as he when, after only half an hour's cross-examination, Williams turned to the judge and said, "No further questions, your honour." Steve had been looking forward to a cosy afternoon putting us all right on the gaps in the fossil record, and he was finished before he had begun: rather than talking of gaps in the record, he was condemned forever to be one.

With the scientists cowing the defence, for us the tension broke. That night, nearly everyone on the plaintiff's team—lawyers, assistants, witnesses, assorted friends on the fringe—went out to eat in a restaurant—to talk, to drink, to play. Toward the end of the evening, someone started singing, and before long we were all joined in chorus. Inevitably we launched into hymns. My experience is that liberals almost always have a good church background, and that under the influence this comes to the fore. An angelic member of the Skadden, Arps contingent led us in that beautiful hymn "Amazing Grace." We all held forth: paleontologist, philosopher, leading counsel for the ACLU. We came to the line which talks of worshiping God for 10,000 years. That sort of time span was a little too close for comfort. We broke off and looked at each other in embarrassment. Then we broke into uncontrollable laughter.

It was a good moment.

NOTES

1. London: John Murray.
2. Darwin, *Origin,* pp. 80–81.

3. The classic account of the reception of the *Origin,* showing how rapidly people did accept evolution, is A. Ellegard, *Darwin and the General Reader* (Goteborg: Goteborgs Universitets Arsskrift, 1958).
4. This is from an unpublished letter, written and sent by the philosopher/scientist John F. W. Herschel to the geologist Charles Lyell, April 14, 1863. I discuss the various proposed supplements and alternatives to Darwinian selection in my book *The Darwinian Revolution: Science Red in Tooth and Claw* (Chicago: University of Chicago Press, 1979).
5. For full details of modern evolutionary controversies, see my *Darwinism Defended: A Guide to the Evolution Controversies* (Reading, Mass.: Addison-Wesley, 1982).
6. L. Huxley, *The Life and Letters of Thomas H. Huxley* (New York: Appleton, 1900), 1, 366.
7. The Scopes Trial has been written about extensively. See, for instance, L. S. de Camp, *The Great Monkey Trial* (New York: Doubleday, 1968).
8. D. Nelkin, *Science Textbook Controversies and the Politics of Equal Time* (Cambridge, Mass.: M.I.T. Press, 1977). Nelkin points out that although Scopes may have been a moral victory for evolution, at another level evolution suffered badly. Textbook publishers simply took evolution out of school texts. Even if it was not always illegal to teach evolution, as a matter of fact it was often not taught.
9. The "classic" work, which seems acknowledged as the founding text of the new Creationist movement, is John C. Whitcomb, Jr. and Henry M. Morris, *The Genesis Flood* (Nutley, N.J.: Presbyterian and Reformed Publishing Co., 1961).
10. The standard work, produced by a large team of Creation-scientists, is H. M. Morris (ed.), *Scientific Creationism* (San Diego: Creation-Life Publishers, 1974). A very popular work (over 130,000 copies sold) is Duane T. Gish, *Evolution?—The Fossils Say No!* (San Diego: Creation-Life Publishers, 1972). A recent book is Gary E. Parker, *Creation: The Facts of Life* (San Diego: Creation-Life Publishers, 1980).
11. Ark. Stat. Ann. §80-1663, *et seq.* (1981 Supp.).
12. These were facts that were to emerge at the trial of Act 590. They are sketched in the memorandum opinion drawn up by the presiding judge, William R. Overton.
13. Canadian newspeople had utmost difficulty in thinking that perhaps

someone from north of the border could make a useful contribution in Arkansas. Actually, however, my nationality never became an issue. It was of course irrelevant, but I suspect the main reason is that the State wanted to use a Canadian witness also. Incredibly, the witness turned out to be the daughter of a colleague in one of my departments.

14. Had I been really bright, I should have specialised in classics.
15. *Darwinism Defended.* Other pertinent books by me include *The Philosophy of Biology* (London: Hutchinson, 1973); *The Darwinian Revolution;* and *Is Science Sexist? And Other Problems in the Biomedical Sciences* (Dordrecht: Reidel, 1981).
16. Presidential address to the Geological Society. *Pro. Geol. Soc. Lond.*, 1, 281–316.
17. Several witnesses suggested definitions of science. A descriptive definition was said to be that science is what is "accepted by the scientific community" and is "what scientists do." The obvious implication of this description is that, in a free society, knowledge does not require the imprimatur of legislation in order to become science. More precisely, the essential characteristics of science are: (1) it is guided by natural law; (2) it has to be explanatory by reference to natural law; (3) it is testable against the empirical world; (4) its conclusions are tentative, i.e., are not necessarily the final word; and (5) it is falsifiable. Creation-science as described in section 4(A) fails to meet these essential characteristics.
18. London: Dent, 1971.
19. "Belief in evolution is thus parallel to belief in special creation—both are concepts which believers know to be true but neither, up to the present, has been capable of truth" (p. x). This is quoted in Morris, *Scientific Creationism,* p. 6, n.1.
20. *The Philosophy of Biology,* Chapter 6. I was talking of the theory which came together in the 1930's, combining Darwinian selection with Mendelian genetics. I was defending it against a lot of vitalistic arguments, and within the context of the discussion, I think the conclusion still holds.
21. Brief accounts of the trial were written for *Science* by Roger Lewin ("Creationism on the defensive in Arkansas," [1982] 215, 33–34; "Where is the science in Creation science?" *Science* [1982] 215, 142–146.)
22. This is certainly not to claim that no one has any sensible ideas about natural causes for ultimate origins. For more on this subject, see my *Darwinism Defended.*
23. *Evolution from Space* (London: Dent, 1981).

24. Technically speaking, Creationists and the State of Arkansas confuse "contraries" with "contradictories." With contradictories, if one side is right then the other side is wrong, and vice versa. (This pen is red/This pen is not red.) With contraries, both can be wrong but both cannot be right. (This pen is blue/This pen is yellow.) The "two-model approach" assumes that evolution and Creation are contradictories, whereas they are really contraries.
25. As noted, a number of Creationists do now admit that Creation science is not genuine science. They argue that evolution is not genuine science either, and so since the latter may be taught, the former should be taught also. But as also noted, the correct inference is surely that *nothing* should be taught about origins. Since I do not accept all of the premises, I do not accept the conclusion either.
26. See Morris, *Scientific Creationism,* pp. 71–72.
27. The full statement is as follows: "1. The Bible is the written Word of God, and because we believe it to be inspired thruout [sic], all of its assertions are historically and scientifically true in all of the original autographs. To the student of nature, this means that the account of origins in Genesis is a factual presentation of simple historical truths. (2) All basic types of living things, including man, were made by direct creative acts of God during Creation Week as described in Genesis. Whatever biological changes have occurred since Creation have accomplished only changes within the original created kinds. (3) The great Flood described in Genesis, commonly referred to as the Noachian Deluge, was an historical event, world-wide in its extent and effect. (4) Finally, we are an organization of Christian men of science, who accept Jesus Christ as our Lord and Savior. The account of the special creation of Adam and Eve as one man and one woman, and their subsequent Fall into sin, is the basis for our belief in the necessity of a Savior for all mankind. Therefore, salvation can come only thru [sic] accepting Jesus Christ as our Savior." (Px 115)
28. New York: Columbia University Press, 1974.
29. The misquoted article, entitled "Adaptation," is from an excellent issue of *Scientific American* (September 1978, pp. 213–30), devoted entirely to evolutionary thought.
30. I keep harping on the fact, but one thing which sticks in my mind is the extreme youth of all of the lawyers. Novik, Klasfeld, and Clark were all about 35. Several lawyers, including Williams, were under 30. Overton was an old man; like me, he is in his early 40's!
31. The point about Sir Karl Popper is that at one point he thought

Darwinian evolutionary biology fails the test of being a genuine science because it is not falsifiable. (This is Popper's "criterion of demarcation" between science and nonscience. Popper argued that Darwinism is a "metaphysical research programme." Recently, Popper has allowed that he was wrong and that Darwinism is indeed genuine science. See K. R. Popper, "Darwinism as a metaphysical research programme" in P. A. Schilpp, ed., *The Philosophy of Karl Popper* (LaSalle, Ill.: Open Court, 1974); letter to the editor, *New Scientist* (1980), 87, 611. See also my *Is Science Sexist?*

32. J. S. Mill, *On Liberty* (New York: Norton, 1975), p. 18. First published 1859.
33. See Lewin, "Where is the science in Creation science?"
34. For a brief summary, see R. Lewin, "Judge's ruling hits hard at Creationism." *Science* (1982) 215, 381–84. For the full text, see W. R. Overton, Memorandum on Rev. Bill McLean et al. LR C 81 322 (January 5, 1982), reprinted below.
35. *McCollum v. Board of Education,* 333 U.S. 203, 232 (1948).

GENE LYONS

REPEALING THE ENLIGHTENMENT

We must respect the other fellow's religion, but only in the sense and to the extent that we respect his theory that his wife is beautiful and his children smart.

—H. L. Mencken

The subspecies *Homo nesciens arkansas* comprises two distinct varieties: Country, and Country-Come-to-Town. It was ever thus. Back in the bad old days before the invention of polyester suits and communications satellites, however, genuine yokels held all the power in the state we call "the Land of Opportunity." In fact the first Arkansas anti-evolution law was not a product of the legislature. Passed on November 6, 1928, the day of Herbert Hoover's ascension to the presidency, the statute forbidding mention of godless, atheistic Darwinism in public schools was enacted by popular referendum.

A couple of days before the election, advertisements appeared in newspapers across the state. THE BIBLE OR ATHEISM, WHICH? read the headline on one favoring the passage of Act No. 1. But more than a hundred prominent citizens, including two former governors and the editor of the *Arkansas Gazette*—most of them from the sinful metropolis of Little Rock—signed another advertisement, urging common sense. Only three years earlier, after all, in 1925, Arkansas's neighbor to the east had convicted John Scopes for uttering heresies within the hearing of schoolchildren, and in the process had made the word "Tennessee" a synonym for "benighted." The people at large, the second avertisement maintained, were not qualified to pass on the veracity of a theory taught "in every first-class university and college in America, Europe, Asia, Africa and Aus-

From *Harper's Magazine,* April 1982. Copyright © 1982 by Harper's Magazine. Reprinted by special permission.

tralia." It was not the credibility of science that was at stake, but the state's reputation.

Voter turnout was heavy, for not only were science and religion contending on the ballot but Arkansas faced an excruciating presidential choice. Al Smith, the Democrat, represented both Demon Rum and the Pope of Rome. Hoover, though, was a Republican, the party of Lincoln. Hayseeds emerged from every God-intoxicated hollow in the Ozarks; automobile and mule jams clogged the flat dirt roads of the Delta. Al Smith won the Wonder State, but evolution lost. The vote for banning biological science was 108,991 to 63,406. Only Pulaski County (Little Rock) dissented.

Having made a ritual gesture in favor of the Lord, fundamentalists returned to the sleep of ages. Darwin made little headway in the boondocks, but then neither did any other sort of civilized learning. Most persons capable of reason in those districts found out about evolutionary theory anyway. In Little Rock and the other larger towns the law was ignored, albeit with caution. Acquaintances of mine who grew up in country towns tell stories of science teachers' voices dropping into conspiratorial whispers, of books being slipped to them on the sly as if they were racy French novels. A cruder version of the Moral Majority has been regnant in the Arkansas outback for at least 150 years, after all, without having effected a diminution of freelance sin. Alcohol in drinkable form is still forbidden the rustics across vast swatches of the state. While a federal court in Little Rock wrestled recently with creationism, the school board in Paragould voted not to allow a school prom on the grounds that dancin' leads to drinkin' and drinkin' to lust. Even so, Arkansas leads the nation in teenage pregnancies and ranks high in the incidence of venereal disease. By setting up coherent thought as temptation, Arkansas's anti-evolution law has probably lured as many young Arkansans to science over the years as it has prevented from hearing about it.

Initiated Act No. 1, in any event, remained on the books for forty years with nary a prosecution. It was removed in 1968 by the United States Supreme Court after a Little Rock Central High biology teacher made an issue of it. *Epperson* v. *Arkansas* was the second Supreme Court case involving Central High in little more than a decade. The first, of course, concerned racial segregation.

Know-Nothingism in a Lab Coat

Act 590, or the "Balanced Treatment of Creation-Science and Evolution-Science Act," as adepts call it, has a more socially acceptable pedigree than the 1928 monkey law. Reporters who came to Little Rock to cover the recent trial about this one's constitutionality found no snake handlers or fulminating barefoot hillbillies. Yessir, folks, with Act 590, country has done come to town. Dress up an ambitious fraud in a suit made of synthetic fiber, style his hair like a health-spa instructor's, give him a pocketful of credit cards, a push-button phone with a "hold" button, electric windows in his late model car, stick a Bible in his pocket, provide a neatly coiffed wife who knows how to make goo-goo eyes at the back of his head for TV cameras, and that man can play the media like a church organ. The statute's very concept of "balanced treatment" derives from, and therefore appeals to, the idea of journalistic fairness taught in the nation's "Schools of Communication." Are there not, after all, "two sides to every question"? Unfortunately that concept, which is shallow enough when dealing with persons holding a post-Enlightenment world view, ill equips a reporter to get at the truth when confronted with persons who do not. Creationists, you see, do not believe that there is or can be a distinction between the sacred and the secular. All ideas to them are religious ideas. Hence they do not hold themselves to the arbitration of facts, evidence, and logic; they reject the metaphysics of science even while claiming its cultural authority.

Neither do creationists believe, accordingly, in the separation of church and state, although they will prevaricate and squirm like sixteenth-century Jesuits when the question is put to them directly. So if the story of the 1981 Monkey Trial strikes you as ludicrous, which I hope it will, do not therefore be deceived into taking creationism lightly. Theirs is a coherent and internally consistent world view. The "scientists" in the movement do science as one does literary criticism, picking among facts and theories for ones that support a preexisting point of view—which in their case is a literal reading of Genesis—and either twisting whatever does not fit, or simply discarding it. Creationism is no more science than is astrology or palm-reading; it is William Jennings Bryan's know-nothingism in a lab coat. Creationists claim the designation "scientific" partly as propaganda, but, as with most propaganda, they are their own first victims. Oddly, while not believing in real science, which strikes them as pessimistic, European, and anti-Christian—perhaps even "Jewish"—they believe quite heartily, most of them, in technology and progress. Up to the day of Armageddon, that is. Most would also be shocked to hear themselves described as Social Darwinists, but all are free-enterprise zealots whose views are perfectly congruent with that turn-of-the-century philosophy. And there are a whole lot more of them in California, to come to the point, than there are in Arkansas.

> [Evolution is] theory only. In recent years [it] has been challenged in the world of science. If evolutionary theory is going to be taught in the schools, then I would think that also the biblical theory of creation, which is not a theory but the biblical story of creation, should also be taught.
>
> —Ronald Reagan, on the campaign trail in Dallas, 1980

But the Arkansas experience with creationism is instructive. The sponsor of the "Balanced Treatment" Act was one Sen. James L. Holsted of North Little Rock, a tall, handsome graduate of Vanderbilt University who was at the time president of the Providential Life Insurance Company, a family concern. Creationism zipped through the senate on the last day of the 1981 session, with no hearings and only a few comments from the floor. The house of representatives held no hearings either, having scheduled the bill for a period reserved for "noncontroversial" legislation. Debate consumed all of fifteen minutes, some of which was spent refusing to hear Arkansas's Methodist bishop Kenneth Hicks, who had rushed in vain to the capitol when a member of his flock warned him what was up. The tally there was 69–18.

Better than the Circus

Arkansans in general are probably no more ignorant than the American public at large, but all the ignoramuses do agree. Political tradition here pardons a legislator who votes on symbolic issues to soothe the prejudices of the mouth-breathing element in the dirt-road churches. Arkansas is more than 90 percent Protestant, the hard-shell sects predominate, and ambitious youths yearn to be television evangelists as others wish to emulate Reggie Jackson or Donny Osmond. No sense, runs the usual logic, in stirring people up; the federal courts can take care of it. Then everybody can whoop it up in the next campaign about meddlin' judges thwarting the will of the people, can get reelected, and can continue to work on the truly important business of democracy, like exempting farm equipment from the sales tax or allowing the poultry industry to load as many chickens as can be jammed into a semi-trailer regardless of highway weight limits.

If the reader detects bitterness, that is an error of tone. Traditionally, an Arkansas legislature in session is a spectacle more diverting than any circus, and best of all, it is free. The lawmakers hit Little Rock every two years from such rustic venues as Oil Trough, Smackover, and Hogeye like so many sailors just off a six-month cruise. Any curious citizen may venture of an evening to the saloons where the fun lovers among them congregate, and there be treated to a carnival of boozing, lurching, and panting such as one rarely sees so far from salt water. But there have been no fistfights on the floor this year, spitoons have given way to discreetly handled styrofoam cups, and ironists lament that the diverting spectacles of old are probably gone for good.

Indeed, it appears that many of the leglislators mistook the creationist bill for yet another in the series of harmless resolutions in praise of Christianity that they customarily endorse. Others were simply gulled. Had scientists uncovered evidence proving Genesis to be biologically and historically accurate? Who could doubt it? Were atheists and "secular humanists" laboring to suppress the truth? It sounded logical. The legislature was besieged by a well-organized phalanx not of backwoods fulminators but of live-wire "Christian" businessmen and doctors' wives from the newer suburbs of Little Rock. The creationists have laid their traps where the money is: among the semieducated who, by their prosperity, deem themselves members of contemporary Puritanism's visible elect, but who cling to the childish theology of their fathers because contemporary life has flooded them with a confusion of moral values that will not compute unless the Bible is accepted as a rule book. At the time of the "debate," only the Moral Majority and a local organization called FLAG (Family, Life, America, and God) seemed to know that Act 590 had been introduced at all, much less made it to the floor.

In fact, Act 590 was not written in Arkansas, and there is reason to doubt that anybody here read it all the way through until after it was already law. Senator Holsted got it from an employee of his, who in turn took it from a group of fundamentalist ministers who received it by mail from its author, a respiration therapist named Paul Ellwanger of Anderson, South Carolina. Ellwanger, founder and proprietor of an organization he calls Citizens for Fairness in Education, wrote it with the help of an outfit called the Institute for Creation Research in (where else?) San Diego. The "scientific" godfathers of creationism are Henry Morris and one Dr. Duane Gish, a preposterous buncombe artist about whom more later. The principal legal consultant was Wendell Bird, also of the ICR and author, for those readers who may be tempted to dismiss creationism as a mere regional delusion, of a very long article in the *Yale Law Journal* three years ago that not only posited creationism as a science but proposed its inclusion in public school curricula to "balance" and thereby "neutralize" the teaching of evolution, which it equated with atheism. The University of Arkansas law journal, I am confident, would have rejected Bird's vaporizings out of hand. Not only would the people in charge have recognized the theological underpinnings, but they would have feared for their academic reputation.

Governor Frank White certainly did not read the creationism bill. A Little Rock bank executive and a graduate of the U.S. Naval Academy, White ran for office as God's own candidate. The Lord, he said repeatedly during his campaign against incumbent Democrat Bill Clinton, had told him to declare his candidacy. On winning a narrow victory in the Reagan landslide, he declared the deity well pleased. White's equally pious second wife told the press that God had not only introduced her to her second husband but He had even done a turn as celestial realtor, divinely inspiring their choice of a home. After he signed the bill, White boasted to reporters that Arkansas had assumed the scientific leadership of the known world. White asserted that the new law was undoubtedly constitutional. But when asked specifically about the clause forbidding the "establishment of Theologically Liberal, Humanist, Nontheist, or Atheistic religions," the governor confessed that he was ignorant of the text. His office issued a clarification saying he had been thoroughly briefed, but the aide responsible for keeping track of legislation told the *Arkansas Gazette* that to her knowledge nothing of the sort had transpired. Sponsor Holsted told the same newspaper that "of course" his motives were religious, but, he added, "If I'd known people were going to be asking me about the specifics of creation science, I might have gotten scared off because I don't know anything about that stuff." Democratic Attorney General Steve Clark, in a remark that would come back to haunt him, said he had his doubts the law could be defended.

When the educated portion of the citizenry heard about the law, reaction was strong. There was a near unanimous outcry from the universities, teacher organizations, and the Arkansas Academy of Sciences. Editorial scorn was heaped on the perpetrators by virtually every newspaper in the state. So far have we come since 1928 that editorialists in places like Warren, McGehee, Stuttgart, Searcy, and Lonoke felt free to denounce Act 590 without having to fear burning crosses. The prevailing theme was that the thoughtless bozos of the legislature had again made Arkansas a national joke, just when its image had begun to improve after the damage done by Orval Faubus in the 1957 Central High integration crisis. A Little Rock man had lapel pins made with a banana logo and sold them to benefit the monkey house at the zoo; he raised hundreds of dollars. A series of derisively funny editorial cartoons has appeared in the *Arkansas Gazette* in which Governor White always appears holding a half-eaten banana. Only the *Arkansas Democrat,* the capital's second-string newspaper, involved in a circulation war with the *Gazette* and seeking the lowest common denominator, has defended the law. But then the *Democrat* looks at the world through oddly tinted glasses. Recently the paper devoted its editorial column to the proposition that Franklin Delano Roosevelt was a Fascist. When the American Association for the Advancement of Science pledged itself to resist creationism, the *Democrat* determined it to be an organization of "moral idiots" and "intellectual frauds."

Senator Holsted was prevented from reaping what glory there was to be had from creationism by an untimely indictment for embezzling $105,000 from the family business, but Frank White has got himself an issue. The governor's genius consists of a total inability to be embarrassed. His 1980 campaign was a masterpiece of fraudulent innuendo. Besides the usual denunciations of taxes, Big Government, and welfare cheats—Arkansas has the lowest taxes of any of the fifty states, and thus of the industrial world—White spent most of his money on a series of television commercials showing a minor riot by Cuban refugees housed on a former army base near Fort Smith. Most of the Cubans were black. Had Governor Bill Clinton "stood up" to Jimmy Carter, he asserted, this threat to Arkansas's peace and security could have been prevented.

Whether or not White is the crassest religious hypocrite seen in these parts since Billy Sunday seems to me a question not worth pausing over. Americans overrate sincerity. Morally speaking, it matters little whether a person can't think, won't think, or merely feigns the credulity of a child. In any case, creationism has become so volatile an issue that the governor is welcome to it, should he decide to flog it in the 1982 campaign. Creationism cuts unpredictably across party and ideological lines. As always, the imponderable mystery is how the monkey law plays in the country; White couldn't be elected county assessor in Little Rock. But there are no polls to tell us how many Arkansans favor the law, much less whether its proponents care deeply enough to vote on that basis alone. Many legislators got nervous when they began hearing from their educated constituents, particularly from ministers and churchgoers from nonfundamentalist sects, which have long since given up militant opposition to the visible world, and who believe correctly that antics like those of last year degrade religion rather than advance it. Chambers of commerce anxious to lure new industries, especially of the clean, high-tech variety, found themselves facing embarrassing questions. Some even wondered whether creationism might not hamper the Arkansas Razorback football and basketball coaches in their quest for out-of-state talent. If *that* could be proved, only Jehovah himself could save White from popular wrath. Many legislators said they thought they had made a mistake; there was talk of repeal. But that required the cooperation of the governor, and White stood petulantly firm. If anybody was going to save Arkansas's public school students from necromancy, it would have to be a federal judge. Again.

God Takes the Stand

If one wished to understand why the adult forms of Christianity in America seem afflicted with polite senility while the kindergarten churches bulge with sinners, Little Rock's creationism trial offered many clues. When the American Civil Liberties Union first announced that it would challenge the law and presented its twenty-three plaintiffs to the public, creationism looked to be set up for a quick knockout. Of the twenty-three, twelve were clerics. Their number included not only the Methodist, Roman Catholic, Episcopal, and African Methodist Episcopal bishops of Arkansas but also representatives of the Presbyterians, Southern Baptists, and Reform Jews as well. Here was a perfect opportunity to seize the high rhetorical ground from the electronic fundamentalists. In aligning themselves with an easily exposed religious hoax, the Moral Majority and company would seem finally to have gone too far. To require that a sectarian dogma inimical to most churches be taught *as science* in public schools violates virtually everything sixth graders are taught about Americanism.

But the churchmen blew it, locally at least. They allowed themselves to be muzzled by a platoon of lawyers. Perhaps "muzzled" is a bit strong. Although the trial was political in its essence, the ACLU conducted it as if it were a corporate merger. In their own pulpits and newsletters, the clergy expressed themselves forcefully and with some eloquence. Bishop Hicks of the Arkansas Methodist Church delivered himself early on of a well-written letter to the *Gazette* on the vast presumption underlying fundamentalist bibliolatry: that puny man sets himself up to limit God's power to the dimensions of his own mind. But only a small fraction of readers see the editorial page; Hicks was preaching to the converted. If the churchmen had appeared on the evening news bearing such messages, if they had held regular news conferences and distributed press releases at regular intervals commenting on the trial, if they had put together a paid religious telecast on the subject using some of the very erudite and committed scientists and theologians who came to Little Rock on their own time to testify, they might have dealt creationism a crippling blow. They did not conduct such a campaign. But the Moral Majority and the Institute for Creation Research did. So the fundamentalist line that the trial was a contest between atheism and the Lord went unchallenged, at least in volume and stridency.

> The court would never criticize or discredit any person's testimony based on his or her religious beliefs. While anybody is free to approach a scientific inquiry in any fashion he chooses, he cannot properly describe the methodology used as scientific if he starts with a conclusion and refuses to change it, regardless of the evidence developed during the course of the investigation.
>
> —Judge William R. Overton

But that is a cavil next to the brilliant show the ACLU's witnesses made during the trial last December. Arkansans can thank their governor and legislature for provoking a first-rate seminar on science and theology, featuring an array of erudite men and women whose like we would not otherwise have seen in five years of visiting lecturers. University of Chicago theologian Langdon Gilkey made such a forceful witness that he had the fundamentalist preachers who crowded the back of the courtroom nodding and buzzing in agreement when he dissected the language of Act 590 to reveal at every turn the unacknowledged authority of Genesis, the very phrase "sudden creation of the universe, energy, and life from nothing" implying not only God, but the God of the Old Testament alone.

Of course most of those preachers are simple souls, not up to the rapid donning and doffing of hats required to maintain that creationism is "scientific" and Act 590's purpose is secular. Unlike many of the state's witnesses, they have never wandered in the wood of materialism and doubt. Evolution is to them an unholy fairy tale whose premises they have never credited for one moment. As for the U.S. Constitution, why, if the Founding Fathers had meant for us to separate God's word from our government, the word "Creator" would not appear in the Declaration of Independence. Only communists think otherwise. When Cornell sociologist Dorothy Nelkin said in cross-examination that she was an atheist, there was a muted gasp in the back of the courtroom. Several heads bowed in prayer.

As the plaintiff's witnesses went on, the courtroom took on most of the aspects not of a religious, but an academic, camp meeting. Except for Moral Majoritarians and creationists taking notes, most of the militant godly among the spectators disappeared, to be replaced by honors biology classes from the local high schools and professors from the Little Rock campuses of the University of Arkansas. Even had I not recognized many of the latter, style would have told: in our corner of the world, as the British reporters on hand rapidly established, creationists go in for synthetic fabrics, styled hair, or toupees, while evolutionists sport khaki, wool, and facial hair.

As an academic camp meeting, the first week of the trial was most inspiring to this apostate English professor. Having years ago wearied of the posturings of most academic literary types, I suppose I had grown more than half dubious that useful thinking was going on anywhere in the academic world. But to hear philosopher of science Michael Ruse of the University of Guelph explain how science both limits and lays claim to knowledge, and to be able to listen to such literate practitioners as geneticist Francisco Ayala of the University of California, biophysicist Harold Morowitz of Yale, Harvard's versatile paleontologist Stephen Jay Gould, and Brent Dalrymple of the U.S. Geological Survey was a rare privilege. There may be something more to our species after all than the lust for power and things. Thank you, Governor White.

> A gorilla, true enough, cannot write poetry and neither can it grasp such a concept as that of Americanization or that of relativity, but . . . in some ways, indeed, it is measurably more clever than many men. It cannot be fooled as easily; it does not waste so much time doing useless things. If it desires, for example, to get a banana, hung out of reach, it proceeds to the business with a singleness of purpose and a fertility of resource that, in a traffic policeman, would seem almost pathological. There are no fundamentalists among the primates. They believe nothing that is not demonstrable. When they confront a fact they recognize it instantly, and turn it to their uses with admirable readiness. There are liars among them, but no idealists.
>
> —H. L. Mencken

It was hard not to pity Attorney General Steve Clark and his outgunned staff, attempting to show that the creationism law had no religious origins. The record contained letters from the law's author. "I view this whole battle as one between God and anti-God forces," Paul Ellwanger had written. He advised his supporters to conceal their sacred motives, lest the courts catch on; if they could not forbear witnessing for the Lord when petitioning their representatives, they should reserve the apologetics for a separate sheet of paper.

The text of the law itself betrayed its intent at every turn. Here, for example, is the definition of the law of creationism:

> "Creation-science" means the scientific evidences for creation and inferences from those scientific evidences. Creation-science includes the scientific evidences and related inferences that indicate: (1) Sudden creation of the universe, energy, and life from nothing; (2) The insufficiency of mutation and natural selection in bringing about development of all living kinds from a single organism; (3) Changes only within fixed limits of originally created kinds of plants and animals; (4) Separate ancestry for man and apes; (5) Explanation of the earth's geology by catastrophism, including the occurrence of a worldwide flood; and (6) A relatively recent inception of the earth and living kinds.

In nearly two weeks of testimony, no scientist, whether "creation" or otherwise, could enlighten the court as to the exact meaning of "kind." Creationist Wayne A. Friar of King's College, Briarcliff Manor, N.Y., said it could mean "species," "genus," "family," or even "order" in which case number four above stands contradicted, since Adam and Eve, Governor White, and Bonzo the Chimp all belong to the order of primates. Friar, who labors at refuting Darwin by comparing the blood-cell sizes of various turtles, said he was still working on the problem of whether or not the shelled beasts constitute a kind. Neither turtles nor tortoises, of course, are specifically mentioned in Genesis 1:11-12 and 21-25, where the concept of "kinds" originates; some, indeed, are "swimming creatures, with which the waters abound," others "animals that crawl on the earth." It is a difficult problem.

Perhaps no more difficult, though, than the problem the attorney general faced in seeking expert witnesses for creationism. Everybody, both in the movement and outside it, cites Henry Morris and Duane Gish of the Institute for Creation Research as not simply *the* authorities on the subject, but in fact its originators. Both, unfortunately, are prolific authors. Putting them on the stand to prove creationism to be science would be like calling Richard Nixon to testify that politicians never lie. In his treatise *Scientific Creationism,* for example, Morris says:

> A. Creation cannot be proved
> 1. Creation . . . is inaccessible to the scientific method.
> 2. It is impossible to devise a scientific experiment to describe the creation process, or even to ascertain whether such a process *can* take place. The creator does not create at the whim of a scientist.

The learned Dr. Gish—he has a Ph.D. in biochemistry from Berkeley—is similarly honest, at least part of the time. In *Evolution? The Fossils Say No!,* he puts it this way:

> We do not know how the Creator created, what processes He used, for *He used processes which are not now operating anywhere in the natural universe* [his emphasis]. This is why we refer to creation as special creation. We cannot discover by scientific investigations anything about the creative processes used by the Creator.

As an article of faith, of course, Gish's is a perfectly sound position and places creationism exactly where it belongs: outside science's claim to know. In a recent letter to *Discover* magazine, though, Gish went further. He was responding to an article attacking creationism's pretensions:

> Stephen Jay Gould states that creationists claim creation is a scientific theory. *This is a false accusation.* Creationists have repeatedly stated that neither creation nor evolution is a scientific theory (and each is equally religious).

Yet this same eminence was everywhere in Little Rock during the trial, sitting two rows behind the state's lawyers, passing them notes, indulging in heated colloquies during recesses, and making pronouncements about the indubitable scientific merits of creationism for the television reporters. Indeed, the man's creator seems to have blessed him with a tropism for bright lights and camera lenses.

Despite a definitely simian aspect, which made him the butt of many cruel jokes in the press row, Gish is in fact a masterful artist of the televised debate, that bastard form of showmanship first visited on us by presidential politics. During the trial, good old Jerry Falwell, of Moral Majority and "Old Time Gospel Hour" fame, staged just such a confrontation between Gish and Professor Russell Doolittle, a chemist from the University of California, who was naïve enough to think that the winners and losers of such events are determined by evidence and logic. Perhaps in graduate seminars and laboratories they are, but for all of his earnest learning, Doolittle might as well have been trading insults with Johnny Carson. Gish's presentation was timed to the minute and consisted of a premium assortment of half-truths, semifacts, quasi-logic, outright falsehoods, and simple balderdash. All replete, of course, with scriptural authority. In front of the audience in Falwell's Lynchburg church, cheering and whistling to see the infidel routed, Gish was triumphant.

Gish argued, for example, that the Second Law of Thermodynamics renders evolution impossible. How childish of "evolution scientists" to imagine, he implied, that they could push this ludicrous hoax past such a learned and reverent authority as himself. What the cheering faithful do not know, however, is that the Second Law of Thermodynamics states almost the opposite of what Gish says it does. In a closed system, it is true, greater organization of heat energy cannot occur. A closed system is one that energy is not entering from the outside. In an open system, into which energy does flow, increased organization of energy can and does occur. Until very recently, when scientists simply ignored the creationists, Gish and his followers did not trouble to make the distinction, although if the Second Law meant what they said it did, not only evolution but life itself would be impossible. On the "Old Time Gospel Hour" debate Gish even slipped for a brief moment into the Old Time Second Law of Thermodynamics, telling the audience that "on the hypothetical primordial earth, you did not have an energy conversion machine." This is heresy. According to my Bible, God created the sun on the first day. Perhaps Gish reads a different translation.

Soon enough, though, he was back on course, acknowledging his critics by maintaining that the very "universe itself is an isolated system." This is why scientists should not debate the man. By the time they got through explaining that a "closed system" in thermodynamics and the "closed universe" hypothesized by some astrophysicists had no more bearing on the matter than a closed checking account or a closed mind, Gish's audience would have swallowed his pun whole and be uttering hosannahs.

With creationism's chief apologists eliminated as potential witnesses by reason of their own past words, Attorney General Clark had no recourse but to call creationists who had published little or nothing. What began in the first week as a fine seminar degenerated into a boring farce with overtones of pathos. That the state's case was incoherent was no fault of the lawyers: it was Act 590 that bequeathed to them the "two-model approach," in turn taken from creationist authors, who in turn plagiarized the notion, as I have suggested earlier, from the "equal time" doctrine that allots television coverage to political candidates in our imperfect world of Republicans and Democrats. Briefly stated, the argument runs like this: "evolution science" posits atheism. "Creation science," while not religious, of course, posits theism. There are no other possibilities. Either there is a God, in which case "evolution science" is falsified, or there is not, in which case . . . But let us not get into that thicket. Suffice it to say, though, that the theory of evolution does not posit atheism. Science agrees to exclude the supernatural, yes. But so do accounting, law, and the rules of baseball. Are we now to have Bowie Kuhn denounced as a godless purveyor of materialistic satanism? Perhaps a creation baseball league will be next.

It was by such incantation that Clark, a handsome man so cleft of chin that he could have been sent down from Central Casting to play the role of up-and-coming Southern moderate, hoped to avoid political disaster. Because a luncheon with Clark had once been auctioned off at an ACLU fund-raiser, the Attorney General was denounced as "crooked" by Pat Robertson, the host of the Christian Broadcast Network's "700 Club," and accused of deliberately throwing the trial. The Rev. Dr. Falwell followed with a similar charge, although the Arkansas Moral Majority confined itself to attacking Judge Overton's impartiality. Even so, Clark is probably safe. Historically, Arkansans have resented outside criticism of their own politicians, and Clark had never sought the hardcore fundamentalist vote anyway.

The "two-model approach" allowed Clark to pretend what creationists pretend: that all evidence against any aspect of any scientific theory tending to support evolution constitutes proof of creationism. Logically, of course, this is like saying that evidence I was not in Little Rock last Wednesday establishes that I was in fact golfing on Mars. Hence scientists were easily convicted of doing science. Does Stephen Jay Gould's theory of "punctuated equilibrium"—i.e., of evolutionary change in relatively rapid bursts, with cataclysms altering the environment—disagree with those of more orthodox theorists who think the process has been more gradual? Very well. Both are refuted and creationism proved. One of the funnier moments in the trial's first week came when one young barrister tried to ensnare the wickedly articulate Francisco Ayala, a former priest with scientific training *and* the equivalent of a doctorate in theology, into admitting the validity of the two-model approach. "Your name," the scientist told the expectant young lawyer, "is Mr. Williams. But my name isn't not-Mr. Williams. The courtroom is full of people whose name isn't not-Mr. Williams." The real Mr. Williams changed his line of questioning. At another point, a state's attorney asked Professor Morowitz of Yale: "Can you tell me the name of one Ivy League university that has a creation scientist on its staff?" Morowitz could not. Neither could he name any other prestigious graduate school or journal that employed creationists. Morowitz added, "I can't give you the name of an Ivy League school, graduate school or journal which houses a flat earth theorist either."

The state's most coherent witness by far was Dr. Norman Geisler of the Dallas Theological Seminary. It was Geisler who admitted, under cross-examination, that besides the two-model theory, he also believed that UFOs were "a satanic manifestation in the world for the purpose of deception." As nearly as I can work it out, Geisler believes that any abstract idea held strongly by any number of people constitutes what he calls "transcendence," and answers his definition of religion. He quoted, as all good fundamentalists do, from something called *The Humanist Manifesto,* and intimated that because there is such an organization, and because a footnote in a Supreme Court decision once classified that organization as a religious one, that all persons who are "humanists" are acolytes of that faith. I shall refrain from insulting *Harper's* readers by letting them work out the syllogism themselves. It is by such arguments that fundamentalist "intellectuals" propose to render the First Amendment tautological and thus useless: if all intellectual positions are equally "religious" in nature, then why bother?

The most profound part of Geisler's testimony was his attempt to prove that the "Creator" of the universe and life mentioned in Act 590 was not an inherently religious concept. After citing Aristotle, Plato, and one or two other classical philosophers who supposedly believed in a God or gods without worshiping them—albeit not as creators of the world "from nothing"—Geisler offered his most thundering proof: the Epistle of James. He cited a line of Scripture to the effect that Satan acknowledges God, but chooses not to worship Him. "The Devil," he said, "believes that there is a God." Whee! If Geisler has not yet squared the circle in his meditations, he has at least, well, circled it. Who would have thought one could prove the Creator a nonreligious idea by means of hearsay evidence from Beelzebub? After unloading that bombshell, Geisler, too, hastened to face the cameras in the courtroom hallway. "We don't rule out stones from a geology class just because some people have worshiped stones, and we don't rule God out of science class because some believe in him." As I listened to Geisler I could not help but recall the words of the Rev. C. O. Magee, a Presbyterian minister who is a member of the Little Rock School Board. "Any time religion gets involved in science," Magee told the *Gazette,* "religion comes off looking like a bunch of nerds. . . . The Book of Genesis told who created the world and why it was created and science tells how it was done." Amen.

After Geisler, the state's case went straight downhill. These witnesses were supposedly learned men, possessing advanced degrees, most of them resident in institutions that purported to be colleges and universities. Some of my own prejudices against academia would have revived, except that this collection of sad sacks, flub-a-dubs, and third-rate hobbyists had been gleaned mostly from the kinds of schools where the faculty must sign pledges certifying their literal belief in the factual inerrancy of the Bible, and were not, in the post-Enlightenment sense, really academic institutions at all. (The Institute for Creation Research requires such a pledge.) Most were like Donald Chittick, a physical chemist from George Fox College in Newberg, Oregon. Chittick spent hours telling the court how fuel could actually be made very rapidly from "biomass" materials. (In the Ozarks, of course, a good deal of biomass fuel has been distilled and drunk over the years.) To Chittick's mind, this proves that the world does have to be 4.5 billion years old at all. Chittick's most telling point was that the amount of helium present in the earth's atmosphere indicates that radioactive decay has been taking place on earth for about 10,000 years only. That is just about how old creationists say the earth is. Either Chittick did not mention, or does not know, that helium is too light to be held by earth's gravity and disperses constantly into space.

The trial's only poignant moment came during cross-examination of Harold Coffin, a dreadfully earnest Seventh-Day Adventist who spends his time floating horsetail ferns in tanks of water to demonstrate that their fossilized ancestors found standing upright in coal seams hundreds of feet thick could have floated to that position during Noah's flood. Coffin was asked to say how old the earth would seem to a person unaided by Scripture, and considering only the available scientific evidence. Coffin paused for what seemed five minutes before answering, so it must have been at least fifteen seconds. As old as evolutionists claim, he said, about 4.5 billion years.

To his credit, Judge Overton kept his patience throughout, although he did seem to be losing it once with a pompous faculty lounge-lizard type from Wofford College in South Carolina, one W. Scott Morrow, a chemist who claimed to be an "evolutionist," but took it upon himself to testify to the closed-mindedness of "my fellow evolutionists." After more than an hour's worth of plausible generalities about how scientists are slow to accept new ideas, Overton asked Morrow if scientific papers were ever rightly rejected. He said he couldn't answer, as he'd never been an editor. Pressed by Overton for one specific example of a scientifically valid creationist paper's having been rejected, Morrow could not provide one. (Indeed, in the course of the trial the state could not produce a single creationist paper that had been published in a refereed scientific journal anywhere in the world, nor even one that had been submitted.) "Are you saying," the judge challenged, "that the entire national and international scientific community is engaged in a conspiracy?" Morrow replied that he knew a lot of his colleagues in science and "I know a closed mind when I see one." Afterward Morrow, too, hustled in the direction of the cameras, and told the press that the judge wasn't paying attention and was obviously biased. Then he beat it back to South Carolina, which is welcome to him. Have I mentioned that there was only one Arkansan among the creationist witnesses?

The pro-creationist witness who traveled farthest for the trial, however, was one Dr. Chandra Wickramasinghe, a native of Sri Lanka who teaches mathematics in Wales. Having allied himself several years ago with Sir Fred Hoyle, the notable English astronomer, who seems to have slipped into scientific dotage, Wickramasinghe has collaborated with his mentor on two books that have done very well on best-seller lists in England, *Life Cloud* and *Diseases from Space.* The first book is an elaboration of a science-fiction novel by Hoyle which I read about twenty-five years ago. It posits that life originated in swirling clouds of intergalactic dust and was brought to earth by a comet. So far the hypothesis has not been falsified, but at the moment it cannot be seriously tested either. Wickramasinghe seemed astonished that he would be cross-examined, and devolved from condescension to giggles when forced after protesting that he was being taken out of context to read three paragraphs of *Evolution from Space.* Insects, the passage said, may be more intelligent than human beings, but pretend stupidity because they don't want us to know what a great deal they have on earth. Well, maybe so. Then again, maybe not.

Diseases from Space elaborates on the idea that viral epidemics are in fact visited on us from the great beyond and asserts that viruses cannot be transmitted horizontally from one human being to another. This hypothesis provoked the best joke of the trial. If viruses cannot be transmitted from one person to another, some unknown wag on the ACLU side wondered, then how about the following scenario: a man comes home and tells his wife, "Honey, I've got good news and bad news. The bad news is I've caught herpes. The good news is it came from outer space." As for the creationist notion that the universe is just 10,000 years old, Dr. Wickramasinghe said, "one would have to be crazy to believe that."

A Blow for Theocracies

We are all the poorer for Attorney General Clark's decision not to appeal Judge Overton's ruling that the creationism law is unconstitutional. No rationally consequent adult who sat through Little Rock's creationism trial can have expected another outcome. Even the Moral Majority's fulminations were clearly a reaction to the dismal showing the creationist witnesses made. Examined in the light of reason, with evidence honestly given and logically assessed, creationism cannot prevail. Unlike a televised debate or a local school-board committee meeting, the trial was a fair fight. But nothing said that Overton's opinion—and I hope readers will have patience with my pointing out that he was educated at Malvern (Ark.) High School and the University of Arkansas—would be as cogent and well written as it was. Many of the creationist faithful were privately contacting the attorney general's office here to advise against appeal. They would like to believe they have a better chance in Louisiana, where the local authorities have deputized the Institute for Creation Research's Wendell Bird. Overton dismissed Bird's argument as having no legal merit:

> If creation science is, in fact, science and not religion, as the defendants claim, it is difficult to see how the teaching of such a science could "neutralize" the religious nature of evolution.
>
> Assuming for the purposes of argument, however, that evolution is a religion or religious tenet, the remedy is to stop the teaching of evolution, not establish another religion in opposition to it. Yet it is clearly established in the case law, and perhaps also in common sense, that evolution is not a religion and that teaching evolution does not violate the establishment clause.

It is equally clear that the state has a "compelling interest" in the teaching not only of biological science, of which the theory of evolution is the fundamental organizing principle, but also of chemistry, physics, geology, and even history, all subjects that would have required "balancing" with creationist gibberish if Act 590 had stood. Where that is the case, the Supreme Court has ruled many times, aggrieved fundamentalists who do not wish to have their children hear what offends them, and wish the shelter of the "free exercise" clause of the First Amendment, are permitted to withdraw their children from science classes or from public school.

Ultimately, the creationists cannot prevail in the courts. Now that the scientific community and the educated public are aroused by the Little Rock spectacle, I doubt that a bill in the U.S. Congress of Rep. William Dannemeyer's (Rep.-Calif.), which would limit funding for the Smithsonian Institution if it refuses to put up creationist exhibits, will get anywhere either. So long as current attempts to limit the power of the federal judiciary are fought back—Arkansas's Act 590 controversy being a textbook example of the political cowardice that has led to courts currently having more power than most of us are comfortable with—we will not have a theocracy in this country, fundamentalist or otherwise. Leave it up to the Arkansas legislature, and in five years we would have an Inquisition.

Creationism was mortally damaged by the Little Rock spectacle. That is why the slippery Dr. Duane Gish now says he thinks state laws mandating its teaching are a mistake; he wants to go back to strong-arming local school boards, as in the past. In fact, the Moral Majority and its politico-religious allies, I believe, will soon be muttering only to each other again. One could not observe the Arkansas Moral Majority head, the Rev. Roy McLaughlin, in action in his modernistic pulpit in Vilonia without speculating that his boyish charm—he looks like a sort of cross between Pat Boone and Howdy Doody—might just be wearing a mite thin. Arkansans may be hotter than most citizens for that old-time fundamental religion, but are they really ready to credit McLaughlin when he says, with unmistakable reference to the clergymen on the opposite side of the creationism case, that "a preacher who does not believe the word of God to be the inspired, inerrant, infallible word of God . . . is a crook and he ought to resign his pulpit . . . and quit robbing money from God's people"? Even out on the dirt roads, they know McLaughlin is talking about their friends and neighbors. In the long run, Arkansas folks aren't *mean* enough for that.

Before closing, I should bring readers up to date on the activities of two of creationism's hotter enthusiasts. A month after the trial, Senator James Holsted, having repaid the money he was accused of embezzling from his family's insurance company, pleaded guilty to the lesser charge of filing a false statement. As part of the deal with the Pulaski County prosecutor, he resigned his senate seat. Yet he told reporters not to count him out of politics for good. "It's life," he smiled. "A winner never accepts defeat."

Governor Frank White is a winner too. He had considerable trouble fulfilling his pledge to rid Arkansas of black Cubans. By January 1982, though, the government had succeeded in finding sponsors for all but 300 of them. The Reagan Administration talked of moving them to a camp outside Glasgow, Montana. But January temperatures often reach thirty below zero there. Whether it was the parallels to Siberia or the exorbitant cost of preparing a new site just to fulfill a campaign promise to the governor of a small state Reagan had in his pocket anyway, the idea was dropped. White was getting worried. The awful Cubans had been at Fort Chaffee longer than they were under Democrat Bill Clinton. Far from "standing up to the President," he wrote beseeching letters to Washington, which found their way into the newspaper, reminding him that "my political future" depended upon their removal. Finally Reagan acted. Buses appeared one night before dawn at the gates of Fort Chaffee. The Cubans, all of them impoverished, many physically or mentally ill, some simply retarded, but none convicted of any crimes in American courts of law, were loaded and taken away to Federal prisons. The Arkansas ACLU tried to intervene, but by the time it found out, the Cubans were safely out of state. In the morning paper there was a letter from a lady in Texarkana who said White had committed an obscenity in our names. The *Clarendon Sentinel,* a paper printed so far out in the country it's halfway back to town, reminded its readers of the internment of Japanese-Americans in camps not far from there during World War II. They said White's and Reagan's action filled them with "despair and shame." A few days later White flew off to Washington with a bevy of fundamentalist preachers for a prayer breakfast with the President.

POSTSCRIPT:

Governor Frank White was defeated for reelection in November 1982 by Democrat Bill Clinton, whom he had defeated in 1980. Although Clinton said he would not have signed Act 590 bill and White reaffirmed his stand for it, creationism was not a significant issue in the campaign. Attorney General Steve Clark was reelected, taking 73 percent of the vote against a candidate who criticized him for not winning the case.

DECISION OF THE COURT

Judgment

Pursuant to the Court's Memorandum Opinion filed this date, judgment is hereby entered in favor of the plaintiffs and against the defendants. The relief prayed for is granted.

Dated this January 5, 1982.

Injunction

Pursuant to the Court's Memorandum Opinion filed this date, the defendants and each of them and all their servants and employees are hereby permanently enjoined from implementing in any manner Act 590 of the Acts of Arkansas of 1981.

It is so ordered this January 5, 1982.

Memorandum Opinion

Introduction

On March 1, 1981, the Governor of Arkansas signed into law Act 590 of 1981, entitled the "Balanced Treatment for Creation-Science and Evolution-Science Act." The Act is codified as Ark. Stat. Ann. §80-1663, *et seq.* (1981 Suppl.). Its essential mandate is stated in its first sentence: "Public schools within this State shall give balanced treatment to creation-science and to evolution-science." On May 27, 1981, this suit was filed[1] challenging the constitutional validity of Act 590 on three distinct grounds.

First, it is contended that Act 590 constitutes an establishment of religion prohibited by the First Amendment to the Constitution, which is made applicable to the states by the Fourteenth Amendment. Second, the plaintiffs argue the Act violates a right to academic freedom which they say is guaranteed to students and teachers by the Free Speech Clause of the First Amendment. Third, plaintiffs allege the Act is impermissibly vague and thereby violates the Due Process Clause of the Fourteenth Amendment.

The individual plaintiffs include the resident Arkansas Bishops of the United Methodist, Episcopal, Roman Catholic and African Methodist Episcopal Churches, the principal official of the Presbyterian Churches in Arkansas, other United Methodist, Southern Baptist and Presbyterian clergy, as well as several persons who sue as parents and next friends of minor children attending Arkansas public schools. One plaintiff is a high school biology teacher. All are also Arkansas taxpayers. Among the organizational plaintiffs are the American Jewish Congress, the Union of American Hebrew Congregations, the American Jewish Committee, the Arkansas Education Association, the National Association of Biology Teachers and the National Coalition for Public Education and Religious Liberty, all of which sue on behalf of members living in Arkansas.[2]

The defendants include the Arkansas Board of Education and its members, the Director of the Department of Education, and the State Textbooks and Instructional Materials Selecting Committee.[3] The Pulaski County Special School District and its Directors and Superintendent were voluntarily dismissed by the plaintiffs at the pre-trial conference held October 1, 1981.

The trial commenced December 7, 1981, and continued through December 17, 1981. This Memorandum Opinion constitutes the Court's findings of fact and conclusions of law. Further orders and judgment will be in conformity with this opinion.

I

There is no controversy over the legal standards under which the Establishment Clause portion of this case must be judged. The Supreme Court has on a number of occasions expounded on the meaning of the clause, and the pronouncements are clear. Often the issue has arisen in the context of public education, as it has here. In *Everson v. Board of Education,* 330 U.S. 1, 15–16 (1947), Justice Black stated:

> The "establishment of religion" clause of the First Amendment means at least this: Neither a state nor the Federal Government can set up a church. Neither can pass laws which aid one religion, aid all religions, or prefer one religion over another. Neither can force nor influence a person to go to or to remain away from church against his will or force him to profess a belief or disbelief in any religion. No person can be punished for entertaining or professing religous beliefs or disbeliefs, for church-attendance or non-attendance. No tax, large or small, can be levied to support any religious activities or institutions, whatever they may be called, or whatever form they may adopt to teach or practice religion. Neither a state nor the Federal Government can, openly or secretly, participate in the affairs of any religious organizations or groups and *vice versa.* In the words of Jefferson, the clause . . . was intended to erect "a wall of separation between church and State."

The Establishment Clause thus enshrines two central values: voluntarism and pluralism. And it is in the area of the public schools that these values must be granted most vigilantly.

> Designed to serve as perhaps the most powerful agency for promoting cohesion among a heterogeneous democratic people, the public school must keep scrupulously free from entanglement in the strife of sects. The preservation of the community from divisive conflicts, of Government from irreconcilable pressures by religious groups, of religion from censorship and coercion however subtly exercised, requires strict confinement of the State to instruction other than religious, leaving to the individual's church and home, indoctrination in the faith of his choice. [*McCollum v. Board of Education,* 333 U.S. 203, 216–217 (1948), (Opinion of Frankfurter, J., joined by Jackson, Burton and Rutledge, J.J.).]

The specific formulation of the establishment prohibition has been refined over the years, but its meaning has not varied from the principles articulated by Justice Black in *Everson.* In *Abbington School District v. Schempp,* 374 U.S. 203, 222 (1963), Justice Clark stated that "to withstand the strictures of the Establishment Clause there must be a secular legislative purpose and a primary effect that neither advances nor inhibits religion." The Court found it quite clear that the First Amendment does not permit a state to require the daily reading of the Bible in public schools, for "[s]urely the place of the Bible as an instrument of religion cannot be gainsaid." *Id.* at 224. Similarly, in *Engel v. Vitale,* 370 U.S. 421 (1962), the Court held that the First Amendment prohibited the New York Board of Regents from requiring the daily recitation of a certain prayer in the schools. With characteristic succinctness, Justice Black wrote, "Under [the First] Amendment's prohibition against the governmental establishment of religion, as reinforced by the provisions of the Fourteenth Amendment, government in this country, be it state or federal, is without power to prescribe by law any particular form of prayer which is to be used as an official prayer in carrying on any program of governmentally sponsored religious activity." *Id.* at 430. Black also identified the objective at which the Establishment Clause was aimed: "Its first and most immediate purpose rested on the belief that a union of government and religion tends to destroy government and to degrade religion." *Id.* at 431.

Most recently, the Supreme Court has held that the clause prohibits a state from requiring the posting of the Ten Commandments in public school classrooms for the same reasons that officially imposed daily Bible reading is prohibited. *Stone v. Graham,* 449 U.S. 39 (1980). The opinion in *Stone* relies on the most recent formulation of the Establishment Clause test, that of *Lemon v. Kurtzman,* 403 U.S. 602, 612–613 (1971):

> First, the statute must have a secular legislative purpose; second, its principal or primary effect must be one that neither advances nor inhibits religion . . .; finally, the statute must not foster "an excessive government entanglement with religion." [*Stone v. Graham,* 449 U.S. at 40]

It is under this three part test that the evidence in this case must be judged. Failure on any of these grounds is fatal to the enactment.

II

The religious movement known as Fundamentalism began in nineteenth century America as part of evangelical Protestantism's response to social changes, new religious thought and Darwinism. Fundamentalists viewed these developments as attacks on the Bible and as responsible for a decline in traditional values.

The various manifestations of Fundamentalism have had a number of common characteristics,[4] but a central premise has always been a literal interpretation of the Bible and a belief in the inerrancy of the Scriptures. Following World War I, there was again a perceived decline in traditional morality, and Fundamentalism focused on evolution as responsible for the decline. One aspect of their efforts, particularly in the South, was the promotion of statutes prohibiting the teaching of evolution in public schools. In Arkansas, this resulted in the adoption of Initiated Act 1 of 1929.[5]

Between the 1920's and early 1960's, anti-evolutionary sentiment had a subtle but pervasive inffuence on the teaching of biology in public schools. Generally, textbooks avoided the topic of evolution and did not mention the name of Darwin. Following the launch of the Sputnik satellite by the Soviet Union in 1957, the National Science Foundation funded several programs designed to modernize the teaching of science in the nation's schools. The Biological Sciences Curriculum Study (BSCS), a nonprofit organization, was among those receiving grants for curriculum study and revision. Working with scientists and teachers, BSCS developed a series of biology texts which, although emphasizing different aspects of biology, incorporated the theory of evolution as a major theme. The success of the BSCS effort is shown by the fact that fifty percent of American school children currently use BSCS books directly and the curriculum is incorporated indirectly in virtually all biology texts. (Testimony of Mayer: Nelkin, Px 1)[6]

In the early 1960's, there was again a resurgence of concern among Fundamentalists about the loss of traditional values and a fear of growing secularism in society. The Fundamendalist movement became more active and has steadily grown in numbers and political influence. There is an emphasis among current Fundamentalists on the literal interpretation of the Bible and the Book of Genesis as the sole source of knowledge about origins.

The term "scientific creationism" first gained currency around 1965 following publication of *The Genesis Flood* in 1961 by Whitcomb and Morris. There is undoubtedly some connection between the appearance of the BSCS texts emphasizing evolutionary thought and efforts by Fundamentalists to attack the theory. (Mayer)

In the 1960's and early 1970's, several Fundamentalist organizations were formed to promote the idea that the Book of Genesis was supported by scientific data. The terms "creation science" and "scientific creationism" have been adopted by these Fundamentalists as descriptive of their study of creation and the origins of man. Perhaps the leading creationist organization is the Institute for Creation Research (ICR), which is affiliated with the Christian Heritage College and supported by the Scott Memorial Baptist Church in San Diego, California. The ICR, through the Creation-Life Publishing Company, is the leading publisher of creation science material. Other creation science organizations include the Creation Science Research Center (CSRC) of San Diego and the Bible Science Association of Minneapolis, Minnesota. In 1963, the Creation Research Society (CRS) was formed from a schism in the American Scientific Affiliation (ASA). It is an organization of literal Fundamentalists[7] who have the equivalent of a master's degree in some recognized area of science. A purpose of the organization is "to reach all people with the vital message of the scientific and historic truth about creation." Nelkin, *The Science Textbook Controversies and the Politics of Equal Time,* 66. Similarly, the CSRC was formed in 1970 from a split in the CRS. Its aim has been "to reach the 63 million children of the United States with the scientific teaching of Biblical creationism." *Id.* at 69.

Among creationist writers who are recognized as authorities in the field by other creationists are Henry M. Morris, Duane Gish, G. E. Parker, Harold S. Slusher, Richard B. Bliss, John W. Moore, Martin E. Clark, W. L. Wysong, Robert E. Kofahl and Kelly L. Segraves. Morris is Director of ICR, Gish is Associate Director and Segraves is associated with CSRC.

Creationists view evolution as a source of society's ills, and the writings of Morris and Clark are typical expressions of that view.

> Evolution is thus not only anti-Biblical and anti-Christian, but it is utterly unscientific and impossible as well. But it has served effectively as the pseudo-scientific basis of atheism, agnosticism, socialism, fascism, and numerous other false and dangerous philosophies

> over the past century. [Morris and Clark, *The Bible Has The Answer,* (Px 31 and Pretrial Px 89).[8]]

Creationists have adopted the view of Fundamentalists generally that there are only two positions with respect to the origins of the earth and life: belief in the inerrancy of the Genesis story of creation and of a worldwide flood as fact, or belief in what they call evolution.

Henry Morris has stated, "It is impossible to devise a legitimate means of harmonizing the Bible with evolution." Morris, "Evolution and the Bible," *ICR Impact Series* Number 5 (undated, unpaged), quoted in Mayer, Px 8, at 3. This dualistic approach to the subject of origins permeates the creationist literature.

The creationist organizations consider the introduction of creation science into the public schools part of their ministry. The ICR has published at least two pamphlets[9] containing suggested methods for convincing school boards, administrators and teachers that creationism should be taught in public schools. The ICR has urged its proponents to encourage school officials to voluntarily add creationism to the curriculum.[10]

Citizens For Fairness in Education is an organization based in Anderson, South Carolina, formed by Paul Ellwanger, a respiratory therapist who is trained in neither law nor science. Mr. Ellwanger is of the opinion that evolution is the forerunner of many social ills, including Nazism, racism and abortion. (Ellwanger Depo. at 32–34). About 1977, Ellwanger collected several proposed legislative acts with the idea of preparing a model state act requiring the teaching of creationism as science in opposition to evolution. One of the proposals he collected was prepared by Wendell Bird, who is now a staff attorney for ICR.[11] From these various proposals, Ellwanger prepared a "model act" which calls for "balanced treatment" of "scientific creationism" and "evolution" in public schools. He circulated the proposed act to various people and organizations around the country.

Mr. Ellwanger's views on the nature of creation science are entitled to some weight since he personally drafted the model act which became Act 590. His evidentiary deposition with exhibits and unnumbered attachments (produced in response to a subpoena *duces tecum*) speaks to both the intent of the Act and the scientific merits of creation science. Mr. Ellwanger does not believe creation science is a science. In a letter to Pastor Robert E. Hays he states, "While neither evolution nor creation can qualify as a scientific theory, and since it is virtually impossible at this point to educate the whole world that evolution is not a true scientific theory, we have freely used these terms—the evolution theory and the theory of scientific creationism—in the bill's text." (Unnumbered attachment to Ellwanger Depo. at 2.) He further states in a letter to Mr. Tom Bethell, "As we examine evolution (remember, we're not making any scientific claims for creation, but we are challenging evolution's claim to be scientific) . . ." (Unnumbered attachment to Ellwanger Depo. at 1.)

Ellwanger's correspondence on the subject shows an awareness that Act 590 is a religious crusade, coupled with a desire to conceal this fact. In a letter to State Senator Bill Keith of Louisiana, he says, "I view this whole battle as one between God and anti-God forces, though I know there are a large number of evolutionists who believe in God." And further, ". . . it behooves Satan to do all he can to thwart our efforts and confuse the issue at every turn." Yet Ellwanger suggests to Senator Keith, "If you have a clear choice between having grassroots leaders of this statewide bill promotion effort to be ministerial or non-ministerial, be sure to opt for the non-ministerial. It does the bill effort no good to have ministers out there in the public forum and the adversary will surely pick at this point . . . Ministerial persons can accomplish a tremendous amount of work from behind the scenes, encouraging their congregations to take the organizational and P.R. initiatives. And they can lead their churches in storming Heaven with prayers for help against so tenacious an adversary." (Unnumbered attachment to Ellwanger Depo. at 1.)

Ellwanger shows a remarkable degree of political candor, if not finesse, in a letter to State Senator Joseph Carlucci of Florida:

> 2. It would be very wise, if not actually essential, that all of us who are engaged in this legislative effort be careful not to present our position and our work in a religious framework. For example, in written communications that might somehow be shared with those other persons whom we may be trying to convince, it would

> be well to exclude our own personal testimony and/or witness for Christ, but rather, if we are so moved, to give that testimony on a separate attached note. (Unnumbered attachment to Ellwanger Depo. at 1.)

The same tenor is reflected in a letter by Ellwanger to Mary Ann Miller, a member of FLAG (Family, Life, America under God) who lobbied the Arkansas Legislature in favor of Act 590:

> . . . we'd like to suggest that you and your co-workers be very cautious about mixing creation-science with creation-religion . . . Please urge your co-workers not to allow themselves to get sucked into the 'religion' trap of mixing the two together, for such mixing does incalculable harm to the legislative thrust. It could even bring public opinion to bear adversely upon the higher courts that will eventually have to pass judgment on the constitutionality of this new law." (Ex. 1 to Miller Depo.)

Perhaps most interesting, however, is Mr. Ellwanger's testimony in his deposition as to his strategy for having the model act implemented:

> Q. You're trying to play on other people's religious motives.
> A. I'm trying to play on their emotions, love, hate, their likes, dislikes, because I don't know any other way to involve, to get humans to become involved in human endeavors. I see emotions as being a healthy and legitimate means of getting people's feelings into action, and . . . I believe that the predominance of population in America that represents the greatest potential for taking some kind of action in this area is a Christian community. I see the Jewish community as far less potential in taking action . . . but I've seen a lot of interest among Christians and I feel, why not exploit that to get the bill going if that's what it takes. (Ellwanger Depo. at 146–147.)

Mr. Ellwanger's ultimate purpose is revealed in the closing of his letter to Mr. Tom Bethell: "Perhaps all this is old hat to you, Tom, and if so, I'd appreciate your telling me so and perhaps where you've heard it before—the idea of killing evolution instead of playing these debating games that we've been playing for nigh over a decade already." (Unnumbered attachment to Ellwanger Depo. at 3.)

It was out of this milieu that Act 590 emerged. The Reverend W. A. Blount, a Biblical literalist who is pastor of a church in the Little Rock area and was, in February, 1981, chairman of the Greater Little Rock Evangelical Fellowship, was among those who received a copy of the model act from Ellwanger.[12]

At Reverend Blount's request, the Evangelical Fellowship unanimously adopted a resolution to seek introduction of Ellwanger's act in the Arkansas Legislature. A committee composed of two ministers, Curtis Thomas and W. A. Young, was appointed to implement the resolution. Thomas obtained from Ellwanger a revised copy of the model act which he transmitted to Carl Hunt, a business associate of Senator James L. Holsted, with the request that Hunt prevail upon Holsted to introduce the act.

Holsted, a self-described "born again" Christian Fundamentalist, introduced the act in the Arkansas Senate. He did not consult the State Department of Education, scientists, science educators or the Arkansas Attorney General.[13] The Act was not referred to any Senate committee for hearing and was passed after only a few minutes' discussion on the Senate floor. In the House of Representatives, the bill was referred to the Education Committee which conducted a perfunctory fifteen minute hearing. No scientist testified at the hearing, nor was any representative from the State Department of Education called to testify.

Ellwanger's model act was enacted into law in Arkansas as Act 590 without amendment or modification other than minor typographical changes. The legislative "findings of fact" in Ellwanger's act and Act 590 are identical, although no meaningful fact-finding process was employed by the General Assembly.

Ellwanger's efforts in preparation of the model act and campaign for its adoption in the states were motivated by his opposition to the theory of evolution and his desire to see the Biblical version of creation taught in the public schools. There is no evidence that the pastors, Blount, Thomas, Young or The Greater Little Rock Evangelical Fellowship were motivated by anything other than their religious convictions when proposing its adoption or during their lobbying efforts in its behalf. Senator Holsted's sponsorship and lobbying efforts in behalf of the Act were motivated solely by his religious beliefs and desire to see the Biblical version of creation taught in the public schools.[14]

The State of Arkansas, like a number of states whose citizens have relatively homogeneous religious beliefs, has a long history of official opposition to evolution which is motivated by adherence to Fundamentalist beliefs in the inerrancy of the Book of Genesis. This history is documented in Justice Fortas' opinion in *Epperson v. Arkansas,* 393 U.S. 97 (1968), which struck down Initiated Act 1 of 1929, Ark. Stat. Ann. §§80-1627–1628, prohibiting the teaching of the theory of evolution. To this same tradition may be attributed Initiated Act 1 of 1930, Ark. Stat. Ann. §80-1606 (Repl. 1980), requiring "the reverent daily reading of a portion of the English Bible" in every public school classroom in the State.[15]

It is true, as defendants argue, that courts should look to legislative statements of a statute's purpose in Establishment Clause cases and accord such pronouncements great deference. See, e.g., *Committee for Public Education & Religious Liberty v. Nyquist,* 413 U.S. 756, 773 (1973) and *McGowan v. Maryland,* 366 U.S. 420, 445 (1961). Defendants also correctly state the principle that remarks by the sponsor or author of a bill are not considered controlling in analyzing legislative intent. See, e.g., *United States v. Emmons,* 410 U.S. 396 (1973) and *Chrysler Corp. v. Brown,* 441 U.S. 281 (1979).

Courts are not bound, however, by legislative statements of purpose or legislative disclaimers. *Stone v. Graham,* 449 U.S. 39 (1980); *Abbington School Dist. v. Schempp,* 374 U.S. 203 (1963). In determining the legislative purpose of a statute, courts may consider evidence of the historical context of the Act, *Epperson v. Arkansas,* 393 U.S. 97 (1968), the specific sequence of events leading up to passage of the Act, departures from normal procedural sequences, substantive departures from the normal, *Village of Arlington Heights v. Metropolitan Housing Corp.,* 429 U.S. 252 (1977), and contemporaneous statements of the legislative sponsor, *Fed. Energy Admin. v. Algonquin SNG, Inc.,* 426 U.S. 548, 564 (1976).

The unusual circumstances surrounding the passage of Act 590, as well as the substantive law of the First Amendment, warrant an inquiry into the stated legislative purposes. The author of the Act had publicly proclaimed the sectarian purpose of the proposal. The Arkansas residents who sought legislative sponsorship of the bill did so for a purely sectarian purpose. These circumstances alone may not be particularly persuasive, but when considered with the publicly announced motives of the legislative sponsor made contemporaneously with the legislative process; the lack of any legislative investigation, debate or consultation with any educators or scientists; the unprecedented intrusion in school curriculum;[16] and official history of the State of Arkansas on the subject, it is obvious that the statement of purposes has little, if any, support in fact. The State failed to produce any evidence which would warrant an inference or conclusion that at any point in the process anyone considered the legitimate educational value of the Act. It was simply and purely an effort to introduce the Biblical version of creation into the public school curricula. The only inference which can be drawn from these circumstances is that the Act was passed with the specific purpose by the General Assembly of advancing religion. The Act therefore fails the first prong of the three-pronged test, that of secular legislative purpose, as articulated in *Lemon v. Kurtzman, supra,* and *Stone v. Graham, supra.*

III

If the defendants are correct and the Court is limited to an examination of the language of the Act, the evidence is overwhelming that both the purpose and effect of Act 590 is the advancement of religion in the public schools.

Section 4 of the Act provides:

> Definitions. As used in this Act:
> (a) "Creation-science" means the scientific evidences for creation and inferences from those scientific evidences. Creation-science includes the scientific evidences and related inferences that indicate: (1) Sudden creation of the universe, energy, and life from nothing; (2) The insufficiency of mutation and natural selection in bringing about development of all living kinds from a single organism; (3) Changes only within fixed limits of originally created kinds of plants and animals; (4) Separate ancestry for man and apes; (5) Explanation of the earth's geology by catastrophism, in-

cluding the occurrence of a worldwide flood; and (6) A relatively recent inception of the earth and living kinds.

(b) "Evolution-science" means the scientific evidences for evolution and inferences from those scientific evidences. Evolution-science includes the scientific evidences and related inferences that indicate: (1) Emergence by naturalistic processes of the universe from disordered matter and emergence of life from nonlife; (2) The sufficiency of mutation and natural selection in bringing about development of present living kinds from simple earlier kinds; (3) Emergence by mutation and natural selection of present living kinds from simple earlier kinds; (4) Emergence of man from a common ancestor with apes; (5) Explanation of the earth's geology and the evolutionary sequence by uniformitarianism; and (6) An inception several billion years ago of the earth and somewhat later of life.

(c) "Public schools" mean public secondary and elementary schools.

The evidence establishes that the definition of "creation science" contained in 4(a) has as its unmentioned reference the first 11 chapters of the Book of Genesis. Among the many creation epics in human history, the account of sudden creation from nothing, or *creatio ex nihilo,* and subsequent destruction of the world by flood is unique to Genesis. The concepts of 4(a) are the literal Fundamentalists' view of Genesis. Section 4(a) is unquestionably a statement of religion, with the exception of 4(a)(2) which is a negative thrust aimed at what the creationists understand to be the theory of evolution.[17]

Both the concepts and wording of Section 4(a) convey an inescapable religiosity. Section 4(a)(1) describes "sudden creation of the universe, energy and life from nothing." Every theologian who testified, including defense witnesses, expressed the opinion that the statement referred to a supernatural creation which was performed by God.

Defendants argue that: (1) the fact that 4(a) conveys ideas similar to the literal interpretation of Genesis does not make it conclusively a statement of religion; (2) that reference to a creation from nothing is not necessarily a religious concept since the Act only suggests a creator who has power, intelligence and a sense of design and not necessarily the attributes of love, compassion and justice;[18] and (3) that simply teaching about the concept of a creator is not a religious exercise unless the student is required to make a commitment to the concept of a creator.

The evidence fully answers these arguments. The ideas of 4(a)(1) are not merely similar to the literal interpretation of Genesis; they are identical and parallel to no other story of creation.[19]

The argument that creation from nothing in 4(a)(1) does not involve a supernatural deity has no evidentiary or rational support. To the contrary, "creation out of nothing" is a concept unique to Western religions. In traditional Western religious thought, the conception of a creator of the world is a conception of God. Indeed, creation of the world "out of nothing" is the ultimate religious statement because God is the only actor. As Dr. Langdon Gilkey noted, the Act refers to one who has the power to bring all the universe into existence from nothing. The only "one" who has this power is God.[20]

The leading creationist writers, Morris and Gish, acknowledge that the idea of creation described in 4(a)(1) is the concept of creation by God and make no pretense to the contrary.[21] The idea of sudden creation from nothing, or *creatio ex nihilo,* is an inherently religious concept. (Vawter, Gilkey, Geisler, Ayala, Blount, Hicks.)

The argument advanced by defendants' witness, Dr. Norman Geisler, that teaching the existence of God is not religious unless the teaching seeks a commitment, is contrary to common understanding and contradicts settled case law. *Stone v. Graham,* 449 U.S. 39 (1980); *Abbington School District v. Schempp,* 374 U.S. 203 (1963).

The facts that creation science is inspired by the Book of Genesis and that Section 4(a) is consistent with a literal interpretation of Genesis leave no doubt that a major effect of the Act is the advancement of particular religious beliefs. The legal impact of this conclusion will be discussed further at the conclusion of the Court's evaluation of the scientific merit of creation science.

IV(A)

The approach to teaching "creation science" and "evolution science" found in Act 50 is identical to the two-model approach espoused by the Institute for Creation Research and is taken almost verbatim from ICR writings. It is an extension of Fundamentalists' view that one must either accept the literal interpretation of Genesis or else believe in the godless system of evolution.

The two model approach of the creationists is simply a contrived dualism[22] which has no scientific factual basis or legitimate educational purpose. It assumes only two explanations for the origins of life and existence of man, plants and animals: It was either the work of a creator or it was not. Application of these two models, according to creationists, and the defendants, dictates that all scientific evidence which fails to support the theory of evolution is necessarily scientific evidence in support of creationism and is, therefore, creation science "evidence" in support of Section 4(a).

IV(B)

The emphasis on origins as an aspect of the theory of evolution is peculiar to creationist literature. Although the subject of origins of life is within the province of biology, the scientific community does not consider origins of life a part of evolutionary theory. The theory of evolution assumes the existence of life and is directed to an explanation of *how* life evolved. Evolution does not presuppose the absence of a creator or God and the plain inference conveyed by Section 4 is erroneous.[23]

As a statement of the theory of evolution, Section 4(b) is simply a hodgepodge of limited assertions, many of which are factually inaccurate.

For example, although 4(b)(2) asserts, as a tenet of evolutionary theory, "the sufficiency of mutation and natural selection in bringing about the existence of present living kinds from simple earlier kinds," Drs. Ayala and Gould both stated that biologists know that these two processes do not account for all significant evolutionary change. They testified to such phenomena as recombination, the founder effect, genetic drift and the theory of punctuated equilibrium, which are believed to play important evolutionary roles. Section 4(b) omits any reference to these. Moreover, 4(b) utilizes the term "kinds" which all scientists said is not a word of science and has no fixed meaning. Additionally, the Act presents both evolution and creation science as "package deals." Thus, evidence critical of some aspect of what the creationists define as evolution is taken as support for a theory which includes a worldwide flood and a relatively young earth.[24]

IV(C)

In addition to the fallacious pedagogy of the two model approach, Section 4(a) lacks legitimate educational value because "creation science" as defined in that section is simply not science. Several witnesses suggested definitions of science. A descriptive definition was said to be that science is what is "accepted by the scientific community" and is "what scientists do." The obvious implication of this description is that, in a free society, knowledge does not require the imprimatur of legislation in order to become science.

More precisely, the essential characteristics of science are:

(1) It is guided by natural law;
(2) It has to be explanatory by reference to natural law;
(3) It is testable against the empirical world;
(4) Its conclusion are tentative, i.e., are not necessarily the final word; and
(5) It is falsifiable. (Ruse and other science witnesses)

Creation science as described in Section 4(a) fails to meet these essential characteristics. First, the section revolves around 4(a)(1) which asserts a sudden creation "from nothing." Such a concept is not science because it depends upon a supernatural intervention which is not guided by natural law. It is not explanatory by reference to natural law, is not testable and is not falsifiable.[25]

If the unifying idea of supernatural creation by God is removed from Section 4, the remaining parts of the section explain nothing and are meaningless assertions.

Section 4(a)(2), relating to the "insufficiency of mutation and natural selection in bringing about development of all living kinds from a single organism," is an incomplete negative generalization directed at the theory of evolution.

Section 4(a)(3) which describes "changes only within fixed limits of originally created kinds of plants and animals" fails to conform to the essential characteristics of science for several reasons. First, there is no scientific definition of "kinds" and none of the witnesses was able to point to any scientific authority which recognized the term or knew how many "kinds" existed. One defense witness suggested there may be 100 to 10,000 different "kinds." Another believes there were "about 10,000, give or take a few thousand." Second, the assertion appears to be an effort to establish outer limits of changes within species. There is no scientific explanation for these limits which is guided by natural law and the limitations, whatever they are, cannot be explained by natural law.

The statement in 4(a)(4) of "separate ancestry of man and apes" is a bald assertion. It explains nothing and refers to no scientific fact or theory.[26]

Section 4(a)(5) refers to "explanation of the earth's geology by catastrophism, including the occurrence of a worldwide flood." This assertion completely fails as science. The Act is referring to the Noachian flood described in the Book of Genesis.[27] The creationist writers concede that *any* kind of Genesis Flood depends upon supernatural intervention. A worldwide flood as an explanation of the world's geology is not the product of natural law, nor can its occurrence be explained by natural law.

Section 4(a)(6) equally fails to meet the standards of science. "Relatively recent inception" has no scientific meaning. It can only be given meaning by reference to creationist writings which place the age at between 6,000 and 20,000 years because of the genealogy of the old Testament. See, e.g., Px 78, Gish (6,000 to 10,000); Px 87, Segraves (6,000 to 20,000). Such a reasoning process is not the product of natural law; not explainable by natural law; nor is it tentative.

Creation science, as defined in Section 4(a), not only fails to follow the canons defining scientific theory, it also fails to fit the more general descriptions of "what scientists think" and "what scientists do." The scientific community consists of individuals and groups, nationally and internationally, who work independently in such varied fields as biology, paleontology, geology and astronomy. Their work is published and subject to review and testing by their peers. The journals for publication are both numerous and varied. There is, however, not one recognized scientific journal which has published an article espousing the creation science theory described in Section 4(a). Some of the State's witnesses suggested that the scientific community was "close-minded" on the subject of creationism and that explained the lack of acceptance of the creation science arguments. Yet no witness produced a scientific article for which publication had been refused. Perhaps some members of the scientific community are resistant to new ideas. It is, however, inconceivable that such a loose knit group of independent thinkers in all the varied fields of science could, or would, so effectively censor new scientific thought.

The creationists have difficulty maintaining among their ranks consistency in the claim that creationism is science. The author of Act 59, Ellwanger, said that neither evolution nor creationism was science. He thinks both are religion. Duane Gish recently responded to an article in *Discover* critical of creationism by stating:

> Stephen Jay Gould states that creationists claim creation is a scientific theory. This is a false accusation. Creationists have repeatedly stated that neither creation nor evolution is a scientific theory (and each is equally religious). (Gish, letter to editor of *Discover,* July, 1981, App. 30 to Plaintiffs' Pretrial Brief.)

The methodology employed by creationists is another factor which is indicative that their work is not science. A scientific theory must be tentative and always subject to revision or abandonment in light of facts that are inconsistent with, or falsify, the theory. A theory that is by its own terms dogmatic, absolutist and never subject to revision is not a scientific theory.

The creationists' methods do not take data, weigh it against the opposing scientific data, and thereafter reach the conclusions stated in Section 4(a). Instead, they take the literal wording of the Book of Genesis and attempt to find scientific support for it. The method is best explained in the language of Morris in his book (Px 31) *Studies in The Bible and Science* at page 114:

> . . . it is . . . quite impossible to determine anything about Creation through a study of present processes, because present processes are not creative in character. If man wishes to know anything about Creation (the time of Creation, the duration of Creation, the order of Creation, the methods of Creation, or anything

> else) his sole source of true information is that of divine revelation. God was there when it happened. We were not there . . . Therefore, we are completely limited to what God has seen fit to tell us, and this information is in His written Word. This is our textbook on the science of Creation!

The Creation Research Society employs the same unscientific approach to the issue of creationism. Its applicants for membership must subscribe to the belief that the Book of Genesis is "historically and scientifically true in all of the original autographs."[28] The Court would never criticize or discredit any person's testimony based on his or her religious beliefs. While anybody is free to approach a scientific inquiry in any fashion they choose, they cannot properly describe the methodology used as scientific, if they start with a conclusion and refuse to change it regardless of the evidence developed during the course of the investigation.

IV(D)

In efforts to establish "evidence" in support of creation science, the defendants relied upon the same false premise as the two model approach contained in Section 4, i.e., all evidence which criticized evolutionary theory was proof in support of creation science. For example, the defendants established that the mathematical probability of a chance chemical combination resulting in life from non-life is so remote that such an occurrence is almost beyond imagination. Those mathematical facts, the defendants argue, are scientific evidences that life was the product of a creator. While the statistical figures may be impressive evidence against the theory of chance chemical combinations as an explanation of origins, it requires a leap of faith to interpret those figures so as to support a complex doctrine which includes a sudden creation from nothing, a worldwide flood, separate ancestry of man and apes, and a young earth.

The defendants' argument would be more persuasive if, in fact, there were only two theories or ideas about the origins of life and the world. That there are a number of theories was acknowledged by the State's witnesses, Dr. Wickramasinghe and Dr. Geisler. Dr. Wickramasinghe testified at length in support of a theory that life on earth was "seeded" by comets which delivered genetic material and perhaps organisms to the earth's surface from interstellar dust far outside the solar system. The "seeding" theory further hypothesizes that the earth remains under the continuing influence of genetic material from space which continues to affect life. While Wickramasinghe's theory[29] about the origins of life on earth has not received general acceptance within the scientific community, he has, at least, used scientific methodology to produce a theory of origins which meets the essential characteristics of science.

Perhaps Dr. Wickramasinghe was called as a witness because he was generally critical of the theory of evolution and the scientific community, a tactic consistent with the strategy of the defense. Unfortunately for the defense, he demonstrated that the simplistic approach of the two model analysis of the origins of life is false. Furthermore, he corroborated the plaintiffs' witnesses by concluding that "no rational scientist" would believe the earth's geology could be explained by reference to a worldwide flood or that the earth was less than one million years old.

The proof in support of creation science consisted almost entirely of efforts to discredit the theory of evolution through a rehash of data and theories which have been before the scientific community for decades. The arguments asserted by creationists are not based upon new scientific evidence or laboratory data which has been ignored by the scientific community.

Robert Gentry's discovery of radioactive polonium haloes in granite and coalified woods is, perhaps, the most recent scientific work which the creationists use as argument for a "relatively recent inception" of the earth and a "worldwide flood." The existence of polonium haloes in granite and coalified wood is thought to be inconsistent with radiometric dating methods based upon constant radioactive decay rates. Mr. Gentry's findings were published almost ten years ago and have been the subject of some discussion in the scientific community. The discoveries have not, however, led to the formulation of any scientific hypothesis or theory which would explain a relatively recent inception of the earth or a worldwide flood. Gentry's discovery has been treated as a minor mystery which will eventually be explained. It may deserve further investigation, but the National Science Foundation has not deemed it to be of sufficient import to support further funding.

The testimony of Marianne Wilson was persuasive evidence that creation science is not science. Ms. Wilson is in charge of the science curriculum for Pulaski County Special School District, the largest school district in the State of Arkansas. Prior to the passage of Act 590, Larry Fisher, a science teacher in the District, using materials from the ICR, convinced the School Board that it should voluntarily adopt creation science as part of its science curriculum. The District Superintendent assigned Ms. Wilson the job of producing a creation science curriculum guide. Ms. Wilson's testimony about the project was particularly convincing because she obviously approached the assignment with an open mind and no preconceived notions about the subject. She had not heard of creation science until about a year ago and did not know its meaning before she began her research.

Ms. Wilson worked with a committee of science teachers appointed from the District. They reviewed practically all of the creationist literature. Ms. Wilson and the committee members reached the unanimous conclusion that creationism is not science; it is religion. They so reported to the Board. The Board ignored the recommendation and insisted that a curriculum guide be prepared.

In researching the subject, Ms. Wilson sought the assistance of Mr. Fisher who initiated the Board action and asked professors in the science departments of the University of Arkansas at Little Rock and the University of Central Arkansas[30] for reference material and assistance, and attended a workshop conducted at Central Baptist College by Dr. Richard Bliss of the ICR staff. Act 590 became law during the course of her work so she used Section 4(a) as a format for her curriculum guide.

Ms. Wilson found all available creationists' materials unacceptable because they were permeated with religious references and reliance upon religious beliefs.

It is easy to understand why Ms. Wilson and other educators find the creationists' textbook material and teaching guides unacceptable. The materials misstate the theory of evolution in the same fashion as Section 4(b) of the Act, with emphasis on the alternative mutually exclusive nature of creationism and evolution. Students are constantly encouraged to compare and make a choice between the two models, and the material is not presented in an accurate manner.

A typical example is *Origins* (Px 76) by Richard B. Bliss, Director of Curriculum Development of the ICR. The presentation begins with a chart describing "preconceived ideas about origins" which suggests that some people believe that evolution is atheistic. Concepts of evolution, such as "adaptive radiation," are erroneously presented. At page 11, figure 1.6, of the text, a chart purports to illustrate this "very important" part of the evolution model. The chart conveys the idea that such diverse mammals as a whale, bear, bat and monkey all evolved from a shrew through the process of adaptive radiation. Such a suggestion is, of course, a totally erroneous and misleading application of the theory. Even more objectionable, especially when viewed in light of the emphasis on asking the student to elect one of the models, is the chart presentation at page 17, figure 1.6. That chart purports to illustrate the evolutionists' belief that man evolved from bacteria to fish to reptile to mammals and, thereafter, into man. The illustration indicates, however, that the mammal from which man evolved was *a rat.*

Biology, A Search For Order in Complexity[31] is a high school biology text typical of creationists' materials. The following quotations are illustrative:

> Flowers and roots do not have a mind to have purpose of their own; therefore, this planning must have been done for them by the Creator. (at page 12)
>
> The exquisite beauty of color and shape in flowers exceeds the skill of poet, artist, and king. Jesus said (from Matthew's gospel),
>
> "Consider the lilies of the field, how they grow; they toil not, neither do they spin . . ." (Px 129 at page 363)

The "public school edition" texts written by creationists simply omit Biblical references but the content and message remain the same. For example, *Evolution—The Fossils Say No!,*[32] contains the following:

> Creation. By creation we mean the bringing into being by a supernatural Creator of the basic kinds of plants and animals by the process of sudden, or fiat, creation.
>
> We do not know how the Creator created, what processes He used, *for He used processes which are not now operating anywhere in the natural universe.* This is why we refer to creation as Special

> Creation. We cannot discover by scientific investigation anything about the creative processes used by the Creator." (page 40)

Gish's book also portrays the large majority of evolutionists as "materialistic atheists or agnostics."

Scientific Creationism (Public School Edition) by Morris, is another text reviewed by Ms. Wilson's committee and rejected as unacceptable. The following quotes illustrate the purpose and theme of the text:

> Foreword
>
> Parents and youth leaders today, and even many scientists and educators, have become concerned about the prevalence and influence of evolutionary philosophy in modern curriculum. Not only is this system inimical to orthodox Christianity and Judaism, but also, as many are convinced, to a healthy society and true science as well. (at page iii)
>
> The rationalist of course finds the concept of special creation insufferably naive, even "incredible." Such a judgment, however, is warranted only if one categorically dismisses the existence of an omnipotent God. (at page 17)

Without using creationist literature, Ms. Wilson was unable to locate one genuinely scientific article or work which supported Section 4(a). In order to comply with the mandate of the Board she used such materials as an article from *Readers Digest* about "atomic clocks" which inferentially suggested that the earth was less than 4½ billion years old. She was unable to locate any substantive teaching material for some parts of Section 4 such as the worldwide flood. The curriculum guide which she prepared cannot be taught and has no educational value as science. The defendants did not produce any text or writing in response to this evidence which they claimed was usable in the public school classroom.[33]

The conclusion that creation science has no scientific merit or educational value as science has legal significance in light of the Court's previous conclusion that creation science has, as one major effect, the advancement of religion. The second part of the three-pronged test for establishment reaches only those statutes having as their *primary* effect the advancement of religion. Secondary effects which advance religion are not constitutionally fatal. Since creation science is not science, the conclusion is inescapable that the *only* real effect of Act 590 is the advancement of religion. The Act therefore fails both the first and second portions of the test in *Lemon v. Kurtzman,* 403 U.S. 602 (1971).

IV(E)

Act 590 mandates "balanced treatment" for creation science and evolution science. The Act prohibits instruction in any religious doctrine or references to religious writings. The Act is self-contradictory and compliance is impossible unless the public schools elect to forego significant portions of subjects such as biology, world history, geology, zoology, botany, psychology, anthropology, sociology, philosophy, physics and chemistry. Presently, the concepts of evolutionary theory as described in 4(b) permeate the public school textbooks. There is no way teachers can teach the Genesis account of creation in a secular manner.

The State Department of Education, through its textbook selection committee, school boards and school administrators will be required to constantly monitor materials to avoid using religious references. The school boards, administrators and teachers face an impossible task. How is the teacher to respond to questions about a creation suddenly and out of nothing? How will a teacher explain the occurrence of a worldwide flood? How will a teacher explain the concept of a relatively recent age of the earth? The answer is obvious because the only source of this information is ultimately contained in the Book of Genesis.

References to the pervasive nature of religious concepts in creation science texts amply demonstrate why State entanglement with religion is inevitable under Act 590. Involvement of the State in screening texts for impermissible religious references will require State officials to make delicate religious judgments. The need to monitor classroom discussion in order to uphold the Act's prohibition against religious instruction will necessarily involve administrators in questions concerning religion. These continuing involvements of State officials in questions and issues of religion create an excessive and prohibited entanglement with religion. *Brandon v. Board of Education,* 487 F.Supp. 1219, 1230 (N.D.N.Y.), *aff'd.,* 635 F.2d 971 (2nd Cir. 1980).

V

These conclusions are dispositive of the case and there is no need to reach legal conclusions with respect to the remaining issues. The plaintiffs raised two other issues questioning the constitutionality of the Act and, insofar as the factual findings relevant to these issues are not covered in the preceding discussion, the Court will address these issues. Additionally, the defendants raised two other issues which warrant discussion.

V(A)

First, plaintiff teachers argue the Act is unconstitutionally vague to the extent that they cannot comply with its mandate of "balanced" treatment without jeopardizing their employment. The argument centers around the lack of a precise definition in the Act for the word "balanced." Several witnesses expressed opinions that the word has such meanings as equal time, equal weight, or equal legitimacy. Although the Act could have been more explicit, "balanced" is a word subject to ordinary understanding. The proof is not convincing that a teacher using a reasonably acceptable understanding of the word and making a good faith effort to comply with the Act will be in jeopardy of termination. Other portions of the Act are arguably vague, such as the "relatively recent" inception of the earth and life. The evidence establishes, however, that relatively recent means from 6,000 to 20,000 years, as commonly understood in creation science literature. The meaning of this phrase, like Section 4(a) generally, is, for purposes of the Establishment Clause, all too clear.

V(B)

The plaintiffs' other argument revolves around the alleged infringement by the defendants upon the academic freedom of teachers and students. It is contended this unprecedented intrusion in the curriculum by the State prohibits teachers from teaching what they believe should be taught or requires them to teach that which they do not believe is proper. The evidence reflects that traditionally the State Department of Education, local school boards and administration officials exercise little, if any, influence upon the subject matter taught by classroom teachers. Teachers have been given freedom to teach and emphasize those portions of subjects the individual teacher considered important. The limits to this discretion have generally been derived from the approval of textbooks by the State Department and preparation of curriculum guides by the school districts.

Several witnesses testified that academic freedom for the teacher means, in substance, that the individual teacher should be permitted unlimited discretion subject only to the bounds of professional ethics. The Court is not prepared to adopt such a broad view of academic freedom in the public schools.

In any event, if Act 590 is implemented, many teachers will be required to teach material in support of creation science which they do not consider academically sound. Many teachers will simply forego teaching subjects which might trigger the "balanced treatment" aspects of Act 590 even though they think the subjects are important to a proper presentation of a course.

Implementation of Act 590 will have serious and untoward consequences for students, particularly those planning to attend college. Evolution is the cornerstone of modern biology, and many courses in public schools contain subject matter relating to such varied topics as the age of the earth, geology and relationships among living things. Any student who is deprived of instruction as to the prevailing scientific thought on these topics will be denied a significant part of science education. Such a deprivation through the high school level would undoubtedly have an impact upon the quality of education in the State's colleges and universities, especially including the pre-professional and professional programs in the health sciences.

V(C)

The defendants argue in their brief that evolution is, in effect, a religion, and that by teaching a religion which is contrary to some students' religious views, the State is infringing upon the student's free exercise rights under the First Amendment. Mr. Ellwanger's legislative findings, which were adopted as a finding of fact by the Arkansas Legislature in Act 590, provides:

> Evolution-science is contrary to the religious convictions or moral values or philosophical beliefs of many students and parents, including individuals of many different religious faiths and with diverse moral and philosophical beliefs. (Act 590, §7[d].)

The defendants argue that the teaching of evolution alone presents both a free exercise problem and an establishment problem which can only be redressed by giving balanced treatment to creation science, which is admittedly consistent with some religious beliefs. This argument appears to have its genesis in a student note written by Mr. Wendell Bird, "Freedom of Religion and Science Instruction in Public Schools," 87 Yale L.J. 515 (1978). The argument has no legal merit.

If creation science is, in fact, science and not religion, as the defendants claim, it is difficult to see how the teaching of such a science could "neutralize" the religious nature of evolution.

Assuming for the purposes of argument, however, that evolution is a religion or religious tenet, the remedy is to stop the teaching of evolution; not establish another religion in opposition to it. Yet it is clearly established in the case law, and perhaps also in common sense, that evolution is not a religion and that teaching evolution does not violate the Establishment Clause, *Epperson v. Arkansas, supra, Willoughby v. Stever, No.* 15574-75 (D.D.C. May 18, 1973); *aff'd.* 504 F.2d 271 (D.C. Cir. 1974), *cert. denied,* 420 U.S. 924 (1975); *Wright v. Houston Indep. School Dist.,* 366 F.Supp. (S.D. Tex. 1978), *aff'd.* 486 F.2d 137 (5th Cir. 1973), *cert. denied* 417 U.S. 969 (1974).

V(D)

The defendants presented Dr. Larry Parker, a specialist in devising curricula for public schools. He testified that the public school's curriculum should reflect the subjects the public wants taught in schools. The witness said that polls indicated a significant majority of the American public thought creation science should be taught if evolution was taught. The point of this testimony was never placed in a legal context. No doubt a sizeable majority of Americans believe in the concept of a Creator or, at least, are not opposed to the concept and see nothing wrong with teaching school children about the idea.

The application and content of First Amendment principles are not determined by public opinion polls or by a majority vote. Whether the proponents of Act 590 constitute the majority or the minority is quite irrelevant under a constitutional system of government. No group, no matter how large or small, may use the organs of government, of which the public schools are the most conspicuous and influential, to foist its religious beliefs on others.

The Court closes this opinion with a thought expressed eloquently by the great Justice Frankfurter:

> We renew our conviction that "we have staked the very existence of our country on the faith that complete separation between the state and religion is best for the state and best for religion." [*Everson v. Board of Education,* 330 U.S. at 59.] If nowhere else, in the relation between Church and State, "good fences make good neighbors." [*McCollum v. Board of Education,* 333 U.S. 203, 232 (1948).]

An injunction will be entered permanently prohibiting enforcement of Act 590.

It is so ordered this January 5, 1982.

—WILLIAM R. OVERTON *in United States District Court, Eastern District of Arkansas, Western Division*

NOTES

1. The complaint is based on 42 U.S.C. §1983, which provides a remedy against any person who, acting under color of state law, deprives another of any right, privilege or immunity guaranteed by the United States Constitution or federal law. This Court's jurisdiction arises under 28 U.S.C. §§1331, 1343(3) and 1343(4). The power to issue declaratory judgments is expressed in 28 U.S.C. §§2201 and 2202.
2. The facts necessary to establish the plaintiffs' standing to sue are contained in the joint stipulation of facts, which is hereby adopted and incorporated herein by reference. There is no doubt that the case is ripe for adjudication.
3. The State of Arkansas was dismissed as a defendant because of its immunity from suit under the Eleventh Amendment. *Hans v. Louisiana,* 134 U.S. 1 (1890).

4. The authorities differ as to generalizations which may be made about Fundamentalism. For example, Dr. Geisler testified to the widely held view that there are five beliefs characteristic of all Fundamentalist movements, in addition, of course, to the inerrancy of Scripture: (1) belief in the virgin birth of Christ, (2) belief in the deity of Christ, (3) belief in the substitutional atonement of Christ, (4) belief in the second coming of Christ, and (5) belief in the physical resurrection of all departed souls. Dr. Marsden, however, testified that this generalization, which has been common in religious scholarship, is now thought to be historical error. There is no doubt, however, that all Fundamentalists take the Scriptures as inerrant and probably most take them as literally true.
5. Initiated Act 1 of 1929, Ark. Stat. Ann. §80-1627 *et seq.*, which prohibited the teaching of evolution in Arkansas schools, is discussed *infra* at text accompanying note 26.
6. Subsequent references to the testimony will be made by the last name of the witness only. References to documentary exhibits will be by the name of the author and the exhibit number.
7. Applicants for membership in the CRS must subscribe to the following statement of belief: "(1) The Bible is the written Word of God, and because we believe it to be inspired thruout (sic), all of its assertions are historically and scientifically true in all of the original autographs. To the student of nature, this means that the account of origins in Genesis is a factual presentation of simple historical truths. (2) All basic types of living things, including man, were made by direct creative acts of God during Creation Week as described in Genesis. Whatever biological changes have occurred since Creation have accomplished only changes within the original created kinds. (3) The great Flood described in Genesis, commonly referred to as the Noachian Deluge, was an historical event, worldwide in its extent and effect. (4) Finally, we are an organization of Christian men of science, who accept Jesus Christ as our Lord and Savior. The account of the special creation of Adam and Eve as one man and one woman, and their subsequent Fall into sin, is the basis for our belief in the necessity of a Savior for all mankind. Therefore, salvation can come only thru (sic) accepting Jesus Christ as our Savior." (Px 115)
8. Because of the voluminous nature of the documentary exhibits, the parties were directed by pre-trial order to submit their proposed exhibits for the Court's convenience prior to trial. The numbers assigned to the pre-trial submissions do not correspond with those

assigned to the same documents at trial and, in some instances, the pre-trial submissions are more complete.

9. Px 130, Morris, *Introducing Scientific Creationism Into the Public Schools* (1975), and Bird, "Resolution for Balanced Presentation of Evolution and Scientific Creationism," *ICR Impact Series* No. 71, App. 14 to Plaintiffs' Pretrial Brief.

10. The creationists often show candor in their proselytization. Henry Morris has stated, "Even if a favorable statute or court decision is obtained, it will probably be declared unconstitutional, especially if the legislation or injunction refers to the Bible account of creation." In the same vein he notes, "The only effective way to get creationism taught properly is to have it taught by teachers who are both willing and able to do it. Since most teachers now are neither willing nor able, they must first be both persuaded and instructed themselves." Px 130, Morris, *Introducing Scientific Creationism Into the Public Schools* (1975) (unpaged).

11. Mr. Bird sought to participate in this litigation by representing a number of individuals who wanted to intervene as defendants. The application for intervention was denied by this Court. *McLean v. Arkansas,* ——— F. Supp. ———, (E.D. Ark. 1981), aff'd. *per curiam,* Slip Op. No. 81-2023 (8th Cir. Oct. 16, 1981).

12. The model act had been revised to insert "creation science" in lieu of creationism because Ellwanger had the impression people thought creationism was too religious a term. (Ellwanger Depo. at 79.)

13. The original model act had been introduced in the South Carolina Legislature, but had died without action after the South Carolina Attorney General had opined that the act was unconstitutional.

14. Specifically, Senator Holsted testified that he holds to a literal interpretation of the Bible; that the bill was compatible with his religious beliefs; that the bill does favor the position of literalists; that his religious convictions were a factor in his sponsorship of the bill; and that he stated publicly to the *Arkansas Gazette* (although not on the floor of the Senate) contemporaneously with the legislative debate that the bill does presuppose the existence of a divine creator. There is no doubt that Senator Holsted knew he was sponsoring the teaching of a religious doctrine. His view was that the bill did not violate the First Amendment because, as he saw it, it did not favor one domination over another.

15. This statute is, of course, clearly unconstitutional under the Supreme Court's decision in *Abbington School Dist. v. Schempp,* 374 U.S. 203 (1963).

16. The joint stipulation of facts establishes that the following areas are the only *information* specifically required by statute to be taught in all Arkansas schools: (1) the effects of alcohol and narcotics on the human body, (2) conservation of national resources, (3) Bird Week, (4) Fire Prevention, and (5) Flag etiquette. Additionally, certain specific courses, such as American history and Arkansas history, must be completed by each student before graduation from high school.
17. Paul Ellwanger stated in his deposition that he did not know why Section 4(a)(2) (insufficiency of mutation and natural selection) was included as an evidence supporting creation science. He indicated that he was not a scientist, "but these are the postulates that have been laid down by creation scientists." Ellwanger Depo. at 136.
18. Although defendants must make some effort to cast the concept of creation in non-religious terms, this effort surely causes discomfort to some of the Act's more theologically sophisticated supporters. The concept of a creator God distinct from the God of love and mercy is closely similar to the Marcion and Gnostic heresies, among the deadliest to threaten the early Christian church. These heresies had much to do with development and adoption of the Apostle's Creed as the official creedal statement of the Roman Catholic Church in the West. (Gilkey.)
19. The parallels between Section 4(a) and Genesis are quite specific: (1) "sudden creation from nothing" is taken from Genesis, 1:1–10 (Vawter, Gilkey); (2) destruction of the world by a flood of divine origin is a notion peculiar to Judeo-Christian tradition and is based on Chapters 7 and 8 of Genesis (Vawter); (3) the term "kinds" has no fixed scientific meaning, but appears repeatedly in Genesis (all scientific witnesses); (4) "relatively recent inception" means an age of the earth from 6,000 to 10,000 years and is based on the genealogy of the Old Testament using the rather astronomical ages assigned to the patriarchs (Gilkey and several of defendants' scientific witnesses): (5) separate ancestry of man and ape focuses on the portion of the theory of evolution which Fundamentalists find most offensive, *Epperson v. Arkansas,* 393 U.S. 97 (1968).
20. "[C]oncepts concerning . . . a supreme being of some sort are manifestly religious . . . These concepts do not shed that religiosity merely because they are presented as philosophy or as a science . . ." *Malnak v. Yogi,* 440 F. Supp. 1284, 1322 (D.N.J. 1977); *aff'd per curiam,* 592 F.2d 197 (3d Cir. 1979).

21. See, e.g., Px 76, Morris, *et al., Scientific Creationism,* 203 (1980) ("If creation really is a fact, this means there is a *Creator,* and the universe is His creation.") Numerous other examples of such admissions can be found in the many exhibits which represent creationist literature, but no useful purpose would be served here by a potentially endless listing.

22. Morris, the Director of ICR and one who first advocated the two model approach, insists that a true Christian cannot compromise with the theory of evolution and that the Genesis version of creation and the theory of evolution are mutually exclusive. Px 31, Morris, *Studies in the Bible & Science,* 102–103. The two model approach was the subject of Dr. Richard Bliss's doctoral dissertation. (Dx 35.) It is presented in Bliss, *Origins: Two Models—Evolution, Creation* (1978). Moreover, the two model approach merely casts in educationalist language the dualism which appears in all creationist literature—creation (i.e., God) and evolution are presented as two alternative and mutually exclusive theories. See, e.g., Px 75, Morris, *Scientific Creationism* (1974) (public school edition); Px 59, Fox, *Fossils: Hard Facts from the Earth.* Particularly illustrative is Px 61, Boardman, *et al., Worlds Without End* (1971), a CSRC publication: "One group of scientists, known as creationists, believe that God, in a miraculous manner, created all matter and energy . . .

"Scientists who insist that the universe just grew, by accident, from a mass of hot gases without the direction or help of a Creator are known as evolutionists."

23. The idea that belief in a creator and acceptance of the scientific theory of evolution are mutually exclusive is a false premise and offensive to the religious views of many. (Hicks) Dr. Francisco Ayala, a geneticist of considerable reknown and a former Catholic priest who has the equivalent of a Ph.D. in theology, pointed out that many working scientists who subscribed to the theory of evolution are devoutly religious.

24. This is so despite the fact that some of the defense witnesses do not subscribe to the young earth or flood hypotheses. Dr. Geisler stated his belief that the earth is several billion years old. Dr. Wickramasinghe stated that no rational scientist would believe the earth is less than one million years old or that all the world's geology could be explained by a worldwide flood.

25. "We do not know how God created, what processes He used, for *God used processes which are not now operating anywhere in the*

natural universe. This is why we refer to divine creation as Special Creation. We cannot discover by scientific investigation anything about the creative processes used by God." Px 78, Gish, *Evolution? The Fossils Say No!,* 42 (3d ed. 1979) (emphasis in original).

26. The evolutionary notion that man and some modern apes have a common ancestor somewhere in the distant past has consistently been distorted by anti-evolutionists to say that man descended from modern monkeys. As such, this idea has long been most offensive to Fundamentalists. See, *Epperson v. Arkansas,* 393 U.S. 97 (1968).
27. Not only was this point acknowledged by virtually all the defense witnesses, it is patent in the creationist literature. See, e.g., Px 89, Kofahl & Segraves, *The Creation Explanation,* 40: "The Flood of Noah brought about vast changes in the earth's surface, including vulcanism, mountain building, and the deposition of the major part of sedimentary strata. This principle is called 'Biblical catastrophism.' "
28. See n. 7, *supra,* for the full text of the CRS creed.
29. The theory is detailed in Wickramasinghe's book with Sir Fred Hoyle, *Evolution From Space* (1981), which is Dx 79.
30. Ms. Wilson stated that some professors she spoke with sympathized with her plight and tried to help her find scientific materials to support Section 4(a). Others simply asked her to leave.
31. Px 129, published by Zonderman Publishing House (1974), states that it was "prepared by the Textbook Committee of the Creation Research Society." It has a disclaimer pasted inside the front cover stating that it is not suitable for use in public schools.
32. Px 77, by Duane Gish.
33. The passage of Act 590 apparently caught a number of its supporters off guard as much as it did the school district. The Act's author, Paul Ellwanger, stated in a letter to "Dick," (apparently Dr. Richard Bliss at ICR): "And finally, if you know of any textbooks at any level and for any subjects that you think are acceptable to you and also constitutionally admissible, these are things that would be of *enormous* to these bewildered folks who may be caught, as Arkansas now has been, by the sudden need to implement a whole new ball game with which they are quite unfamiliar." (sic) (Unnumbered attachment to Ellwanger depo.)

SIDNEY RATNER

EVOLUTION AND THE RISE OF THE SCIENTIFIC SPIRIT IN AMERICA

Ten years ago the civilized world was shocked out of an unwarranted confidence in the victory of science over superstition by the news that Tennessee had outlawed the teaching of man's evolutionary origin. Within the next three years both Mississippi and Arkansas joined Tennessee in defense of Biblical truth against the scepticism of science. The furor raised by the dramatic clashes between William Jennings Bryan and Clarence Darrow at the Scopes Trial in Dayton, Tennessee, has now become a faded memory. The struggle between Fundamentalists and Modernists has died down, though the embers still smoulder; and few ever think of the issues then raised, now that war and fascism threaten civilization.

Yet the crucial principle of the controversy over evolution cannot be forgotten; it still lives, in a different form, in the issues that face us now. Shall Reason or Prejudice be the arbiter of our destiny? The answer that will be given to that question depends in part on the extent to which the scientific spirit has become dominant in society. What rôle Evolution, as a revelation of science, played in the creation of the Scientific Temper in America has never been critically evaluated; few have dared to study at first hand how the impact of the theory of evolution upon the Victorian Age brought about a cataclysmic religious and scientific upheaval and readjustment.

Change in human affairs is sometimes gradual and imperceptibly

From *Philosophy of Science,* Vol. 3 (Jan. 1936), pp. 104–22.

cumulative; but sometimes swift, sudden, catastrophic. If men are plastic, eager to learn, and strong in reason, civilization advances; if not, civilization declines or perishes. Now the most significant effect of Darwin's discovery was, not the triumph of certain specific scientific theories, but a revolutionary change in the very process by which men arrived at their convictions. To throw overboard the sanctions of authority, ancient prejudices, and the hallowed rationalizations of *a priori* philosophy, and to embrace the method of science was a far greater achievement than the discrediting of an outmoded cosmology and the shocking incorporation of man into the animal kingdom.

But many who discarded old theological dogmas and accepted the popularized (and distorted) versions of evolution only exchanged one mythology for another. One name, however, stands out as the pre-eminent exponent of intellectual honesty and scientific method in that age of intellectual confusion and emotional stress: Chauncey Wright. Now unknown and unhonored, this philosopher was the truest champion of evolution and science; unlike John Fiske and Asa Gray, he did not make a new religion of evolution or force science to be the handmaiden of theology. He was wise enough to see that "to live by science requires intelligence and faith, but not to live by it is folly." Only when the contribution of Chauncey Wright is placed against the background of his age, can the rise of the scientific spirit in America be seen in proper perspective.

On November 11th, 1859, Charles Darwin sent copies of his latest work, *The Origin of Species,* to two men in America—Louis Agassiz and Asa Gray. Asa Gray was the greatest botanist in America; Louis Agassiz, his colleague at Harvard, the greatest zoologist. Gray, a personal friend and correspondent of Darwin's, had already been won over to his theories. But Agassiz, the brilliant Swiss naturalist, noted as the protégé of Alexander von Humboldt and Cuvier, could not abandon the philosophy of a life-time so easily. Cuvier, his master, fifty years before had ridiculed Lamarck's evolutionary theory into obscurity and had popularized a theory that the species of different geologic epochs had been created after successive cataclysms. Agassiz, in his famous *Essay on Classification* published in 1857, had elaborated on this theme. He declared that species were thoughts of God and immutable and that the diversity of species was the result of repeated interventions on the part of the Creator. A theory so firmly based on the method of intuition would not yield at once to a new scientific doctrine that seemed charged with materialism. Agassiz threw the weight of his immense influence as the foremost zoologist and popular lecturer in America against Darwin's theory.

Soon after an American edition of Darwin's book appeared in January, 1860, an irrepressible conflict of ideas on science and religion began to parallel the struggle over slavery and secession. Heated debates took place at the Boston Society of Natural History and the American Academy of Arts and Sciences. In the spring and fall of 1860, magazines like the *American Journal of Science,* the *North American Review,* the *Christian Examiner,* the *Atlantic Monthly,* the *Methodist Quarterly Review,* and the *American Theological Review* carried long and lively articles attacking or defending this latest of heresies.

To one sufficiently detached by time or philosophy, the attacks on Darwin's theory—even by so great a scientist as Agassiz—are impressive, not as scientific arguments, but as beautiful rationalizations of religious faith in an obsolete metaphysics. True, the scientific evidence bulked large in the discussions, and some sound criticisms were made, but these were the pretexts, not the causes, for rejection. As C. S. Peirce pointed out, most, if not all, reasoning on vitally important topics consists in finding reasons for believing what the heart desires.

Yet intellectual honesty requires that justice be done to some penetrating points made by Agassiz, the foremost opponent of evolution. He stressed, for instance, the fact that not all primitive organisms are simple, and that the most perfect organs are not the result of gradual development.

Francis Bowen, Professor of Moral Philosophy and Natural Theology at Harvard, deserves mention as being next to Agassiz the keenest enemy of evolution. His scorn some fifteen years before for the evolutionary theory of Robert Chamber's *Vestiges of the Natural History of Creation* as a "pure hypothesis which might be summarily dismissed into the region of cloud-land and dreams, where it had its origin" was now visited upon the "ingenious and captivating" theory of Darwin's. He alone called attention to Darwin's evasion of the origin of the human species and asserted Darwin would have to find the means of bridging over, by imperceptibly fine gradations, "the immense gap which now separates man from the animals most nearly allied to him."

Bowen also was the first to declare evolution incompatible with the doctrine of final causes, and a negation of design or purpose in the animate or organic world. In this opinion he was upheld by President Goodwin of Trinity College, Hartford, Conn., who wrote in the *American Theological Review* for May, 1860: "Mr. Darwin does not expressly attack Christianity or Theism. He scarcely notices them at all. But a theory of origins or ends that ignores Theism altogether, is as truly its enemy as one which attacks it openly in front, and must be treated by the friends of Theism accordingly."

The lengths to which friends of Theism would go in defending the deeply cherished proofs of natural theology for the existence of God against the atheistic doctrine of evolution is shown in the fervent assertion by J. A. Lowell in the *Christian Examiner* for May, 1860 that miracles "have never presented to our mind any metaphysical difficulty. The Power that could enact and sustain, must, in our apprehension, of necessity be equally able to suspend or alter, the laws of nature."

Against such sublime faith in orthodoxy and frequent confusion between vituperation and refutation of evolution, rare courage and ability were needed. These were supplied in ample measure first and foremost by Asa Gray, who was an army in himself, secondly by William B. Rogers—an able geologist, and by Professor Theophilus Parsons of Harvard Law School. In public debate and magazine articles they vigorously met their opponents point by point and proved their superiority in sound logic. They admitted that Darwin's theories did not adequately explain all phases of the processes of evolution, but they insisted that his evidence as to the fact of evolution was overwhelming and that Natural Selection was a necessary, although not a sufficient cause of evolution.

Professor Parsons, "almost if not quite a man of genius," according to Mr. Justice Holmes, was one of the first throughout the world to advance a mutation theory of evolution to account for the gap between species in the geologic record (Proc. American Academy Arts and Sciences, April 10, 1860).

It is also only just to Parsons' memory to point out that he stood next to Asa Gray in repelling the attacks upon the atheistic character of evolution and argued cogently that rational religion need not fear the revelations of science. He alone in America at that time, so far as I can discover, dared to declare that since the Bible was no authority on the question of man's origin, neither reason nor religion would be shocked if science considered man descended from the Simiae. Surely this was an advance upon even the bold rationalism of Thomas Cooper, who had so ardently taken Benjamin Silliman, the Yale Geologist, to task in 1829 for distorting geology to prove Biblical assertions about the Flood. The revolt of American geologists in the 1830's against the strict literal interpretation of the Bible as laid down by such prophets as Moses Stuart of Andover Theological Seminary was now bearing fruit.

Unfortunately for civilization, the progress made by Asa Gray, Parsons, and Rogers in breaking down resistance to evolution was halted by man's concern over the fortunes of war. The appeal to arms in the War for Southern Independence so overshadowed all appeals to reason in science and religion that a diligent search through some twenty scientific, religious and popular magazines reveals but four articles on evolution from 1861 to 1865. Only a few scientists found time or strength to shut out the news of daily battle and consider the struggle for existence and Natural Selection *sub specie aeternitatis.*

Asa Gray, at the suggestion of Charles Darwin, had three essays he had written for the *Atlantic Monthly* in July, August and October 1860 on *Natural Selection not inconsistent with Natural Theology* reprinted as a pamphlet in 1861. Darwin, although not in sympathy with this point of view, believed it was of much value in lessening opposition and making converts to evolution and showed his high opinion of it by inserting a special notice of it in a most prominent place in the third edition of the *Origin of Species.*

But Louis Agassiz still clung to his opinion that Darwin's theory was "a scientific mistake, untrue in its facts, unscientific in its method, and mischievous in its tendency." In the preface to a volume on *Methods of Study in Natural History* published in 1863, he wrote with passionate eloquence against the evolutionary theory and said:

> I confess that there seems to me to be a repulsive poverty in this explanation that is contradicted by the intellectual grandeur of the universe; the resources of the Deity cannot be so meager, that in order to create a human being endowed with reason, he must change a monkey into a man. . . .

Realizing this to be only a personal opinion, Agassiz went on to speak in the name of science:

> I insist that this theory is opposed to the processes of Nature, as far as we have been able to apprehend them, that it is contradicted by the facts of Embryology and Paleontology. . . .

Against such ancient prejudice masquerading as scientific fact, Huxley's short but explosive book: *Evidence as to Man's Place in Nature,* was an effective answer. But James Dwight Dana shared Agassiz's intuitions and wrote in his review of Huxley for the *American Journal of Science,* May 1863, that "with regard to Man and the Man-apes, no evidence has been pointed out, derived from Man, or the Apes, proving either the fact, or the probability, or the possibility of a common origin." A critic of Huxley in the *North American Review* for July 1864, admitted that evolution might be of some use as a provisional hypothesis in comparative anatomy, but declared that if evolution were admitted with regard to the physical structure of the human being, something he heartily deprecated, "it could not reach the realm of reason, conscience, and will, it could not cast a ray of doubt on our Divine sonship and immortal birthright on the spiritual side. . . ."

'Eternal truth bears eternal repetition,' but what passes as eternal truth usually is uncritical assumption and hasty generalization erected into unquestionable first principles. Slowly doubt spread and undermined dogmatic faith. Typical of the process of faith and final conversion that many scientists underwent was the experience of Jeffries Wyman, the brilliant comparative anatomist. In the *American Journal of Science* for September 1863 he remarked in passing that scientists had to choose between supernatural creation or evolution. Although he suspended judgment for a time, he finally made the scientific choice.

Even students of Agassiz began to shake off the spell of his personality and free themselves from the method of authority. Men like Nathaniel S. Shaler read books by Darwin and others on evolution and debated the theory in private (it would never do for the faithful to be caught in the careful study of a heresy). They dared to press their master for some conception of how a species first appeared and were no longer satisfied with his mystical answer that a species was a thought of God.

The deep silence of the press on evolution was broken as the conflict over slavery drew to its tragic close. An obscure essayist in the *Evangelical Quarterly Review* for January 1865 upheld the immutability of species and damned evolution as inconsistent with the biblical doctrine of Providence and as an attempt "to substitute a power of nature for the personal, intelligent, and overruling God whom Christians worship."

But that very month the *North American Review* carried the most effective refutation of religious dogmatism on science that could have been written: an essay on *Natural Theology as a Positive Science* by Chauncey Wright, who was later to prove the most brilliant defender of evolution in America. He declared:

> That there is a fundamental distinction between the natures of scientific and religious ideas ought never to be doubted; but that contradiction can arise, except between religious and superstitious ideas, ought not for a moment to be admitted. . . . If the teachings of natural theology are liable to be refuted or corrected by progress in knowledge, it is legitimate to suppose, not that science is irreligious, but that these teachings are superstitious; and whatever evils result from the discoveries of science are attributable to the rashness of the theologian, and not to the supposed irreligious tendencies of science.

In accord with Wright's philosophy, an anonymous reviewer of Huxley's *Lectures on the Elements of Comparative Anatomy,* wrote in the same issue of the *North American Review* that evolution seemed destined to prevail and suggested that perhaps "by accustoming our imagination to contemplate the possibility of our ape descent now and then, as a precautionary measure, the dire prospect, should it ever really burst upon us, will appear shorn of some of its novel horrors, and our humanity appear no less worthy than before."

Soon after the publication of these three interesting discussions, the War ended, but not the sectional antagonisms. During the critical years of Reconstruction philosophic detachment became impossible for all but a few, and little of great moment appeared on evolution until 1868. Yet that little indicates that the forces of evolution were gaining ground against the opposition.

Leo Lesquereux, the distinguished paleo-botanist, wrote Darwin towards the end of 1865 that after repeated readings of Darwin's book, he had become a convert, but to Darwin's astonishment, chiefly because the *Origin of Species* made the Birth of Christ and Redemption by Grace more easily intelligible to him. That same winter J. P. Lesley lectured on *Man's Origin and Destiny* before the Lowell Institute in Boston and championed evolution with much enthusiasm, force, and unscientific dogmatism. The next year Charles J. Sprague contributed a defense of the Darwinian theory to the *Atlantic Monthly,* and in August 1867 Professor J. S. Newberry of Columbia, a friend of Agassiz's and President of the American Association for the Advancement of Science that year, came out as a supporter of evolution, although he stressed the need for more research to clarify the origin of life and the mechanism of evolution.

This encouraging trend received a definite impetus in 1868. Darwin's monumental work: *Variation of Animals and Plants under Domestication,* published in January, met with high praise from the leading periodicals and buttressed the evidence of evolution. In April, a writer for the *Southern Review* defended evolution as a scientifically sound, though still not adequately proved, and as not destructive of belief in God. Francis E. Abbot, a penetrating thinker now unjustly neglected, wrote in the *North American Review* for October a strong attack on the hypothesis of special creations as "the delusive substitute of words for thoughts" and declared evolution to be the only intelligible hypothesis.

A new and powerful defender of evolution against the charges of atheism was James McCosh, the noted Scotch theologian who came to America in 1868 to become President of the College of New Jersey (now Princeton). At Harvard Charles W. Eliot became President in 1869 and inaugurated a reign of liberalism by having John Fiske deliver a series of lectures on evolution and positivism. His audiences were small, but the lectures received wide circulation by being published in full in the *New York World.* A storm of orthodox protest against this new infidelity fell upon Fiske, Eliot, and the *New York World,* but Eliot was not the man to be daunted by public fury and reengaged Fiske for another series of lectures on evolution to be given in 1871.

Another champion of religious liberalism was James Freeman Clarke, the great Unitarian minister of Boston. In 1870 he boldly declared in his volume *Steps of Belief* that evolution had not, in itself, the slightest atheistic tendency.

Nevertheless, the progress of evolution during these years was not free from opposition. Articles against evolution appeared in the *Christian Examiner* for January 1866 and *Harper's New Monthly Magazine* for December 1867. Louis Agassiz still warred against evolution with his old fervor. In a lecture before the National Academy of Science in 1866 and in a book on Brazil published in 1868, he presented new evidence that he felt sure would kill what he called "this mania" for evolution; but he was doomed to disappointment.

The gathering strength of the evolutionists brought new allies to Agassiz. Horace Bushnell wrote a vehement criticism of evolution in *Putnam's Magazine* for March 1868 and pontifically declared that if evolution were true, then both science and religion were impossible. Parke Godwin at a dinner in honor of Charles Dickens in April 1868 made fun of the new doctrine of evolution which, as he put it, announced that long ago a lot of dust began twisting and twisting until it got twisted into Parthenons and Shakesperes. This revolt against science was continued in equally scornful essays in the *Baptist Quarterly,* the *Biblical Repertory, the Catholic World,* and the *Congregational Review* from 1868 to 1870.

But a new phase of the struggle over evolution opened with the publication of Darwin's *Descent of Man* in the Spring of 1871. All the objections to evolution that appeared before 1871, no matter how impassioned or widespread, were mild and insignificant in comparison with the hurricane of outraged feeling and religious indignation aroused by this book. Edward L. Youmans wrote Herbert Spencer on April 21st.:

> Things are going here furiously. I have never known anything like it. Ten thousand *Descent of Man* have been printed, and I guess they are nearly all gone. Five or six thousand of [Huxley's] *Lay Sermons* have been printed . . . the progress of liberal thought is remarkable. Everybody is asking for explanations. The clergy are in a flutter.

We can understand the intense excitement and interest in 1871 only when we remember that Darwin had very cautiously and carefully avoided all discussion of man's origin in *The Origin of Species* in order to concentrate attention first on the general problem. True, he had promised "light would be thrown on the origin of man and his history," but despite Huxleys explicit exposition on *Man's Place in Nature* in 1863, comparatively few had worried about man's animal origin.

This explains why a liberal in religion like Joseph P. Thompson, as late as 1870, could say in his book: *Man in Genesis and in Geology,* that "in the present stage of science, I may safely lay down the postulate, that Man had a beginning, and that Nature is not proved adequate to have caused that beginning. He appeared upon the plane of nature with an organism that Nature fails to account for, and with powers for which Nature furnishes no precedents." C. L. Brace, a champion of evolution, could attack certain German scientists for believing that men were descended from apes and denying that man's origin and mind required special intervention by God. Even Alfred Russel Wallace, the co-discoverer with Darwin of natural selection, felt that miracles were needed to explain man's consciousness and moral sense.

Hence, it is not astonishing that practically the entire religious press in 1871 lunited in denouncing Darwin's most terrible of libels against man's divine origin. The *Methodist Quarterly Review,* the *Congregational Review,* the *American Presbyterian Review,* and the *Christian Quarterly* affirmed with great vigor and much vehemence that man belonged to a separate kingdom from all the rest of creation and had been specially blest by God with consciousness and morality. Similar views appeared in non-sectarian magazines like the *Eclectic, Littel's Living Age,* the *National Quarterly Review,* and the *Overland Monthly.*

What anguish and turmoil of spirit advanced liberals in theology underwent is illustrated by the case of James McCosh. He felt and preached that no conflict existed between evolution and religion and declared in a noted series of lectures on *Christianity and Positivism* in 1871 that evolution, "properly understood," invoked supernatural intervention to account for life, intelligence, and morality. He lacked the courage to extend the logic of his half-way liberalism to man. He shrank with horror from the picture drawn by Darwin of man descending from a hairy quadruped, and with the desperate courage of his confusion used logic most illogically to prove that Scriptures were a weightier authority than Science on the origin of man.

Amazing as McCosh's use of the method of authority was, his irrationalism could not compare in boldness or assurance with the *ex cathedra* dicta of Charles Hodge, Professor of Theology at Princeton and valiant champion of rigid, uncompromising orthodoxy. Strong in conceit, terrible in scorn, Hodge used every method other than that of science to show that Darwin was thoroughly atheistic and therefore could not be scientifically sound. In three noted works that he published between 1871 and 1875 he spurned the mere naturalist, "the man devoted so exclusively to the study of nature as to believe in nothing but natural causes," and proclaimed with an enviable confidence in his own infallibility: "God has revealed his existence and his government of the world so clearly and authoritatively that any philosophical or scientific speculations inconsistent with those truths are like cobwebs in the track of a tornado. They offer no sensible resistance."

Certitude is not the test of certainty, as Mr. Justice Holmes has said. Yet many were so swayed by emotion and blinded by prejudice as to confuse the two and jump to the unwarranted conclusion that evolution was incompatible with true spirituality. The *Southern Review* for January 1872 featured an essay on *Modern Atheism,* in which the author exclaimed with great emotional intensity:

> Is man, in other words, merely a monkey, minus the tail, or is he in his most amorphous, monstrous condition, merely a man minus religion and faith—minus all that is greatest and most God-like in human nature? . . . if in this life only we have hope, then are we of all men, and of all animals, the most miserable. Tell us not then, O ye mighty prophets of science! that the soul itself is

merely a "mode of motion," a bubble blown upon the bosom of a shoreless eternity of life, . . . and then (to) pass away, to make room for other bubbles.

As noted a scientist, educator, and liberal as President Barnard of Columbia College declared in 1873 that if spontaneous generation, organic evolution, and the correlation of mental and physical forces were true, then the existence of God and the immortality of the soul were impossible. So terrible a prospect was more than he could endure; he renounced without shame his intellectual integrity to preserve his dream of heaven:

> Much as I love truth in the abstract I love my sense of immortality still more; and if the final outcome of all the boasted discoveries of modern science is to disclose to men that they are more evanescent than the shadow of the swallow's wing upon the lake . . . if this, after all, is the best that science can give me, give me then, I pray, no more science. I will live on in my simple ignorance, as my fathers did before me; and when I shall at length be sent to my final repose, let me . . . lie down to pleasant, even though they may be deceitful dreams.

Extraordinary as Barnard's renunciation of science and his advocacy of the Will to Believe was, his position was not unique. The presidents of such leading colleges as Yale, Princeton, Williams, Amherst, Brown, Hamilton, Lafayette, and Rochester denied that they taught "such unverified hypotheses" as the evolution of man from irrational animals. As President Seelye of Amherst put it, they did not "teach groundless guesses for ascertained truths of science."

At a time when the higher Biblical criticism and the science of comparative religion were weakening the traditional dogmatic beliefs, the impetus given to a naturalistic philosophy by evolution stirred the opponents of modern science to white heat. The 70's and 80's brought a tragic wave of academic persecution. The first martyr in the cause of evolution was Alexander Winchell, Professor of Geology at Vanderbilt University, which was controlled by the Southern Methodists. Upon Winchell's expulsion in 1878, the Tennessee Methodist Conference commented as follows:

> The unthinking masses have been sadly deluded. But our university alone has had the courage to lay its young but vigorous hands upon the mane of untamed Speculation and say, "We will have no more of this." Science we want, no crude, undigested theories for the sons of our patrons—science we must have, science we intend to have, but we want only science clearly demonstrated. . . .

The next year Professor C. H. Toy of the Southern Baptist Seminary at Louisville, Kentucky, resigned at request because he used modern science in interpreting the Old Testament. In 1884 Professor James Woodrow, Woodrow Wilson's uncle, paid the price for his courageous defense of evolution at the Presbyterian Theological Seminary at Columbia, South Carolina, by being forced from his position. In 1891 Professor Alexander at the University of South Carolina dared to voice doubts concerning the divinity of Christ and lost his position in consequence. Two years later Professor Biggs of Union Theological Seminary of New York City was tried by the General Presbyterian Assembly for heresy and was expelled from the ministry. Economic determinists might well reflect on the fact that property rights are not the only vested interests that inspire persecution; it is the deeply-rooted love of illusion that makes the pursuit of truth so radical a test of courage.

The irony of this spiritual conflict and irrational persecution is that few understood how needless the struggle was. Surely the best part of religion, if not its essence, is its poetic interpretation of moral experience. Science cannot conflict with rational religion, which is based upon the moral adequacy of naturalism. Such a religion recognizes that faith in the supernatural and a future life is a desperate wager made by men eager to be deceived by a play on words; the supernatural, in the last analysis, is an idealization of man's natural situation and a promise to satisfy his earthly interests. But learning, as Santayana has said, does not liberate men from superstition when their souls are cowed or perplexed. They defend supernatural systems although they have only natural interests at heart because they lack the courage or candour "to distinguish the edge of truth from the might of imagination" and "to conceive other moulds for morality and happiness than those to which a respectable tradition has accustomed them."

The tenacity, authoritarianism, and mystical intuition relied upon to support exploded myths refute the facile identification of progress and evolution. No greater obstacle in the advancement of civilization exists than the tendency of men to distort discussions on matters of vital import into apologetics; even professional scientists and philosophers became so absorbed in defending some vested illusion or eloquent idea that they do not welcome truth, but seek victory and the dispelling of their own doubts. Their failure to see the ultimate futility of all romantic self-deception and the classic need for self-understanding if peace with the world and oneself is to be attained, harms not only themselves, but, what is even more deplorable, those who have the strength to envisage the truth and rebel against myth posing as science. Those in revolt against tradition usually pay a terrible tax to the errors of the past: they are led to stress the sharp contrasts between their new philosophy and that in authority, and hence parallel the old false patterns and emphases. Only a very great and generous mind can champion truth and point out the invalidity of established tenets without being carried away by a crusading zeal into injustice to some sound insight underlying the more patent absurdities.

That rare union of vision and balance distinguished Chauncey Wright and made all he wrote the enduring foundation of rational naturalism in America. His name is now unknown to all but a few; yet he was the philosophic master of three American immortals: Charles Sanders Peirce, the greatest logician America has produced; William James, its greatest psychologist, and Justice Holmes, its greatest jurist. Wright's merits as a champion of evolution have been unjustly obscured by the éclat of John Fiske's fame. Future historians, however, will redress the balance; Wright will then come into his own as the foremost philosophic exponent of evolution in the United States and the founder of rational naturalism, pragmatism, and pluralism. His *Philosophic Discussions* (1877) and *Letters* (1878) will then take their place among the philosophic classics of America.

For the ten years after 1865, and particularly between 1870 and 1875, Wright towered above all his American contemporaries as the philosophic champion of evolutionary naturalism. The essays he wrote for the *North American Review* on *The Limits of Natural Selection* (October, 1870), *The Genesis of Species* (July, 1871),

Evolution by Natural Selection (July, 1872), and the *Evolution of Self-Consciousness* (April, 1873) were of such extraordinary brilliance, originality and force that they attracted international attention. Darwin, who had been seriously worried by a powerful attack on his theory of Natural Selection by St. George Mivart, was so impressed by Wright's refutation of Mivart in the essay on the *Genesis of Species* that he had it reprinted at his own expense as a separate pamphlet for distribution in England.

In these essays, his book reviews for the *Nation,* and his private letters, Wright brought a most masterly use of the scientific method to lay bare all the imposing fallacies underlying the opposition to evolution and naturalism, show that rational religion could have no conflict with science, prove that the emergence of self-consciousness and morality were capable of explanation as natural events, and refute all the baseless fears that hope for humanity and reverence for spiritual excellence would disappear once man's animal origin was recognized. He regarded it as most unfortunate that the prepossessions of religious sentiments in favor of antiquated metaphysical theories should make the progress of science always seem like an indignity to religion, or a detraction from what is held as most sacred; yet the responsibility for this, he felt, belonged neither to the progress of science nor to true religious sentiment, but a false conservatism, and irrational respect for the ideas and motives of an unscientific philosophy. He realized men's love of pedigrees and their attribution of moral dignity to lofty origins and therefore exposed the fallacy of confusing the two distinct categories of moral worth and causal dependence, value and history. Men were what they were, whether descended from gods or the ancestors of monkeys. In fact, if the latter theory were true, humanity had cause for hope since men had made some progress.

The sanity that prevented Wright from turning history into mythology and science into mythic cosmology led him to develop the humanistic implications of naturalism. Convinced that most, if not all, of the puzzles of metaphysics might be reduced to unconscious puns, or unseen ambiguities in terms, he exposed the emptiness of supernaturalism and other-worldliness by showing that all the ends of life are within the sphere of life and are, in the last analysis, to be found in the preservation, continuance, and increase of life itself. Questions concerning the uses of life he suggested could be solved by asking: To what ascertainable phase of life is this or that other phase of life valuable or serviceable? All forms of movement in life have some value in themselves; yet life in the widest sense is neither good nor evil, but the theatre of possible goods and evils. Hence, if any particular way of life becomes unbearable, not suicide, but the rational reorganization of the ends of life, is the solution.

Here was the basis for the rational ethics that American society needed most in the moral chaos of the euphemistically-called reconstruction period. But the death of Wright in 1875, while still in his prime, left no peer to develop the system of naturalistic ethics that he had sketched. Yet his influence lived on among a few choice spirits. His work had shown that to be a great thinker one must have a heroic soul, that no man could boast of being civilized until he had questioned his own first principles, and that human life could be brought to perfection only when illuminated by intelligence and based on a morality frankly relative to man's nature and freed from the bonds of the pretentious moral absolutism of a supernatural creed.

So profound and noble a philosophy was condemned by its very excellence from becoming the dominant force in American thought at that time. Wright's evolutionary naturalism was the achievement of a great mind; to be understood it required minds equally great and courageous in applying scientific method to the most dearly held beliefs. The sacrifice of all illusions on the altar of truth was a price that no other champion of evolution in America was willing or able to pay. Even Benjamin Peirce, the foremost American mathematician of the nineteenth century, surrendered reason to faith and though he acknowledged the truth of evolution, said in 1879 that science was erecting the brute and unconscious force of evolution into an original divine power and honoring it as a god. In his eyes, Nature was "God's messenger," "the poem of an infinite imagination, signifying Immortality," and evolution "but the mode in which he is present on whom mortal cannot look with physical eyes and live."

A similar philosophy animated John Fiske, the great apostle of Herbert Spencer in America. Incomparably inferior to Wright in logical acumen, scientific training, and philosophic vision, Fiske nevertheless attained a national popularity and success that Wright never hoped for or dreamed of. Blessed with a lively and vivid style, a remarkable breadth, if not depth, of scholarship, and a contagious moral enthusiasm, he impressed his generation as a very great man and wielded immense influence. From 1869 on he crusaded for evolution, and after 1875 succeeded Wright as its most noted defender. His *Outlines of Cosmic Philosophy,* published in 1874, ran through sixteen editions and helped to make evolution the fashionable creed of the 1890's. Yet the victory for science was a hollow one. Fiske and his associates: Asa Gray, Joseph Le Conte, Henry Ward Beecher, and Lyman Abbott, won a wide currency for evolution by denaturalizing it into a new mythology.

Darwin had conceived of evolution in purely naturalistic terms; all his work implicitly denied the Aristotelian doctrine of evolution by ideal attraction and was a triumph for naturalism as the basis of life and morals. But Herbert Spencer was seduced by the dramatic charm and glamour of evolution and converted good science into bad poetry. He identified evolution with progress, with betterment—a tribute to his humanitarian sympatheties, but completely unwarranted by sober science. He changed a biologic truism, the survival of the fittest, into an unsound moral dogma that enabled many to rationalize their indifference to social injustice, their insensibility to human ideals, and their support of vested interests under pretense of respect for historic necessity and the dictates of science.

Comforting as Spencer's vision of a steadily improving universe was, his cold agnosticism could not satisfy the emotional needs of many. Fiske, speaking in the name of science, most unscientifically abandoned the logical rigor that would have led to an evolutionary naturalism, and promulgated a cosmic theism that made man's spiritual development the goal of the whole evolutionary process. This optimistic mythology appealed to many and became the religious dogma of the day. Men were given a cosmic importance and assurance that made them feel: This is the best of all possible worlds. Only a sceptic like F. H. Bradley could add: "and everything in it is a necessary evil."

Fiske's fall from naturalism was condemned by few. His use of the method of intuition to justify his Will to Believe was confused by the multitude with the method of science because he based his cosmic myth on the results of science. They did not realize how unscientific a synthesis of science might be that was prompted by aesthetic and religious motives and were not too critical of philosophies that answered their desires.

Only in the early part of this century did a decided revolt develop against the last defense of the Genteel Tradition expressed in the "Will to Believe" movement. Naturalism received a new birth when Santayana, Dewey, Singer, Morris R. Cohen and others repudiated all apologetics and asserted the supreme claim of reason in judging all cherished beliefs. To-day, for various social and economic reasons, we are witnessing in philosophy and science an unfortunate return to mysticism and supernaturalism. The defenders of naturalism in philosophy are facing a situation very similar to that which Chauncey Wright encountered in the 1870's. The progress of civilization in America will depend on a reassertion of scientific method as the only sound means of attaining a solid basis for security and happiness. All other methods are a snare and a delusion.

Contributors

Isaac Asimov
BOSTON UNIVERSITY

Kenneth E. Boulding
UNIVERSITY OF COLORADO

Roger J. Cuffey
PENNSYLVANIA STATE UNIVERSITY

Sidney W. Fox
UNIVERSITY OF MIAMI

Roy A. Gallant
UNIVERSITY OF SOUTHERN MAINE

Laurie R. Godfrey
UNIVERSITY OF MASSACHUSETTS, AMHERST

Stephen Jay Gould
HARVARD UNIVERSITY

L. Beverly Halstead
UNIVERSITY OF READING, UNITED KINGDOM

Garrett Hardin
UNIVERSITY OF CALIFORNIA, SANTA BARBARA

Roger Lewin

Gene Lyons

George M. Marsden
CALVIN COLLEGE

Robert M. May
PRINCETON UNIVERSITY

Kenneth R. Miller
BROWN UNIVERSITY

Ashley Montagu
PRINCETON UNIVERSITY

William R. Overton
UNITED STATES DISTRICT COURT, EASTERN DISTRICT OF ARKANSAS, WESTERN DIVISION

Sidney Ratner
RUTGERS UNIVERSITY

Robert Root-Bernstein
SALK INSTITUTE FOR BIOLOGICAL STUDIES

Michael Ruse
UNIVERSITY OF GUELPH, GUELPH ONTARIO CANADA

Gunther S. Stent
UNIVERSITY OF CALIFORNIA, BERKELEY